Lecture Notes in Computer Science   10509

Commenced Publication in 1973
Founding and Former Series Editors:
Gerhard Goos, Juris Hartmanis, and Jan van Leeuwen

More information about this series at http://www.springer.com/series/7409

Mārīte Kirikova · Kjetil Nørvåg
George A. Papadopoulos (Eds.)

Advances in Databases and Information Systems

21st European Conference, ADBIS 2017
Nicosia, Cyprus, September 24–27, 2017
Proceedings

 Springer

Editors
Mārīte Kirikova ⓘ
Riga Technical University
Riga
Latvia

George A. Papadopoulos ⓘ
University of Cyprus
Nicosia
Cyprus

Kjetil Nørvåg
Norwegian University of Science
 and Technology
Trondheim
Norway

ISSN 0302-9743   ISSN 1611-3349   (electronic)
Lecture Notes in Computer Science
ISBN 978-3-319-66916-8   ISBN 978-3-319-66917-5   (eBook)
DOI 10.1007/978-3-319-66917-5

Library of Congress Control Number: 2017951433

LNCS Sublibrary: SL3 – Information Systems and Applications, incl. Internet/Web, and HCI

Printed on acid-free paper

This Springer imprint is published by Springer Nature
The registered company is Springer International Publishing AG
The registered company address is: Gewerbestrasse 11, 6330 Cham, Switzerland

Preface

The 21st European Conference on Advances in Databases and Information Systems (ADBIS 2017) took place in Nicosia, Cyprus, during September 24–27, 2017. The ADBIS series of conferences aims at providing a forum for the dissemination of research accomplishments and at promoting interaction and collaboration between the database and information systems research communities from European countries and the rest of the world. The ADBIS conferences provide an international platform for the presentation of research on database theory, development of advanced DBMS technologies, and their advanced applications. As such, ADBIS has created a tradition: its 21st anniversary edition in 2017 continued the series held in St. Petersburg (1997), Poznan (1998), Maribor (1999), Prague (2000), Vilnius (2001), Bratislava (2002), Dresden (2003), Budapest (2004), Tallinn (2005), Thessaloniki (2006), Varna (2007), Pori (2008), Riga (2009), Novi Sad (2010), Vienna (2011), Poznan (2012), Genoa (2013), Ohrid (2014), Poitiers (2015), and Prague (2016). The conferences are initiated and supervised by an international Steering Committee consisting of representatives from Armenia, Austria, Bulgaria, the Czech Republic, Cyprus, Estonia, Finland, France, Germany, Greece, Hungary, Israel, Italy, Latvia, Lithuania, the FYR of Macedonia, Poland, Russia, Serbia, Slovakia, Slovenia, and the Ukraine.

The program of ADBIS 2017 included keynotes, research papers, thematic workshops, and a doctoral consortium. The conference attracted 107 paper submissions from 38 countries from many continents. After rigorous reviewing by the Program Committee (88 reviewers from 33 countries in the Program Committee and additionally by 26 external reviewers), the 26 papers included in this LNCS proceedings volume were accepted as full contributions, making an acceptance rate of 24%. Springer sponsored the ADBIS 2017 best paper award. Furthermore, the Program Committee selected 12 more papers as short contributions. The authors of the ADBIS 2017 papers come from 38 countries. The four workshop organizations acted on their own and accepted 25 papers for the BigNovelTI (50% acceptance rate), AMSD (50% acceptance rate), SW4CH (54% acceptance rate), and DaS 2017 (55% acceptance rate) workshops, as well as 3 for the doctoral consortium. Short papers, workshop papers, and the paper on Contributions from ADBIS 2017 workshops are published in a companion volume entitled New Trends in Databases and Information Systems in the Springer series Communications in Computer and Information Science. All papers were evaluated by at least three reviewers. The selected papers span a wide spectrum of topics in databases and related technologies, tackling challenging problems and presenting inventive and efficient solutions. In this volume, these papers are organized according to the nine sessions: (1) Conceptual modeling and human factors; (2) Subsequence matching and streaming data; (3) OLAP; (4) Graph databases; (5) Spatial data management; (6) Parallel and distributed data processing; (7) Query optimization, recovery, and databases on modern hardware; (8) Semantic data processing; and (9) Additional database and information systems topics. For this edition of ADBIS 2017, we had two keynote talks: the first one by Ernesto Damiani from the Khalifa University of Science, Technology & Research, United Arab Emirates, on "Model-Based Big Data as a Service," and the second one by Demetris Zeinalipour from the University of Cyprus, Cyprus, on "Indoor Navigation Services from Mobile Data."

The best papers of the main conference and workshops were invited to be submitted to special issues of the following journals: *Information Systems* and *Informatica*.

We would like to express our gratitude to every individual who contributed to the success of ADBIS 2017. Firstly, we thank all authors for submitting their research papers to the conference. We are also indebted to the members of the community who offered their precious time and expertise in performing various roles ranging from organizational to reviewing roles – their efforts, energy, and degree of professionalism deserve the highest commendations. Special thanks to the Program Committee members and the external reviewers for their support in evaluating the papers submitted to ADBIS 2017, ensuring the quality of the scientific program. We also offer thanks to all the colleagues, secretaries, and engineers involved in the conference and workshop organization. We acknowledge the assistance and guidance of the Steering Committee, especially the Vice Chair Yannis Manolopoulos.

The conference would not have been possible without our supporters and sponsors: Austrian Airlines, Cyprus Tourism Organization, and the University of Cyprus. Finally, we thank Springer for publishing the proceedings containing invited and research papers in the LNCS series. The Program Committee work relied on EasyChair, and we thank its development team for creating and maintaining it; it offered great support throughout the different phases of the reviewing process.

July 2017   Mārīte Kirikova
                                                                   Kjetil Nørvåg
                                                          George A. Papadopoulos

Organization

Program Committee

Bader Albdaiwi	Kuwait University, Kuwait
Bernd Amann	LIP6-UPMC, France
Grigoris Antoniou	University of Huddersfield, UK
Ladjel Bellatreche	LIAS/ENSMA, France
Klaus Berberich	Max Planck Institute for Informatics, Germany
Maria Bielikova	Slovak University of Technology in Bratislava, Slovakia
Doulkifli Boukraa	Université de Jijel, Algeria
Drazen Brdjanin	University of Banja Luka, Bosnia and Herzegovina
Stephane Bressan	National University of Singapore, Singapore
Bostjan Brumen	University of Maribor, Slovenia
Albertas Caplinskas	Vilnius University, Lithuania
Barbara Catania	DIBRIS-University of Genoa, Italy
Marek Ciglan	Institute of informatics, Slovak Academy of Sciences, Slovakia
Isabelle Comyn-Wattiau	ESSEC Business School, France
Alfredo Cuzzocrea	ICAR-CNR and University of Calabria, Italy
Ajantha Dahanayake	Georgia College & State University, USA
Christos Doulkeridis	University of Piraeus, Greece
Johann Eder	Alpen Adria Universität Klagenfurt, Austria
Erki Eessaar	Tallinn University of Technology, Estonia
Markus Endres	University of Augsburg, Germany
Werner Esswein	Technische Universität Dresden, Germany
Georgios Evangelidis	University of Macedonia, Greece
Flavio Ferrarotti	Software Competence Centre Hagenberg, Austria
Peter Forbrig	University of Rostock, Germany
Flavius Frasincar	Erasmus University Rotterdam, Netherlands
Jan Genci	Technical University of Kosice, Slovakia
Jānis Grabis	Riga Technical University, Latvia
Gunter Graefe	HTW Dresden, Germany
Franceso Guerra	University of Modena and Reggio Emilia, Italy
Hele-Mai Haav	Institute of Cybernetics at Tallinn University of Technology, Estonia
Theo Härder	TU Kaiserslautern, Germany
Katja Hose	Aalborg University, Denmark
Ekaterini Ioannou	Technical University of Crete, Greece
Mirjana Ivanovic	University of Novi Sad, Serbia
Hannu Jaakkola	Tampere University of Technology, Finland
Lili Jiang	Univeristy of Umeå, Sweden

Ahto Kalja	Tallinn University of Technology, Estonia
Dimitris Karagiannis	University of Vienna, Austria
Randi Karlsen	University of Tromsø, Norway
Panagiotis Karras	Aalborg University, Denmark
Zoubida Kedad	University of Versailles, France
Marite Kirikova	Riga Technical University, Latvia
Margita Kon-Popovska	Ss. Cyril and Methods University, Macedonia
Michal Kopecký	Charles University, Czech Republic
Michal Kratky	VSB-Technical University of Ostrava, Czech Republic
John Krogstie	NTNU, Norway
Ulf Leser	Humboldt-Universität zu Berlin, Germany
Sebastian Link	The University of Auckland, New Zealand
Audrone Lupeikiene	Vilnius University, Lithuania
Hui Ma	Victoria University of Wellington, New Zealand
Leszek Maciaszek	Wrocław University of Economics, Poland
Federica Mandreoli	DII - University of Modena, Italy
Yannis Manolopoulos	Aristotle University of Thessaloniki, Greece
Tadeusz Morzy	Poznan University of Technology, Poland
Martin Nečaský	Charles University, Czech Republic
Kjetil Nørvåg	Norwegian University of Science and Technology, Norway
Boris Novikov	St. Petersburg University, Russia
Eirini Ntoutsi	Gottfried Wilhelm Leibniz Universität Hannover, Germany
Andreas Oberweis	Karlsruhe Institute of Technology (KIT), Germany
Andreas L. Opdahl	University of Bergen, Norway
Odysseas Papapetrou	EPFL, Switzerland
Jaroslav Pokorný	Charles University in Prague, Czech Republic
Giuseppe Polese	University of Salerno, Italy
Boris Rachev	Technical University of Varna, Bulgaria
Milos Radovanovic	University of Novi Sad, Serbia
Heri Ramampiaro	Norwegian University of Science and Technology (NTNU), Norway
Tore Risch	University of Uppsala, Sweden
Gunter Saake	University of Magdeburg, Germany
Petr Saloun	VSB-TU Ostrava, Czech Republic
Kai-Uwe Sattler	TU Ilmenau, Germany
Ingo Schmitt	Technical University Cottbus, Germany
Tomas Skopal	Charles University in Prague, Czech Republic
Bela Stantic	Griffith University, Australia
Kostas Stefanidis	University of Tampere, Finland
Panagiotis Symeonidis	Free University of Bolzano, Italy
James Terwilliger	Microsoft Corporation
Goce Trajcevski	Northwestern University, USA
Christoph Trattner	MODUL University Vienna, Austria
Raquel Trillo-Lado	Universidad de Zaragoza, Spain
Yannis Velegrakis	University of Trento, Italy

Goran Velinov   Ss. Cyril and Methods University, Macedonia
Akrivi Vlachou   University of Piraeus, Greece
Gottfried Vossen   ERCIS Muenster, Germany
Robert Wrembel   Poznan University of Technology, Poland
Anna Yarygina   St. Petersburg University Russia
Weihai Yu   University of Tromsø, Norway
Arkady Zaslavsky   CSIRO, Australia

Additional Reviewers

Hosam Aboelfotoh   Jens Lechtenbörger
Dionysis Athanasopoulos   Jevgeni Marenkov
George Baryannis   Denis Martins
Sotiris Batsakis   Robert Moro
Panagiotis Bozanis   Ludovit Niepel
Loredana Caruccio   Wilma Penzo
Vincenzo Deufemia   Horst Pichler
Senén González   Benedikt Pittl
Sven Hartmann   Tarmo Robal
Zaid Hussain   Eliezer Souza Silva
Pavlos Kefalas   Nikolaos Tantouris
Julius Köpke   Eleftherios Tiakas
Vimal Kunnummel   H. Yahyaoui

Abstracts

Toward Model-Based Big Data-as-a-Service: The TOREADOR Approach

Ernesto Damiani[1,2], Claudio Ardagna[1,3], Paolo Ceravolo[1,3],
and Nello Scarabottolo[1,3]

[1] Consorzio Interuniversitario Nazionale per l'Informatica, Italy
[2] EBTIC/Khalifa University of Science and Technology, UAE
[3] Università degli Studi di Milano, Italy

Abstract. The full potential of Big Data Analytics (BDA) can be unleashed only by overcoming hurdles like the high architectural complexity and lack of transparency of Big Data toolkits, as well as the high cost and lack of legal clearance of data collection, access and processing procedures. We first discuss the notion of *Big Data Analytics-as-a-Service* (BDAaaS) to help potential users of BDA in overcoming such hurdles. We then present TOREADOR, a first approach to BDAaaS.

Short Biography: Ernesto Damiani is the Director of the Information Security Research Center at Khalifa University, Abu Dhabi, and the leader of the Big Data Initiative at the Etisalat British Telecom Innovation Center (EBTIC). Ernesto is on extended leave from the Department of Computer Science, Università degli Studi di Milano, Italy, where he leads the SESAR research lab and coordinates several large scale research projects funded by the European Commission, the Italian Ministry of Research and by private companies such as British Telecom, Cisco Systems, SAP, Telecom Italia and many others. Ernesto's research interests include business process analysis and privacy-preserving Big Data analytics. Ernesto is the Principal Investigator of the TOREADOR H2020 project on models and tools for Big data-as-a-service.

Indoor Navigation Services from Mobile Data

Demetrios Zeinalipour-Yazti

Department of Computer Science, University of Cyprus, 1678 Nicosia, Cyprus
dzeina@cs.ucy.ac.cy

Abstract. People spend 80–90% of their time in indoor environments such as offices, undergrounds, libraries, shopping malls and airports. On the other hand, the uptake of interesting applications in indoor spaces has so far been hampered by the lack of technologies that can provide indoor location (position) accurately, in real-time, in an energy-efficient manner and without expensive additional hardware. In this scope, the pervasiveness of smartphones is leading to the uptake of a new class of Internet-based Indoor Navigation (IIN) services, which rely on geo-location databases to store spatial indoor models along with wireless, light and magnetic signals used to localize users and provide better power efficiency and wider coverage than predominant approaches. In this talk, I will overview the research behind the building blocks of the Anyplace IIN, an open, modular, extensible and scalable navigation architecture that exploits crowd-sourced Wi-Fi data to develop a novel navigation service that won several international research awards for its utility and accuracy (i.e., less than 2 meters). Our MIT-licenced open-source software stack has to this date been used by hundreds of researchers and practitioners around the globe, with the public Anyplace service reaching over 100,000 real user interactions. In the second part of this talk, I will focus on an algorithm we developed for protecting users from location tracking by the IIN service, without hindering the provisioning of fine-grained location updates on a continuous basis. Our algorithm exploits a k-Anonymity Bloom filter and a generator of camouflaged localization requests, both of which are shown to be resilient to a variety of privacy attacks. My talk will conclude with a summary of related research challenges and results.

Short Biography: Demetrios Zeinalipour-Yazti (PhD, University of California, Riverside, 2005) is an Assistant Professor of Computer Science at the University of Cyprus, where he founded and directs the Data Management Systems Laboratory (DMSL). Before his current appointment, he served the University of Cyprus and the Open University of Cyprus as a Lecturer of Computer Science. He has also been a short-term Visiting Researcher at the Network Intelligence Lab of Akamai Technologies, Cambridge, MA, USA (2004), a Marie-Curie Fellow at the University of Athens, Greece (2007) and a Visiting Researcher at the University of Pittsburgh, PA, USA (2015). During 2016–2017, he is a Humboldt Fellow at the Max Planck Institute for Informatics, Saarbrücken, Germany. His research interests include Data Management in Computer Systems and Networks, in particular Mobile and Sensor Data Management;

Big Data Management in Parallel and Distributed Architectures; Spatio-Temporal Data Management; Network and Web 2.0 Data Management; Crowd and Indoor Data Management as well as Data Privacy Management. He is an ACM Distinguished Speaker (2017–2020), a Senior Member of ACM, a Senior Member of IEEE, and Member of USENIX. More Info: https://www.cs.ucy.ac.cy/~dzeina/.

Contents

ADBIS 2017 - Keynote Papers

Toward Model-Based Big Data-as-a-Service: The TOREADOR Approach

Ernesto Damiani[1,2]([✉]), Claudio Ardagna[1,3], Paolo Ceravolo[1,3], and Nello Scarabottolo[1,3]

[1] Consorzio Interuniversitario Nazionale per l'Informatica, Rome, Italy
[2] EBTIC/Khalifa University of Science and Technology, Abu Dhabi, UAE
ernesto.damiani@kustar.ac.ae
[3] Università degli Studi di Milano, Milan, Italy

Abstract. The full potential of Big Data Analytics (BDA) can be unleashed only by overcoming hurdles like the high architectural complexity and lack of transparency of Big Data toolkits, as well as the high cost and lack of legal clearance of data collection, access and processing procedures. We first discuss the notion of *Big Data Analytics-as-a-Service* (BDAaaS) to help potential users of BDA in overcoming such hurdles. We then present TOREADOR, a first approach to BDAaaS.

1 Introduction

Big Data technology has recently become a major market estimated to reach $203 billion in 2020, growing at a CAGR of 11.7% [11]. According to [9], every human in the world is producing over 6 megabytes for minute, a total of 1.7 million billion bytes of data. Also, the Compliance, Governance and Oversight Council claimed that the information volume doubles every 18–24 months for most organizations [5]. Many organizations have discovered that, in order to remain competitive, they have to deal with business cases where the volume of data reaches terabytes and even petabytes, and whose requirements include low latency and handling a variety of datatypes [2]. Still, Big Data applications are complex systems whose design and deployment poses challenges at multiple levels, ranging from data representation and storage issues to choice and adaptation of analytics, parallelisation and deployment strategies as well as display and interpretation of results. ICT companies propose to their customers to tackle Big Data application development using a mix of technologies going from NoSQL ("notonlySQL") databases like Cassandra or HBase, data preparation utilities like Paxata, and distributed, parallel computing systems like Apache Hadoop, Stark or Flink. However, the high architectural complexity and lack of transparency of Big Data toolkits leads many customers to use them as blackboxes, with little or no insight on how analytics are actually executed. Another major factor hindering Big Data Analytics (BDA) adoption is the "regulatory barrier:" concerns about violating data access, sharing and custody regulations when using BDA, and the high cost of obtaining legal clearance for their specific scenario are discouraging for many organizations. Finally, the limited size of

© Springer International Publishing AG 2017
M. Kirikova et al. (Eds.): ADBIS 2017, LNCS 10509, pp. 3–9, 2017.
DOI: 10.1007/978-3-319-66917-5_1

the BDA talent pool makes Big Data scientists, architects, and developers costly and in high demand internationally. Even outsourcing BDA to a service provider and/or engaging consultants does not eliminate the need of costly in-house skills. Having to tackle these challenges from scratch creates an entry barrier for small organisations, particularly SMEs, that cannot access the data science and data technology competence pools. This paper presents TOREADOR, a model-based approach for fast specification and roll-out of Big Data applications, fostering reuse via a Software Product Line approach. The TOREADOR methodology and toolkit support collecting user requirements and preferences in a declarative format, converting them into a procedural model of the Big Data computation, and compiling the latter into low-level specification directly deployable on a number of execution platforms and technologies. In this paper, we first give an overview of Big Data concepts (Sect. 2); we then define and compare BDA and BDAaaS providing some relevant application scenarios (Sect. 3); we finally discuss the TOREADOR approach and design to BDAaaS (Sect. 4).

2   Big Data: Overview

In the rest of the paper, we shall use the term *Big Data* to refer to data sets and flows whose size and update frequency cannot be handled by traditional database systems [15]. Big Data have been defined using the so called 5 V model: Volume, Variety, Velocity, Value, Veracity [13]. The term *Big Data Analytics* refers to the implementation of analytics over architectures that automatically adapt to the volume, variety and velocity of data, making it possible to extract valuable results within strict deadlines. Following [4], we now describe some major hurdles that Big Data Analytics is facing today.

The technology opacity hurdle. While Big Data analytics can in principle support existing or new value propositions in a number of business domains, choosing and deploying the "right" analytics on the "right" computational infrastructure is still more an art than an engineering practice [8,12]. Today, only large organizations with deep pockets can afford going trial-and-error for weeks on failure-prone, resource intensive Big Data Analytics projects. If SMEs and other limited budget actors, like start-ups and no-profit organization, have to join the Big Data ecosystem, provisioning a Big Data analysis process must become fast, transparent, affordable, repeatable and robust.

The data diversity hurdle. According to the current Big Data hype, the world is awash in readily accessible Big Data having common time, location, and identity references. Reality is very different. A few Over-The-Top (OTT) operators, like Google, have proprietary, semantically rich, and homogeneous data sources; they can conceivably expand their scope, adding uniform location and identity metadata. For others, data diversity is much higher [1,16,21], as data are independently collected and supplied by multiple actors: utility companies own sensor, management and billing information, telecommunication operators offer location and identity data, while public administrations supply open data on their territory and urban environments. With respect to relatively uniform OTT-style data, these multi-owner data sources are highly diverse: they differ in volume (involving small giga-scale and large peta-scale data sizes), granularity and veracity.

The compliance hurdle. Vertical domains where BDA can make a real difference (healthcare, transportation and energy) are highly regulated [7,14]. Regulatory peculiarities cannot be addressed on a project-by-project basis. Rather, certified compliance of each BDA (e.g., in the form of a Privacy Impact Assessment) should be made available from the outset to all actors that use BDA in their business model. Also, BDA comes with legal issues that may trigger unwanted litigation. How to account intellectual property and how to shape the economical exploitation of BDA in multi-party environments [23]? How to provide evidence that data processing is compliant to ethics, beyond norms and directives [18]? Those are among the questions that still require mature and reliable solutions.

We believe these hurdles to have played a major role in hindering BDA acceptance [20,22]. For example, IDC [10] reports that 60% of organizations are hampered by too little business intelligence and only 10% of employees are satisfied with the Big Data technology resources available [10]. Big Data Analytics-as-a-Service can play a role in bringing Big Data to the mass, representing the entry point also for companies lacking Big Data skills and competences.

3 Big Data Analytics-as-a-Service

The BDAaaS paradigm [4] represents the next evolution step of Big Data to accomplish the hurdles discussed in Sect. 2. It consists of a set of automatic tools and methodologies that allow customers lacking Big Data expertise to manage BDA and deploy a full Big Data pipeline addressing their goals. BDAaaS can be seen as a function that takes as input users' Big Data goals and preferences, and returns as output a ready-to-be-executed Big Data pipeline.

Users with different skills and expertise can benefit by using a BDAaaS paradigm. Users lacking expertise proper of data scientists (e.g., modeling, analysis, problem solving) can use a BDAaaS solution for preparing the real analytics, reason on data to find out hidden patterns and information, and solve business problems. Users lacking data engineering expertise (e.g., build a robust data pipeline, install a Big Data toolkit) can use a BDAaaS to automatically identify and deploy the proper set of technologies that accomplish their requirements. Users lacking both expertise can still use BDAaaS solutions for a proper initiation in the Big Data realm.

Users' requirements are in the form of platform-independent declarative goals, which are then transformed in low-level platform-dependent configurations of the Big Data pipeline. Requirements can be defined in five different conceptual areas as follows:

– **Data preparation** specifies all activities aimed to prepare data for analytics. For instance, it defines how to guarantee data owner privacy.

- **Data representation** specifies how data are represented and expresses representation choices for each analysis process. For instance, it defines the data model and data structure.
- **Data analytics** specifies the analytics to be computed. For instance, it defines the expected outcome and the type of analytics.
- **Data processing** specifies how data are routed and parallelized. For instance, it defines the processing type and the parameters driving a map-reduce processing.
- **Data visualization and reporting** specifies an abstract representation of how the results of analytics are organized for display and reporting. For instance, it defines visualization type and visual density.

BDAaaS paradigm applies to Big Data scenarios involving enterprises that, for different reason, cannot rely on the adequate level of Big Data competences and/or on skilled data scientists and engineers. In the following, we discuss the issues and challenges introduced by the BDAaaS paradigm.

4   The TOREADOR Methodology

Let us consider a Big Data Analytics application from the point of view of the final user. Most of the times, some or all the following activities are needed (not necessarily in this order):

- *Define a business value proposition.* What income increase or cost reduction will the BDA results enable?
- *Identify the data.* Which inputs are needed to feed the BDA ?
- *Define the ingestion data flows (and the data lake where data will be ingested).* Where are (and who supplies) the data ingestion hooks (e.g., URLs or callbacks)? Are the ingestion flows stream or batch-like? Are the data flows to be processed immediately or will they feed a "lake" from which the BDA will periodically take its inputs?
- *Select and apply data preparation filters.* Will the flows be filtered (e.g., for anonymity/obfuscation or density) before ingestions?
- *Select and apply data protection measures.* Will after-ingestion access control be applied on the data lake, so each BDA will be able to access only the information it is entitled to?
- *Select analytics.* Given the available data and a description, classification or prediction goal, which algorithm or model should be employed to achieve the best results?
- *Define the analytics processing.* Will the BDA code be executed in a parallel way? Which HPC parallelization paradigm will be employed?
- *Define visualization, reporting and interaction.* Should the results be presented in a visual fashion, will they be stored in the data lake or will they feed other BDAs?

The questions above can be easily mapped into the five areas discussed in Sect. 3, which are then implemented by the TOREADOR methodology in five steps as follows:

1. A *declarative model* is used to describe the desired functional and non-functional goals of the BDA. The model is generated by a form where the questions listed above are answered by choosing among a closed list of answers, expressed in TOREADOR controlled vocabulary. In a nutshell, the declarative model is expressed as a list of lists, each list corresponding to one the five areas of the BDA. The atoms within each list are {*property, value*} pairs where *property* expresses functional or non-functional indicators, while *value* can be either 0 or 1, for Boolean properties, or a value in an ordinal scale.
2. The TOREADOR declarative model is checked for consistency to eliminate conflicting requirements [3].
3. The declarative model is used to instantiate a platform-independent *procedural model* that describes the BDA computation. This model is a linear composition represented in OWL-S and consists of 5 pipeline stages,[1] one for each area of the BDA pipeline (i.e., preparation, representation, analytics, processing, reporting/visualization). For each stage, the TOREADOR toolkit generates an *intra-area procedural model*, specifying the composition of internal services within the stage. More in detail, the toolkit feeds each list of {*property, value*} pairs into a SPARQL query pattern [19]. The query is then applied to a pre-defined OWL-S service ontology that lists the available services for each area (e.g., k-anonymity obfuscation service for area data preparation). The result is a list of services compatible with the preferences expressed in the declarative model.
4. The user is then called in to specify how the abstract services should be composed. These compositions are not necessarily pipelines; for instance, the analytics stage may involve disjunction, parallel execution of services and even loops [17]. To simplify this operation, the user can choose among a number of pre-defined composition patterns. Once the abstract service interface and their composition have been specified, the TOREADOR toolkit adds to the model the service interface's *grounding*, that is, the corresponding URLs in a target execution environment (e.g., Apache Flink, Spark) selected by the user.
5. Each intra-area procedural model is then compiled by the TOREADOR toolkit into a platform-dependent *deployment model*. The deployment model can be either executed via SaaS or PaaS service interface on the target deployment environment, according to the user preferences. In the former case, the composition workflow is translated in an XML incarnation (e.g., Apache Oozie), whose execution is supported by the target environment; in the latter case, an executable image (a Python bundle or a Docker image [6]) is generated ready for execution on a service provider's infrastructure.

5 Conclusions

In this paper, we defined the Big Data Analytics-as-a-Service (BDAaaS) paradigm as the next evolution step of Big Data domain. We then briefly outlined the

---

[1] Again, the order of the stages depends on the specific BDA and is decided by the user.

TOREADOR approach to BDAaaS as a suitable driver bringing BDA to those organizations and SMEs lacking sufficient in-house competences. The TORE-ADOR toolkit and the project open deliverables describing the toolkit in detail are available on the site www.toreador-project.eu.

Acknowledgements. This work has received funding from the European Union's Horizon 2020 research and innovation programme under the TOREADOR project, grant agreement No. 688797. It was also partly supported by the program "piano sostegno alla ricerca 2016" funded by Università degli Studi di Milano.

References

1. Abadi, D., Agrawal, R., Ailamaki, A., Balazinska, M., Bernstein, P.A., Carey, M.J., Chaudhuri, S., Dean, J., Doan, A., Franklin, M.J., Gehrke, J., Haas, L.M., Halevy, A.Y., Hellerstein, J.M., Ioannidis, Y.E., Jagadish, H.V., Kossmann, D., Madden, S., Mehrotra, S., Milo, T., Naughton, J.F., Ramakrishnan, R., Markl, V., Olston, C., Ooi, B.C., Ré, C., Suciu, D., Stonebraker, M., Walter, T., Widom, J.: The beckman report on database research. ACM SIGMOD Rec. **43**(3), 61–70 (2014)
2. Ardagna, C., Damiani, E.: Network and storage latency attacks to online trading protocols in the cloud. In: Proceedings of the International Conference on Cloud Computing, Trusted Computing and Secure Virtual Infrastructures, Amantea, Italy, October 2014
3. Ardagna, C.A., Bellandi, V., Bezzi, M., Ceravolo, P., Damiani, E.: Model-driven methodology for big data analytics-as-a-service. In: Proceedings of the 6th IEEE International Congress on Big Data (BigData Congress 2017), Honolulu, HI, USA, June 2017
4. Ardagna, C.A., Ceravolo, P., Damiani, E.: Big data analytics as-a-service: issues and challenges. In: Proceedings of the IEEE International Conference on Big Data (Big Data 2016), Washington, DC, USA, December 2016
5. Austin, D.: eDiscovery Trends: CGOCs Information Lifecycle Governance Leader Reference Guide. http://www.ediscoverydaily.com
6. Boettiger, C.: An introduction to docker for reproducible research. ACM SIGOPS Oper. Syst. Rev. **49**(1), 71–79 (2015)
7. Eckhoff, D., Sommer, C.: Driving for big data? privacy concerns in vehicular networking. IEEE Secur. Priv. **12**(1), 77–79 (2014)
8. Ekbia, H., Mattioli, M., Kouper, I., Arave, G., Ghazinejad, A., Bowman, T., Suri, V.R., Tsou, A., Weingart, S., Sugimoto, C.R.: Big data, bigger dilemmas: a critical review. J. Assoc. Inf. Sci. Technol. **66**(8), 1523–1545 (2015)
9. Commission, E.: Helping SMEs Fish the Big Data Ocean. http://ec.europa.eu/digital-agenda/en/news/helping-smes-fish-big-data-ocean
10. IDC: Six patterns of big data and analytics adoption, March 2016. http://www.oracle.com/us/technologies/big-data/six-patterns-big-data-infographic-2956541.pdf
11. IDC: Worldwide Semiannual Big Data and Analytics Spending Guide, October 2016. http://www.idc.com/getdoc.jsp?containerId=prUS41826116
12. Jagadish, H.V., Gehrke, J., Labrinidis, A., Papakonstantinou, Y., Patel, J.M., Ramakrishnan, R., Shahabi, C.: Big data and its technical challenges. Commun. ACM **57**(7), 86–94 (2014)

13. Lomotey, R.K., Deters, R.: Analytics-as-a-service framework for terms association mining in unstructured data. Int. J. Bus. Process Integr. Manage. (IJBPIM) **7**(1), 49–61 (2014)
14. Lu, R., Zhu, H., Liu, X., Liu, J.K., Shao, J.: Toward efficient and privacy-preserving computing in big data era. IEEE Netw. **28**(4), 46–50 (2014)
15. Manyika, J., Chui, M., Brown, B., Bughin, J., Dobbs, R., Roxburgh, C., Byers, A.H.: Big data: the next frontier for innovation, competition, and productivity (2011). http://tinyurl.com/z9wjhuw
16. Markl, V.: Breaking the chains: On declarative data analysis and data independence in the big data era. Proc. VLDB Endow. **7**(13), 1730–1733 (2014)
17. Martin, D., Paolucci, M., McIlraith, S., Burstein, M., McDermott, D., McGuinness, D., Parsia, B., Payne, T., Sabou, M., Solanki, M., et al.: Bringing semantics to web services: the owl-s approach. In: Proceedings of the International Workshop on Semantic Web Services and Web Process Composition (SWSWPC 2004), San Diego, CA, USA, July 2004
18. Martin, K.E.: Ethical issues in the big data industry. MIS Q. Execut. **14**, 2 (2015)
19. Prud, E., Seaborne, A., et al.: SPARQL query language for RDF (2006)
20. Rahman, N.: Factors affecting big data technology adoption (2016). http://pdxscholar.library.pdx.edu/cgi/viewcontent.cgi?article=1099
21. Russom, P.: Big Data Analytics. TDWI best practices report, TDWI Research (2014). http://www.iso.org/iso/home/news_index/news_archive/news.htm?refid=Ref1821
22. Salleh, K.A., Janczewski, L.: Adoption of big data solutions: a study on its security determinants using sec-toe framework. In: Proceedings of the International Conference on Information Resources Management (CONF-IRM 2016), Cape Town, South Africa, May 2016
23. Wu, D., Greer, M.J., Rosen, D.W., Schaefer, D.: Cloud manufacturing: strategic vision and state-of-the-art. J. Manufact. Syst. **32**(4), 564–579 (2013)

Conceptual Modeling
and Human Factors

General and Specific Model Notions

Bernhard Thalheim[✉]

Department of Computer Science, Christian Albrechts University Kiel,
Olshausenstr. 40, 24098 Kiel, Germany
thalheim@is.informatik.uni-kiel.de

Abstract. Models are a universal and widely used instrument in Computer Science and Computer Engineering. There is a large variety of notions of models. A model functions in a utilisation scenario as an instrument. It is well-formed, adequate and dependable. It represents or deputes origins. This conception of the model is a very general one. Based on the notion of a stereotype as a starting point we show that specific or particular model notions are specialisations of the general notion.

1 Models in Computer Engineering and Computer Science

Models are principle instruments in modern computer engineering (CE), in teaching any kind of computer technology, and also modern computer science (CS). They are built, applied, revised and manufactured in many CE&CS subdisciplines in a large variety of application cases with different purposes and context for different communities of practice.

1.1 The Omnipresence of Models in CE&CS

The wide deployment of models is supported by an expansive scientific literature on model usages. There are many different model notions, e.g. [30] discussed more than 50 different definitions of models used in CE&CS programs. All subdisciplines in CE&CS use models such as phenomenological models, computational models, developmental models, explanatory models, didactic models, imaginary models, mathematical models, substitute models, iconic or diagrammatic models, formal models, and analogue models. There is no branch in CE&CS that does not widely use models as instruments.

It is now well understood that models are something different from theories. They are often intuitive, visualisable, and ideally capture the essence of an understanding within some community of practice and some context. At the same time, they are limited in scope, context and the applicability.

We realised also that models become an research issue on their own. Models are expressions, descriptions, icons, statements, etc. from one side and desiderata, representations, deputies, instruments, designs, products etc. from the other side. They might suggest something that we might later be able to explain or to construct. Models also help us to explain a system, help us to deal with more realistic situations, and tell us which intuition and understand is a good one. *How we handle such variety of deployments, understandings, and approaches?*

© Springer International Publishing AG 2017
M. Kirikova et al. (Eds.): ADBIS 2017, LNCS 10509, pp. 13–27, 2017.
DOI: 10.1007/978-3-319-66917-5_2

1.2   The General Notion of the Model

There is however a general notion of a model and of a conception of the model:

> A **model** *is a well-formed, adequate, and dependable instrument that represents origins* [6, 24–26].

Its criteria of well-formedness, adequacy, and dependability must be commonly accepted by its community of practice within some context and correspond to the functions that a model fulfills in utilisation scenarios.

The model should be well-formed according to some well-formedness criterion. As an instrument or more specifically an artifact a model comes with its *background*, e.g. paradigms, assumptions, postulates, language, thought community, etc. The background its often given only in an implicit form.

The background is often implicit and hidden. *Is there any approach to consider the background in a simpler form?*

An well-formed instrument is *adequate* for a collection of origins if it is *analogous* to the origins to be represented according to some analogy criterion, it is more *focused* (e.g. simpler, truncated, more abstract or reduced) than the origins being modelled, and it sufficiently satisfies its *purpose*.

So far, the adequateness notion is far too fuzzy and too wide. *Can be develop a simpler notion of adequateness that still covers the approaches we are used in our subdiscipline?*

Well-formedness enables an instrument to be *justified* by an empirical corroboration according to its objectives, by rational coherence and conformity explicitly stated through conformity formulas or statements, by falsifiability, and by stability and plasticity within a collection of origins.

The instrument is *sufficient* by its *quality* characterisation for internal quality, external quality and quality in use or through quality characteristics [20] such as correctness, generality, usefulness, comprehensibility, parsimony, robustness, novelty etc. Sufficiency is typically combined with some assurance evaluation (tolerance, modality, confidence, and restrictions).

A well-formed instrument is called *dependable* if it is sufficient and is justified for some of the justification properties and some of the sufficiency characteristics.

Again, dependability is a wide field. *Do we need this broad coverage for models? Or is there any specific treatment of dependability for subdisciplines or specific deployment scenarios?*

If there are many specific and particular notions of the model: *Can we relate different notions of models with each other? Can be define interfaces among models? Is there any standard notion for a sub-discipline? Are specific or particular notions derivable from the general notion of the model?*

And finally, this notion is a very general one. *How does the general notion match with other understandings and approaches to modelling in CE&CS?* Or more generally for sciences based on Occam's razor principle: *Are there specific or particular notions of the model within specific constellations that sufficiently represent all relevant aspects requested and nothing more?*

1.3   Generality versus Specificity

The general notion of a model covers all aspects of adequateness, dependability, well-formedness, scenario, functions and purposes, backgrounds (grounding and basis), and outer directives (context and community of practice). It covers all known so far notions in agriculture, archeology, arts, biology, chemistry, computer science, economics, electrotechnics, environmental sciences, farming, geosciences, historical sciences, languages, mathematics, medicine, ocean sciences, pedagogical science, philosophy, physics, political sciences, sociology, and sports. The models used in these disciplines are instruments used in certain scenarios.

Sciences distinguish between general, particular and specific things. Particular things are specific for general things and general for specific things. The same abstraction may be used for modelling. We may start with a general model. So far, nobody knows what is such general model for most utilisation scenarios. Models *function* as *instruments* or tools. Typically, instruments come in a variety of forms and fulfill many different functions. Instruments are partially independent or autonomous of the thing they operate on.

Models are however special instruments. They are used with a specific intention within a utilisation scenario. The quality of a model becomes apparent in the context of this scenario.

It might thus be better to start with generic models. A *generic model* [3, 15] is a model which broadly satisfies the purpose and broadly functions in the given utilisation scenario. It is later tailored to suit the particular purpose and function. It generally represents origins under interest, provides means to establish adequacy and dependability of the model, and establishes focus and scope of the model. Generic models should satisfy at least five properties: (1) they must be accurate; (2) the quality of the generic model allows that it is used consciously; (3) they should be descriptive, not evaluative; (4) they should be flexible so that they can be modified from time to time; (5) they can be used as a first "best guess".

Generic models might also be an abstraction of other models that are used as an inspiration for development of the new model and that are based on the experience of the modeller. Generic models can be calibrated to specific models through a process of data or situation calibration, refinement, concretisation, context enhancement, or instantiation.

Generic models [29] are typically specialised to more specific ones in a development process. Generic models are widely used under different names or development approaches such as inverse modelling, model-driven architectures and development, universal applications, data mining and analysis, pattern-based development, reference models, inductive learning, and model forensics. All these approaches develop models by stepwise refinement of the root or initial model, by selection and integration of model variations, and by mutation and recombination of the model where the root model is a generic model with parameters (also structures and operations as parameters as well as the architecture).

Instead, we also may start with general models. Typically, we prefer however *particular* or idealised *models* as a starting point for a specific community of practice with a specific background, within a specific context, and for representation of a specific world of origins under consideration. Generic models can be calibrated to specific models through a process of data or situation calibration, refinement, concretisation, context enhancement, or instantiation. *Lightweight models* [28] typically cut off background and context. They assume per default some utilisation scenario and reduce the functions of the model to the main function. The purpose is then driven by this function. Often the community of practice is set to some standard community that uses a specific kind of justification.

Therefore, we face the problem: *What is the best starting point for development of a model.* This paper answers this question by introducing stereotypes of particular models in Sects. 3 and 4. For this we use the separation of abstraction into stereotypes, pattern, and templates [1].

1.4   The Storyline and Objectives of This Paper

Since the model notion is too broad we might ask ourselves whether more specific notions can be used in subdisciplines of CE&CS. We might also consider whether some of the proposed notions are simpler and better to use. We might start with the main properties of models (mapping or analogy, truncation or abstraction or focus, pragmatic, amplification, distortion, idealisation, carrier, added value, purpose [12,16,18,21]) and specialise them. We might also discuss the variety of notions [23,30] and compare them with the general one. The main question is however whether these different notions are sufficient within their environment, i.e. which specific notion of the model is sufficient for which utilisation, for which community, within which context, under which general conditions and within which understanding.

Models are used as perception models, mental models, situation models, experimentation models, formal model, mathematical models, conceptual models, computational models, inspiration models, physical models, visualisation models, representation models, diagrammatic models, exploration models, heuristic models, etc. Although this categorisation provides an entry point for a discussion of model properties, the phenomenon of being a model can be properly investigated. Each category is too broad and combines too many different aspects at the same time.

We thus first discuss notions which are commonly accepted and discover that these notions are laden by background, community, context, and utilisation scenarios. This ladenness can be represented by definitional frames for the model notion. These frames may now be used for defining stereotypes of model notions.

2   Specialising and Refining the Model Notion

2.1   Stereotypes for Models and Particular Notions of Models

Modelling stereotypes describe the general modelling situation. Generic models are typically a general modelling solution in a certain utilisation scenario, context, background, and community of practice. For instance, a *structure stereotype* describes data structuring environment within a certain modelling situation. The corresponding generic models can be refined and used during model development. They can be considered to be classes or collections of potential models.

2.2   Two Model Notions and Their Specific Approaches

Let us consider two of the 49 model notions we collected [27] for CE&CS. We will show that these notions are applicable but are heavily biased and thus paradigmatically use a lot of latent semantics behind.

Models for Model-Based Development. The Scandinavian and Dutch schools of (conceptual) modelling have developed a sophisticated approach to modelling since the late 60ies. One result is the famous FRISCO report [7]. More recently, J. Krogstie [11] states:

> "Model: A model is an abstraction externalised in a professional language. A model is assumed to be simpler than, resemble, and have the same structure and way of functioning as the phenomena it represents.
> Phenomenon. A phenomenon is something as it appears in the mind of a person. The world is perceived by persons to consist of phenomena. ...
> Property. A property is an aspect of a phenomena that can be described and given a value. A phenomenon will have a set of potentially relevant properties. ...
> Constitutive rule. A deontic rule that applies to phenomena that exist only because a rule exist. ...
> Professional language. A professional language is a language used by set of persons working in certain kind of area or in a scienctrific discipline. Usually such a language is not learned before the person has been active in the area for a while.
> Language model. The model of a language. Within conceptual modelling, this is often termed 'meta-model', which is only a proper term when looking upon it from the point of view of repository-management for a modelling tool where the instantiation of the model is another model in the same or a different modelling language.
> Conceptual model. A model of a domain made in a formal language or semi-formal language with a limited vocabulary.
> System. A system is a set of correlated phenomena, which is itself a phenomenon. ...
> System model. A model of a system."

Analysing these notions and more specifically the notion of the model, we realise that there must exist an origin that we can call *matured perception model*. At the same time, the modelling approach is entirely biased by its discipline, its school of thought, it context, and - as a partially explicit component - its community of thought. At the same time, we consider only phenomena in a set-based fashion and not within a conception/conception network. So, the modelling approach is using a rather restricted world view.

This restricted world view is however entirely sufficient since the model is used in one very specific utilisation scenario: system construction. We observe the *import of latent* paradigmatic (computing-oriented, function-backed, economic, ...) *models* with predefined meaning, specific context and background concepts (space, time, settlement, environment, ...) within this scenario. The main function of the model is that of a *mediator* that describes the (augmented and perceived) model and that prescribes a system to be investigated or perceived. The adequateness property uses homomorphisms.

This approach is typical for model-based (software) development [4, 10, 11, 17] within the specific consideration of specific platform-independent models such as conceptual models and of platform-dependent models as refinements of the generic ones. This approach uses latent hidden generic models as community knowledge. Beside the community dependence, the development biases are also latent in this model notion.

Model Notions with Justification. Extending and revisiting the model notion with its mapping, truncation and pragmatic properties by H. Stachowiak [16], R. Kaschek [9] introduces a *model* as a material or virtual *artifact* (1) that is called a model within a community of practice (2) based on a judgement (3) of appropriateness for representation of other artifacts (things in reality, systems, ...) and serving a *purpose* (4) within this community.

Already [9] discussed the forgetful development of software products. Classically we observe that (i) developers base their design decisions on a "partial reality", i.e. on a number of observed properties within a part of the application, (ii) developers are developing the information system within a certain context, (iii) developers reuse their experience gained in former projects and solutions known for their reference models, and (iv) developers use a number of theories with a certain exactness and rigidity.

The design decisions made during the design process are deeply influenced by these four hidden factors. In some approaches revisions made during the information systems development are recorded. However, since the background knowledge is not recorded the documentation of the information systems development is fragmentary.

The justification of models [9] is here explicit. It should however be combined with a statement of quality that has been achieved so far. The quality criteria are implicit. The model notion [9] is based on the community of practice behind the model. Forgetful development is one of the specific properties. The community of practice drives the context of the model and of modelling. At the same time, appropriateness is more general than (homomorphic) mapping and truncation.

2.3   The Background as the Hidden Component of Models

The two cases show that the model notion is often laden by its specific background. The background consists of undisputable elements (grounding: paradigms, postulates, restrictions, theories, culture, foundations, commonsense) and disputable one (basis: concepts, foundations, language as carrier, assumptions, thought community, thought style, conventions, practices). *Background laden* models are already using the grounding and the basis without making it explicit.

2.4   The Particular Notion of a Conceptual Model

Conceptual models are nothing else as models that incorporate concepts and conceptions which are denoted by names in a given name space. A concept space[1] consists of concepts [13] as basic elements, constructors for inductive construction of complex elements called conceptions, a number of relations among elements that satisfy a number of axioms, and functions defined on elements.

The general Sapir-Whorf hypothesis [33, 36] states the principles of language determinism (the language governs thinking) and language relativity (coded distinctions made in one language might not be expressible in another language). The weak form refers to the dependence of perception, remembering and simplicity on language. We may transfer this hypothesis to *concept-ladenness* of languages. Some languages might have richer concepts and conceptions than others[2]. Therefore, concepts and conceptions that are expressed in certain language heavily influence semiotics of models since the basis of models is also concerned with concepts and conceptions to be used and thus related to the (discipline's context).

They use a specific background: a concept space that clarifies the meaning of the elements of the model. The concept space is often application dependent and based on the understanding of notions in the application area. The linguistic meaning of designators and annotations is an inherent but hidden element of the
So, we notice: the conceptual model is *concept space laden*.

2.5   The Ladenness of Model Notions

In a similar way we observe also other kinds of ladenness:

Context-ladenness: The application domain and disciplinary context is often already given due to the introduction of the model. It is often enhanced by focus and scope depending on the concrete deployment of the model. The time and space issues are typically implicit.

Community-ladenness: A community of practice tries to be efficient. Such kind of efficiency includes an agreement of the way how thing are considered, i.e. a "school of thought" and commonly accepted practices, conventions, and assumptions.

Development- and utilisation-ladenness: Models must function effectively within the utilisation scenarios. For this reason, a number of biases are inherited by the the model notion due to the orientation and function of the model.

---

[1] We follow R.T. White [24, 35] and distinguish between concepts, conceptual, conceptional, and conceptions.

[2] Think for instance about the finer notions for whole in Aborigine language: yarla, pirti, pirnki, kartalpa, yulpilpa, mutara, nyarrkalpa, pulpa, makarnpa, and katarta.

Utilisation also determines most of the quality characteristics, the assessment of the model, and the tolerance that might be applied.

2.6 Lessons Learning: Towards a General Approach to Modelling

We observe that modelling mainly consists of three macro-steps, two intentional and implicit and one extensional and explicit:

(I) *Setting the definitional frame* with priming, language, and actor setting: Priming defines the undisputable decisions (called grounding), the concept space, and the context. Actors within a community of practice act in certain roles while fulfilling a task. They are biased by their disputable but somehow accepted background or basis.

(II) *Choice of a model stereotype* consisting of accepting the definitional frame, of agenda setting, and of initialisation pattern: Agenda setting restrict potential utilisation scenarios of models. It thus results in a clarification of the model functions and thus also purpose and goal. Initialisation may be based on generic models or modelling experience, e.g. on the basis of reference models.

(III) *Model development and deployment* is the classical macro-step and well investigated for many modelling problems.

The two first intentional macro-steps are hardly often explicitly mentioned. We often use already existing models (generic, reference, perception, situation, documentation, etc.) as a starting point without making a reference to it.

3   Definitional Frames for Model Notions

Definitional frames are often somehow agreed practice and commonsense within a context and within a community of practice. They are somehow implicit. Without knowing and managing them we might however come-up with models that drive us to spurious results or pitfalls. This paradox is well known for natural sciences or economics. Disputes in the past on whether semantical modelling, object-role modelling, relational modelling etc. are based on a misunderstanding of the definitional frames that have been used.

3.1   Priming and Orientation

The model is mostly developed within some *context* of a discipline, an application area, and an environment such as an infrastructure. Context may also incorporate certain foci and scopes for the model. Context may also be concerned with time. The context is taken as granted and not questioned.

Models are instruments and therefore design for utilisation. That means they are also set into the existing world. This world is based on some fundament or *grounding*. The grounding consists of the commonly accepted and not disputed postulates, paradigms, restrictions, theories, culture, foundations, and common-sense. Models thus inherit this grounding and do not explicitly refer to this grounding.

Models represent origins. These origins bring in their own world view, their own concepts and conceptions. The *concept(ion) space* is therefore for models some referred background. It is used for assigning a meaning to the constructs of the model, for consideration of properties of the model, and for validation of the model. Therefore, models often use concepts either in an explicit form (becoming thus conceptual models) or in a reference form as abstract formal notations which provide potentially an explanation of the model and its elements (most often for formal or mathematical models). In the first case, the concept space is given and not disputed whereas in the second case the concept space is hidden but available upon demand.

The fourth component of priming is the *context* agreement. It integrates the application domain, the specific thoughts in this application and thus the disciplinary context, scope, focus, infrastructure and time. We answer the when, whereat, whereabout, wherein, where, for what, wherefrom, and whence questions, may be partially also the what question.

3.2 Actors

The community of practice is far more influential than typically assumed. Community members play their specific roles, have their task portfolio, responsibilities and obligations during a development process. They have however also their interests which are injected into the modelling decisions. They have their preferred method spectrum and neglect others [2]. So, they choose also the modelling language [32] with all the limitations and potential of the language.

A community of practice is typically not interested in revision of the grounding. The community agrees typically also on the basis, i.e. on assumptions, on the thought style and understanding, on practices, and on conventions within the setted definitional frame. That means the community of practice determines the background meaning of a model and adequateness and dependability. The community also has a hidden raw understanding what means that a model is well-defined, analogous, focused and purposeful. A similar raw agreement is already made on dependability, i.e. on justification and sufficiency. The corroboration, rational coherence, validation, stability and plasticity is somehow already generically set and taken as commonly agreed.

So, we need to question the influence of the social and professional community: whom (to whom, by whom), whichever. These questions answer to the presetting of the model. The message of the model is the same within the community.

3.3 Languages and Basics

Languages enable and restrict at the same time [33,36]. They have their own obstinacy and thus restrict representability. From the other side, the provide rules for well-formedness, especially for syntactic ones. Professional languages additionally provide rules for semantic well-formedness. The community of practice also introduces its rules of pragmatic well-formedness. So, the language supports 'beauty' of models due to the inherent phonetics. We known from audience theory that representation determines later thinking, usage, and understanding of a model.

For instance, ER modelling supports well a global-as-design procedure on the basis of a global conceptual schema. If we follow the approach that syntactics also determines the operations and the algebra [19] then the different viewpoints which are required by the business user can be expressed via view collections defined on top of the conceptual schema.

Due to the carrier property, the language enables also to adjust practices, methodologies, pattern, typical routines, and commonsense. These elements of the basis complete the background. The language has a symbolic level which formes the culture of its users and provides a meaning. Professional languages use denotations and connotations. They provide a code that professionals learned to read.

So, the language answers the wherewith question. The imposed basics answer the question with what means. The language and the background form together some kind of 'gatekeeper' since we implicitly decide what to represent.

4   Stereotypes of Models in Utilisation Scenarios

Stereotypes of modelling have already been considered in discussions on methodologies, e.g. [5,8,14]. Typically, a methodology is bound to one stereotype and one kind of model within one utilisation scenario. We can however be more flexible. Stereotypes are governing, conditioning, steering and guiding the model development. They determine the model kind, the background and way of modelling activities. They persuade the activities of modelling. They provide a means for considering the economics of modelling.

4.1   Starting with Completing the Definitional Frame

The potential definitional frames are either selected on convenience or after consideration of appropriateness. Often one frame is taken for granted in most IT modelling approaches. The definitional frame sets up the acceptable background of a model. It is typically implicit.

4.2   Model Utilisation Scenario

Models are used as instruments in some utilisation scenarios. They have a number of functions in these scenarios. Based on an understanding of these functions we know what is the goal and purpose of such models. Therefore, we can now define the profile (goals, purposes, functions) that a model must fulfill. Due to the instrument property we also know which tasks are going to be solved with instruments. That means we know the task portfolio.

The profile and the portfolio create the 'spin' of the model since they convey a value judgement that might be immediately apparent and they create inherent bias by setting of the modelling task. The spin attempts to steer the way a model becomes useful to others.

4.3 Agenda Setting

Finally, we can define what is the agenda of the modelling tasks and of the model deployment. The agenda setting answers the why, for which reason, wherefore, worthiness, and whither away questions. This agenda can be formalised as a protocols setting and an orientation behind the model.

Based on the agenda, we sketch also adequateness and dependability. We can determine what means that a model is well-formed, which analogy or similarity is going the be used, which kind of focus allows to restrict the modelling task, and what means to be purposeful for a model.

At the same time, we have set up the main justification approaches. We already know explanatory statements and viability for the elements of the model based on the profile and background. We can sketch the arguments that support the model. We reflect norms and standards accepted by the community of practice, e.g. common practices for achieving inner coherence. The validation procedure is already set up for the model. We also may use which kind of robustness the model must have in order not to be over-fitted.

The model must not represent anything what might be representable. We know in this pre-setting which quality characteristics for quality in use, external and internal quality must be observed and which ones can be neglected. The quality characteristics are enhanced by evaluation procedures. So, we already define which discrimination is tolerated, which modality (necessity, contingency or possibility, relativity) can applied within the context, and which confidence of the evaluation is necessary.

Justification and sufficiency form our criteria for dependability. We can define for the model that is intended to build what means to be admissible, rigid, right, and fit.

4.4 Initial Model Setting

Models represent their origins. We might start from scratch, explore origins, discover essential and relevant elements, decompose them and explore then the modelling task. A first (nominal) model is the result of a composition or amalgamation step. Model formulation results then in development of a model. We might also base modelling on already existing models either for a given system or on the basis of referential models. We might also start with a generic model. In all these cases, we are already conditioned by the definitional frame. Additionally, we selected a modelling workflow or development strategy [19,22].

The initial setting also inherits *latent models* that come with the grounding, the context, and the basis.

After setting the stereotype, we start with model development according to the chosen strategy within the agenda and the definitional frame. The typical questions answered in this step are: whereof, how, what, with which restrictions. Additional questions are concerned with adequateness and dependability of the model especially with quality characteristics.

4.5   A Test Case for the Approach

We might consider all notions in [27, 31]. Let us only consider the *construction scenario* for IT systems. The *stereotype* we shall use incorporates (1) the typical and also specific *IT grounding* with all its paradigms, postulates, theories, foundations, culture, commonsense, and restrictions, (2) the *mediator function* of models in the construction scenario, (3) the *IT community of practice* with its obligations, interests, tasks portfolio from one side, and the biases accepted in the community such as school of thought, practices, commonsense, and assumptions, and (4) the selection of the *languages* and *concept space* that might be used. It also provides a collection of *reference models* as their basis for opportunities. These reference models are *latent models*.

So, in this case, the modelling case is based on the needs and the functions a model might play in system construction. The context is given by current IT systems, current infrastructures and by system development foci and scopes. Therefore, IT grounded is not reconsidered. The choice of the concept space is determined by the notion of the system. The community of practice determines the language and the biases the community likes. The agenda is a mediating one. The model is used either for description of a development idea and for prescription of a forthcoming system or for documentation of an existing system. Initialisation might be based on generic models, on reference models or on already existing models.

Then we arrive with some model definition as [34]:

> *"A model is a simplified reproduction of a planned or real existing system with its processes on the basis of a notational and concrete concept space. According to the represented purpose-governed relevant properties, it deviates from its origin only due to the tolerance frame for the purpose."*

5   Concluding: Stereotyping as the Spinning Principle

Models are one of the instrument in sciences, engineering and every life. They are not yet properly understood for their way of functioning, their impact, their potential, their capacity, and their anti-profile (not-supported utilisations). We do not want to overload the notion. Models should be used and understood. Therefore, we need a notion that is as simple as possible in the given scenario and given situation. At the same time, we should not loose the specific agreements we have made for models. Models must be effective, efficient, user-friendly, economic, and well-organised. Otherwise, nobody can properly use the conclusions and results that have been generated by the help of models. Sometimes, models may mis-orientate, condition, biase or persuade [23] users in their understanding and must be corrected after paradigmatic revision and synthesis.

So, we need a general specification of the model kind that allows from one side to reason on the potential, capacity, adequacy, and dependability of the given model and from the other side to be aware of the anti-profile and the cases in which the model is not promising, not adequate, may direct to wrong conclusions, and has its pitfalls.

This paper uses definitional frames and stereotypes for a holistic treatment of models. From one side, the model notion covers all what is necessary. From the other side, the specific agreements have to be explicitly given and must not be guessed. So providing the stereotype allows to understand the model, its quality characteristics, its capacity and its potential. It also allows to understand in which cases the model is not useful or more explicitly to know in which cases the model should not been used.

This paper does not claim that existing models or model notions are bad. We cannot handle here the large variety of modelling techniques. Model management is out of scope of this paper. Instead, we contribute to general model theory and harmonise notions of models by development of an approach that allows to derive specific notions of a model from the general one and thus to inherit investigations made for one model notion by other approaches to modelling.

References

1. AlBdaiwi, B., Noack, R., Thalheim, B.: Pattern-based conceptual data modelling. In: Information Modelling and Knowledge Bases, volume XXVI. Frontiers in Artificial Intelligence and Applications, vol. 272, pp. 1–20. IOS Press (2014)
2. Berghammer, R., Thalheim, B.: Methodenbasierte mathematische Modellierung mit Relationenalgebren. In: Wissenschaft und Kunst der Modellierung: Modelle, Modellieren, Modellierung, pp. 67–106. De Gryuter, Boston (2015)
3. Bienemann, A., Schewe, K.-D., Thalheim, B.: Towards a theory of genericity based on government and binding. In: Embley, D.W., Olivé, A., Ram, S. (eds.) ER 2006. LNCS, vol. 4215, pp. 311–324. Springer, Heidelberg (2006). doi:10.1007/11901181_24
4. Bjørner, D.: Domain Engineering. COE Research Monographs, vol. 4. Japan Advanced Institute of Science and Technolgy Press, Ishikawa (2009)
5. Brackett, M.H.: Data Resource Quality. Addison-Wesley, Boston (2000)
6. Embley, D., Thalheim, B. (eds.): The Handbook of Conceptual Modeling: Its Usage and Its Challenges. Springer, Heidelberg (2011)
7. Falkenberg, E.D., Hesse, W., Lindgren, P., Nilsson, B.E., Han, O.J.L., Rolland, C., Stamper, R.K., Van Assche, F.J.M., Verrijn-Stewart, A.A., Voss, K.: A Framework of Information System Concepts, The FRISCO Report (Web Edition). IFIP, ifip@ifip.or.at (1998)
8. Fleming, C.C., von Halle, B.: Handbook of Relational Database Design. Addison-Wesley, Reading, MA (1989)
9. Kaschek, R.: Konzeptionelle Modellierung. Ph.D. thesis, University Klagenfurt, Habilitationsschrift (2003)

10. Kleppe, A., Warmer, J., Bast, W., Explained, M.D.A.: The Model Driven Architecture - Practice and Promise. Addison Wesley, Boston (2006)
11. Krogstie, J.: Model-Based Development and Evolution of Information Systems. Springer, London (2012)
12. Mahr, B.: Information science and the logic of models. Softw. Syst. Model. **8**(3), 365–383 (2009)
13. Murphy, G.L.: The Big Book of Concepts. MIT Press, Cambridge (2001)
14. Simsion, G.: Data Modeling - Theory and Practice. Technics Publications, LLC, New Jersey (2007)
15. Simsion, G., Witt, G.C.: Data Modeling Essentials. Morgan Kaufmann, San Francisco (2005)
16. Stachowiak, H.: Modell. In: Seiffert, H., Radnitzky, G. (eds.) Handlexikon zur Wissenschaftstheorie, pp. 219–222. Deutscher Taschenbuch Verlag GmbH & Co. KG, München (1992)
17. Stahl, T., Völter, M.: Model-Driven Software Architectures. dPunkt, Heidelberg (2005). (in German)
18. Steinmüller, W.: Informationstechnologie und Gesellschaft: Einführung in die Angewandte Informatik. Wissenschaftliche Buchgesellschaft, Darmstadt (1993)
19. Thalheim, B.: Entity-Relationship Modeling - Foundations of Database Technology. Springer, Berlin (2000)
20. Thalheim, B.: Towards a theory of conceptual modelling. J. Univers. Comput. Sci. **16**(20), 3102–3137 (2010). http://www.jucs.org/jucs_16_20/towards_a_theory_of
21. Thalheim, B.: The theory of conceptual models, the theory of conceptual modelling and foundations of conceptual modelling. In: Embley, D., Thalheim, B. (eds.) The Handbook of Conceptual Modeling: Its Usage and Its Challenges, chap. 17, pp. 547–580. Springer, Berlin (2011). doi:10.1007/978-3-642-15865-0_17
22. Thalheim, B.: The art of conceptual modelling. In: Information Modelling and Knowledge Bases XXII. Frontiers in Artificial Intelligence and Applications, vol. 237, pp. 149–168. IOS Press (2012)
23. Thalheim, B.: The conception of the model. In: Abramowicz, W. (ed.) BIS 2013. LNBIP, vol. 157, pp. 113–124. Springer, Heidelberg (2013). doi:10.1007/978-3-642-38366-3_10
24. Thalheim, B.: The conceptual model ≡ an adequate and dependable artifact enhanced by concepts. In: Information Modelling and Knowledge Bases, volume XXV. Frontiers in Artificial Intelligence and Applications, vol. 260, pp. 241–254. IOS Press (2014)
25. Thalheim, B.: Models, to model, and modelling - towards a theory of conceptual models and modelling - towards a notion of the model. Collection of recent papers (2014). http://www.is.informatik.uni-kiel.de/~thalheim/indexkollektionen.htm
26. Thalheim, B.: Conceptual modeling foundations: the notion of a model in conceptual modeling. In: Encyclopedia of Database Systems (2017)
27. Thalheim, B., Nissen, I. (eds.): Wissenschaft und Kunst der Modellierung: Modelle, Modellieren, Modellierung. De Gruyter, Boston (2015)
28. Thalheim, B., Tropmann-Frick, M.: Wherefore models are used and accepted? The model functions as a quality instrument in utilisation scenarios. In: Comyn-Wattiau, I., du Mouza, C., Prat, N. (eds.) Ingenierie Management des Systemes D'Information (2016)
29. Thalheim, B., Tropmann-Frick, M., Ziebermayr, T.: Application of generic workflows for disaster management. In: Information Modelling and Knowledge Bases, volume XXV. Frontiers in Artificial Intelligence and Applications, vol. 260, pp. 64–81. IOS Press (2014)

30. Thomas, M.: Modelle in der Fachsprache der Informatik. Untersuchung von Vorlesungsskripten aus der Kerninformatik. In: DDI. LNI, vol. 22, pp. 99–108. GI (2002)
31. Thomas, O.: Das Modellverständnis in der Wirtschaftsinformatik: Historie, Literaturanalyse und Begriffsexplikation. Technical report Heft 184, Institut für Wirtschaftsinformatik, DFKI, Saarbrücken, Mai (2005)
32. von Dresky, C., Gasser, I., Ortlieb, C.P., Günzel, S.: Mathematische Modellierung: Eine Einführung in zwölf Fallstudien. Vieweg (2009)
33. Wanner, P. (ed.): The Cambridge Encyclopedia of Language. Cambridge University Press, New York (1987)
34. Wenzel, S.: Referenzmodell fr die simulation in produktion und logistik. ASIM Nachr. 4(3), 13–17 (2000)
35. White, R.T.: Commentary: conceptual and conceptional change. Learn. Instr. 4, 117–121 (1994)
36. Whorf, B.L.: Lost generation theories of mind, language, and religion. Popular Culture Association, University Microfilms International, Ann Arbor, Mich (1980)

"Is It a Fleet or a Collection of Ships?": Ontological Anti-patterns in the Modeling of Part-Whole Relations

Tiago Prince Sales[1,2] and Giancarlo Guizzardi[3,4(✉)]

[1] Department of Information Engineering and Computer Science,
University of Trento, Trento, Italy
tiago.princesales@unitn.it
[2] Laboratory for Applied Ontology, ISTC-CNR, Trento, Italy
[3] NEMO Group, Federal University of Espírito Santo, Vitória, Brazil
gguizzardi@unibz.it
[4] Faculty of Computer Science,
Free University of Bozen-Bolzano, Bolzano, Italy

Abstract. Over the years, there is a growing interest in employing theories from philosophical ontology, cognitive science and linguistics to devise theoretical, methodological and computational tools for information systems engineering, in general, and for conceptual modeling, in particular. In this paper, we discuss one particular kind of such tools, namely, ontological anti-patterns. Ontological anti-patterns are error-problem modeling structures that can create a deviation between the possible and the intended interpretations of a model. In this paper, we present two empirically elicited ontological anti-patterns related to the modeling of part-whole relations. In particular, these anti-patterns identify possible mistakes in the modeling of collectives (complex entities that have a uniform role-based structure) and functional complexes (complex entities composed of functional parts). Besides identifying these anti-patterns, the paper presents a series of rectification plans that can be used to eliminate their occurrence in models. Finally, we present a model-based computational tool that supports the automated detection, analysis and elimination of these anti-patterns.

Keywords: Ontology-based conceptual modeling · Anti-patterns · Parthood

1 Introduction

In recent years, there has been an increasing interest in the application of ontologies in conceptual modeling, including the use of foundational ontological theories to improve the theory and practice of this discipline [1, 2]. In these scenarios, ontological theories can play a fundamental role in improving the quality of enterprise-wide conceptual models, improving their quality as artifacts supporting communication, problem-solving, meaning negotiation and, chiefly, semantic interoperability in its various manifestations (e.g., enterprise application integration) [3].

Given the increasing complexity of ontology-driven conceptual modeling, there is an urging need for developing a new generation of complexity management tools for

© Springer International Publishing AG 2017
M. Kirikova et al. (Eds.): ADBIS 2017, LNCS 10509, pp. 28–41, 2017.
DOI: 10.1007/978-3-319-66917-5_3

this discipline [1, 4]. These include a number of methodological and computational tools that are grounded on sound ontological foundations. In particular, as defended in [1], we should advance in these disciplines a well-tested body of knowledge in terms of Ontology Patterns, Ontology Pattern Languages and Ontological Anti-Patterns. This article focuses on the latter.

An anti-pattern is a recurrent error-prone modeling decision [5]. In this paper, we are interested in one specific sort of anti-patterns, namely, model structures that, albeit producing syntactically valid conceptual models, are prone to result in unintended domain representations. In other words, we are interested in configurations that, when used in a model, will typically cause the set of valid (possible) instances of that model to differ from the set of instances representing intended state of affairs in that domain [1, 6]. Such a difference occurs either because the model allows unintended model in-stances or because it forbids intended ones. We name these configurations Onto-logical Anti-Patterns.

In this article, we focus on the study of Ontological Anti-Patterns in a particular conceptual modeling language named OntoUML [7]. OntoUML is a language whose meta-model has been designed to comply with the ontological distinctions and axiomatization of a theoretically well-grounded foundational ontology named UFO (Unified Foundational Ontology) [7, 8]. UFO is an axiomatic formal theory based on theories from Formal Ontology in Philosophy, Philosophical Logics, Cognitive Psychology and Linguistics. OntoUML has been successfully employed in several industrial projects in different domains, such as petroleum and gas, digital journalism, complex digital media management, off-shore software engineering, telecommunications, retail product recommendation, and government [8]. A recent study shows that UFO is the second-most used foundational ontology in conceptual modeling and the one with the fastest adoption rate [2]. Moreover, the study also shows that OntoUML is among the most used languages in ontology-driven conceptual modeling (together with UML, (E)ER, OWL and BPMN).

This article can be seen as complementary to our earlier work in [9, 10]. In [9], we have focused on anti-patterns that are connected to the modeling of material relations (roughly domain associations) and, in [10], on those connected to the modeling of roles. Here, in contrast, we focus on some anti-patterns that emerge when modeling parthood (part-whole) relations. In particular, we focus on anti-patterns that emerge when modelers confuse the ontological unity criteria involved in the modeling of two particular type of parthood relations, namely, the *member of* relation and the *component of* relation.

The contributions of this paper are three-fold. Firstly, we contribute to the identifi-cation of two new Ontological Anti-Patterns for conceptual modeling, in general, and for OntoUML, in particular. Secondly, after precisely characterizing these anti-patterns, we propose a set of refactoring plans that can be adopted to eliminate the possible unin-tended consequences induced by the presence of each of these anti-patterns. Finally, we present an extension for the Menthor Editor[1], an open-source OntoUML model-based editor that: (i) automatically detects anti-patterns in their models; (ii) supports users in

---

[1] https://github.com/menthortools/menthor-editor.

exploring whether the presence of an anti-pattern indeed characterizes a modeling error; (iii) automatically executes refactoring plans to rectify the model.

The remainder of this article is organized as follows: in Sect. 2, we briefly elaborate on the modeling language OntoUML and some of its underlying ontological categories, with a particular focus on the modeling of parthood relations; In Sect. 3, we first briefly present the anti-pattern elicitation method employed here and characterize the model benchmark used in this research; In Sect. 4, we present the newly elicited Ontological Anti-Patterns with their unintended consequences, as well as possible solutions for their rectification in terms of model refactoring plans; Sect. 5 elaborates on the extensions implemented in the OntoUML editor taking into account these anti-patterns; Finally, Sect. 6 presents some final considerations.

2   Ontological and Cognitive Aspects

Parthood is a relation of fundamental importance in conceptual modeling. As such, it is present as a primitive in practically all major conceptual modeling languages and as a micro-theory in all foundational ontologies used in conceptual modeling [7].

Being a cognitively-oriented descriptive ontology, UFO includes micro-theories to address the four kinds of parthood relations generally recognized in Cognitive Science [11, 12], namely, the relations of *subquantity-quantity, subcollective-collective, member-collective* and *component-functional complex*. In UFO, these relations are fully axiomatized and, in their corresponding formal axiomatizations, these relations are characterized w.r.t. formal theories in classical and non-classical mereology [13–17]. In particular, all these relations are shown to conform to the following standard mereological principles: non-reflexivity, asymmetry and the so-called Weak Supplementation Principle (WSP). WSP mandates that if an individual X is part of an individual Y then there must exist at least another individual Z that is mereological disjoint from X and that is also part of Y. In other words, if an individual is mereologically non-atomic (i.e., if it has parts), then it must have at least two disjoint parts.

Being based on UFO, OntoUML has modeling primitives termed *subQuantityOf, subCollectiveOf, memberOf* and *componentOf* representing these four types of parthood relations, respectively. Moreover, it includes in its metamodel formal constraints representing the axiomatization of these relations according to UFO.

The *subquantity-quantity* is focused on modeling parts of an amount of matter (e.g., alcohol-wine, gin-Martini, ice cream-milkshake) and has been discussed in depth in [17]. This paper focuses on part-whole relations of the three remaining kinds, i.e., the ones involving collectives and their parts (*subCollectiveOf* and *memberOf*) the one involving functional complexes and their parts (*componentOf*).

There is an important difference between an ontological account of parthood such as the one included in UFO and classical mereological theories [13], namely, whilst the latter theories are about a binary relation between the part and the whole, the former theories also address the relations that have to hold between the parts in order for them to form a whole. In other words, we must address the question of what kind of *unity principle* binds the parts together such that they can form a particular whole.

An important ontological distinction between collectives and functional complexes is related exactly to the differentiation between the types of unity principles that form these two types of wholes. In the case of collectives, this unity principle is a uniform relationship (i.e., a relation instance) that holds between all parts and only those parts [14, 15]. Because of the uniformity of this relationship, the collective has a uniform structure, i.e., all its *members* are undifferentiated w.r.t. to the whole. In other words, they can be said to play the same role w.r.t. the whole. Take for example collectives such as a forest, a crowd, a pack of lions or a deck of cards with their corresponding instances of the *memberOf* relation (i.e., tree-forest, person-crowd, lion-pack, card-deck). In all of these cases, the wholes have a uniform structure provided by a uniform unity principle (e.g., a crowd is a collective of person all which are positioned in a particular topologically self-connected spatial location) and their parts are all considered to play the same role w.r.t. the whole (e.g., all persons are equally considered to be *membersOf* the crowd). Having uniform criteria regarding their *membership* does not entail that collectives cannot have differentiated *parts*. However, these parts are of a different kind, namely, they further structure collectives in terms of *subcollectives*. For example, the male portion of the crowd and the female portion of the crowd are *subcollectivesOf* the crowd. Likewise, the teenager portion of the crowd is another *subcollectiveOf* the crowd that *mereologically overlaps* with at least one the former sub-collectives.

In contrast to collectives, functional complexes are unified by a *functional architecture* formed by a chain of *functional dependence* relations [16]. In a functional complex, there is a differentiation of the roles played by the different parts. Moreover, by playing these different functional roles, the parts contribute in complementary manners to the functionality of the whole. Take for example functional complexes such as a circulatory system, a car, a computer network, or an organization and their corresponding *componentOf* relations (i.e., heart-circulatory system, engine-car, router-computer network, presidency-organization). In all these examples, the parts play particular functional roles contributing in specific ways to the functionality of the whole (e.g., the heart plays the functional role of pumping blood w.r.t. to the circulatory system such that the circulatory system cannot function as such without having a component play that particular role of blood pump) [16].

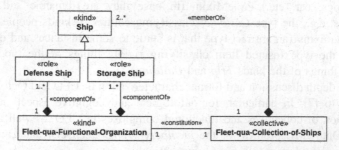

Fig. 1. Fleet as a functional complex *versus* fleet as a collection.

Finally, it is important to highlight that many natural language terms exhibit a case of *systematic polysemy* [18] in referring both to collectives and functional complexes. For example, take the case of a fleet, as discussed by [11, 12]. In the case that all ships of a fleet are conceptualized as playing solely the role of a *memberOf* a fleet, then the term fleet can be said to refer to a collection. In contrast, if a fleet is conceptualized from a functional perspective in which roles are further specialized in *leading ship*, *defense ship*, *storage ship* and so forth, the fleet term refers to a functional complex. In other words, the term fleet seems to refer in a polysemic manner to two different entities: one that is an organizational entity/functional complex that has a functional architecture in which parts play a number of differentiated roles; another that is just a collection of ships. On one hand, these two entities are distinct, following different identity and unity principles (e.g., while replacing an individual ship creates a different *collective of ships* it does not alter the identity of the *fleet-qua-functional-organization*). On the other hand, they bear a particular relation to each other, namely, a relation of *constitution* [7], i.e., the *fleet-qua-functional-organization* is constituted by the *fleet-qua-collection-of-ships*, as depicted in Fig. 1.

Following the ontological distinctions put forth by UFO, OntoUML countenances three different stereotypes that can be applied to types of substantial entities in the domain, depending on the nature of their unity criteria: «kind» for types of functional complexes, «collective» for types of collectives, and «quantity» for types of quantities. The types marked with these three stereotypes represent what the modeler deems to be the *kinds* of entities in the domain (in the ontological sense). As such, these types aggregate essential properties for their instances. For this reason, they are static (i.e., modally rigid types), meaning that they classify their instances in all possible situations. Rigid types that specialize those former three types are stereotyped as «subkind» (e.g., the «subkind» Man specializes the «kind» Person); dynamic types specializing them are either stereotyped as «role», when their dynamic classification condition is a relational one (e.g., student, husband, father), or «phase», in case their dynamic classification condition is an intrinsic one (e.g., teenager, puppy or living person); abstract types (aka *dispersive types* [7]) that classify instances of more than one *kind* (i.e., more than one type stereotyped as «kind», «collective» or «quantity», or any combination of these) are stereotyped as «category» (in case they are rigid abstract types, e.g., the type *Physical Object* rigidly classifying entities of kinds people, buildings, dogs, car, etc.), «roleMixin» (in case they are dynamic and relational abstract types, e.g., the type *Customer* classifying entities of kinds people and organization) or «mixin» (an abstract type that is static to some instances and dynamic to others, e.g., the type Insured Item classifying rigidly things of the type *Car* and dynamically things of the kinds *Trip* and *Building*).

For an in depth discussion and formal characterization of UFO and OntoUML, one should refer to [7]. In particular, for ontological and cognitive aspects and formal characterization of collectives and functional complexes in UFO, as well as the corresponding OntoUML profiles for the *memberOf/subCollectiveOf* and *componentOf* relations, one should refer to [15, 16], respectively.

3 Methods and Materials

Our approach to identify ontological anti-patterns is an empirical qualitative analysis. It starts with the selection of a model for analysis, which is followed by the identification of relevant model fragments for analysis. Such fragments can consist of a whole diagram, a subset of a diagram or even a new "artificial" diagram produced for the sake of analysis (model inspection). Step three is to inspect the selected portion of the model in order to uncover possible problems. We conduct this activity using visual model simulation [1, 6]. This simulation consists in converting OntoUML models into Alloy [19] specifications, generating possible model instances and contrasting these instances with the set of intended instances of the model. The set of intended instances correspond to those that represent intended state of affairs [1, 6] according the creators of the models. Upon the identification of a mismatch, we register it as a potential problem. After detecting a possible problem, we analyze the model in order to identify which structures (i.e., combination of language constructs) caused that problem. In the sequence, we interact with the modelers (when available) or inspect the documentation accompanying the model to define whether the identified structure is indeed problematic. If that is the case, we propose a possible solution to rectify the model and register it as a problem-solution pair. With a modified model, we go back to step three. This iteration is repeated until no more problems can be identified in that fragment and then, another fragment is selected. The analysis stops whenever we inspect all relevant model fragments. After inspecting each model, we analyze the generated problem-solution pairs in order to generalize them into pairs of anti-patterns/refactoring plans.

Our empirical analysis for uncovering anti-patterns was performed using a repository of 54 models[2]. Out of these, 11 models were developed in the context of academic research without industry collaboration. An example is The Configuration Management Task Ontology [20], a product of a Masters dissertation. Furthermore, 7 models had total or partial participation of private companies and/or governmental organizations, the most significant being the MGIC Ontology [21], developed within a re-search project with a regulatory agency responsible for controlling ground transportation services in Brazil.

Concerning the purpose for which the models have been created, the repository contains 10 models (16%) that are intended to serve as a reference domain or core ontologies (e.g. UFO-S [22] for the domain of services). Another 10 models (16%) have been developed in order to perform ontological analysis on existing formalizations, databases or modeling languages. An example is the refactoring of the Conceptual Schema of Human Genome presented in [23]. The repository also contains 8 models (13%) designed for knowledge-based applications, 6 (10%) whose main intention was to support semantic interoperability between systems and/or organizations, and only 2 (3%) for the purpose of enterprise modeling. For the remainder 26 models (42%), there is no information w.r.t. this aspect of classification.

---

[2] The models we used in our research, with an exception of a few (due to non-disclosure agreements), are available at http://www.menthor.net/model-repository.html.

Regarding the modeler's overall expertise in OntoUML, 22 models (41%) have been developed by beginners (18 of these models are also graduate course assignments) and 32 (59%) developed by experienced modelers. Finally, we look into the total number of modelers involved in the model construction. Most models (35 out of 54) were developed individually, whilst 15 were the product of a collaboration between 2–4 people, and 4 involved 7–10 people.

The two anti-patterns discussed in next section appeared in 37,04% of the models with 142 occurrences (*HomoFunc*, see Sect. 4.2) and in 12,96% of the models with 60 occurrences (*HetColl*, see Sect. 4.1).

4 Ontological Anti-patterns

4.1 Heterogeneous Collective (HetColl)

As we discussed in Sect. 2, a collective is an entity whose parts (members) play the same role w.r.t. whole. If we say that a troupe is a collection of artists, we are implying that all artists play merely the generic role of being part of the troupe. Conversely, functional complexes are entities whose parts play different roles w.r.t. whole, thus making different contributions to the behavior of the whole. For instance, the CPU is a functional part of a computer, as well as the hard-drive, since the former is responsible for processing operations, whilst the latter is responsible for storing non-volatile data. As discussed in [7], sometimes, different conceptualizations can articulate a notion in reality as a functional complex or as a collective. As previously mentioned, one conceptualization can articulate a fleet as a functional complex, in which different ships play different functional roles, while another conceptualization can articulate it as merely a collection of ships.

In OntoUML, collectives can be further refined into sub-collections. Again, although defining collective parts of super-collective provides further structure to the whole, it still does not differentiate roles played by their members. For instance, the troupe could be refined into sub-collections of singers, dancers and actors, whose members are the artist who can sing, dance and act, respectively. In this case, the principle unifying these sub-collections is just a strengthen of the common principle that unifies the collection in the first place [15] and, thus, all the members of the collection are still undifferentiated w.r.t. the whole. In other words, the whole regards them merely as *members*.

The Heterogeneous Collective (HetColl) anti-pattern identifies collectives that are connected via membership relation to *members* classified under different types. This can be an indication that the modeler either confused the collection and functional complex interpretations of the same notion or confused the membership and sub-collection relations. By analyzing the models in the OntoUML repository and by discussing their intended semantics with their respective modelers, we have notice that whenever a collective noun (like fleet, group, pack) is used, modelers are most likely to represent it as a collective, without fully analyzing the subtleties of the particular conceptualization at hand.

The key aspect to successfully analyze this anti-pattern is to identify the nature of the unity criteria connecting the parts that form the whole. If one concludes that the parts in fact play different roles w.r.t the whole, the refactoring plan is to change the ontological category of the whole to a functional complex (if necessary, also change the

Table 1. Characterization of the *HetColl* anti-pattern.

Acronym	Name
HetColl	Heterogeneous Collective

Description

A collective connected to multiple types of member through «memberOf» relations suggests a misrepresentation of the concepts w.r.t. to the ontological nature of the whole and the meronymic relations.

Pattern Roles

#	Name	Allowed Sterotypes
1	Whole	«collective», «subkind», «phase», «role», «category», «roleMixin», «mixin»
2..*	$partOf_n$	«memberOf»
2..*	$Part_n$	«kind», «collective», «subkind», «phase», «role», «category», roleMixin», «mixin»

Additional Constraints

The *Whole* should be: (i) stereotyped as collective; (ii) stereotyped as subkind, role or phase and be a direct or indirect subtype of another type stereotyped as collective; or (iii) be stereotyped as mixin, category or roleMixin and have all its direct or indirect subtypes meeting conditions (i) or (ii)

Generic Example

Refactoring Plans

1. Change parts to functional parts: Change the nature of *Whole* to functional complex and change the stereotype of every $partOf_n$ to «componentOf».

2. Change parts to sub-collections: Change the stereotype of every $partOf_n$ to «subCollectionOf». If a $Part_n$ is not a collection, this should be rectified in the model.

3. Set a generic membership: Create *GeneralPart*, a common direct supertype of every $Part_n$; remove every $partOf_n$ from the model; and create *generalPartOf*, a new «memberOf» from *Whole* to *GeneralPart*. The stereotype ascribed to *GeneralPart* can be derived from the stereotype of $Part_n$.

ontological category of parts) and then to change the stereotype of the meronymic relations to *componentOf*, instead of *memberOf*. Alternatively, if one concludes that the members indeed play the same role regarding the whole, we propose to make this position explicit by creating a type as the direct parent of all current part types and then merge all *memberOf* relations into a new that is connected to this newly created supertype. Yet a third alternative is presented when the modeler concludes that the types representing the members are in fact sub-collections, i.e., they are refinements of internal structure of the collective whole [15]. To rectify this case, one must change the stereotypes of the *memberOf* relations to *subCollectionOf* and, if necessary, adjust the ontological category of the parts to be that of a collection (Table 1).

Note that in both the first and third refactoring plans described, whenever there is a need to ontological category of either a part type or a whole type, the following strategy can be adopted. If (1) a type A is erroneously stereotyped as a «kind» or a «quantity», then A should be represented as a «collective». However, if (2) A is stereotyped as «subkind», «role» or «phase» and (directly or indirectly) specializes a «kind» or a «quantity» type S, one can: (2.1) change the stereotype of S to «collective»; (2.2) select another «collective» in the model and make A specialize it; or (2.3) create a new «collective» be the supertype of A. Finally, if (3) P is stereotyped as a «category», «mixin» or «roleMixin», strategies (1) and (2) should be adopted for every subtype of P.

An example of the occurrence of *HetColl* is depicted in Fig. 2 below. This model is an adaptation of a fragment extracted from a governmental conceptual model in the domain of agricultural protection. The fragment describes a particular type of work group, named Technical Administrative Support Group, which has employees that play the roles of technical and/or administrative support. The cardinalities constraints defined in the association ends of parts show that this type of work group requires employees performing different duties. The requirement of the presence of people playing these different roles indicates that the work group should be really modeled a functional complex instead of a collective. As it is typical in these cases, there is another implicit entity, namely, the staff of the work group, which at each point in time constitutes the Work Group as a functional complex. However, the Work Group itself and its staff are ontologically distinct entities as these are associated with different identity criteria (e.g., while changing a member of the staff creates a different staff, it can still be the same Work Group, just then constituted by a different staff) [7].

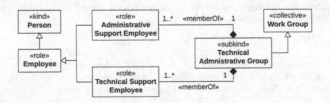

Fig. 2. HetColl occurrence in a fragment of a governmental model.

4.2 Homogeneous Functional Complex (HomoFunc)

The Homogeneous Functional Complex (*HomoFunc*) anti-pattern is the counterpart of the *HetColl* anti-pattern. As discussed in Sect. 2, functional complexes have heterogeneous structures, such that its parts play different functional roles w.r.t. the whole. Therefore, when a modeler describes a functional complex, it is usually expected that she would represent multiple *componentOf* relations connected to such type to account for the diversity of such functional roles.

In OntoUML, a type represents a functional complex when: (i) it is stereotyped as a «kind»; (ii) it is a subtype of kind (i.e., as a «subkind», «role» or «phase»); (iii) it represents a non-sortal type (i.e., «category», «mixin» or «roleMixin») and all its (direct or indirect) sortal subtypes satisfy conditions (i) or (ii). An occurrence of the *HomoFunc* anti-pattern is observed in a model when there is a model fragment representing a homogenous structure of a functional complex, i.e., there is a «kind» type connected through a single *componentOf* relation to one single type of part.

In our empirical investigation, we observed that a common reason for an occurrence of *HomoFunc* is when a modeler mistakenly represents a functional complex while actually intending to represent a collective. This can happen because alternative conceptualizations can ascribe different interpretations for the same term in the domain (see discussion on Sect. 4.1). For this case, we propose the following refactoring plan (see Table 2): one should transform the functional parthood relation at hand (a *componentOf*) in a membership relation (a *memberOf*). Then, one should change the ontological category of the whole type to that of a collection.

In a second situation, we have that a modeler actually intended to represent a heterogeneous structure for the whole. In this case, the modeler should refine the model to include additional types of part. This can be accomplished in two different albeit non-exclusive ways. First, through the specification of new types of functional parts, i.e., types that bear no taxonomic relations to the type already present in the model representing the part (see refactoring plan 2 in Table 2). Second, through the creation of subtypes of the single functional part, alongside with the additional corresponding *componentOf* relations (see refactoring plan 3 in Table 2).

In Table 2, the constraint number 2 for the characterization of this anti-pattern expresses that the parthood relation represented should satisfy the weak supplementation axiom (one of the most fundamental axioms of parthood, see [13–17]). Otherwise, the situation would indicate simply an incomplete model and would not exemplify an occurrence of this anti-pattern. Constraint number 3, instead, exclude the situation in which role differentiation is guarantee by additional parthood relations inherited from a possible supertype of the type representing the whole.

An example of the occurrence of *HomoFunc* is depicted in Fig. 3. The model fragment is extracted from the PAS 77 ontology [25], a model in the domain of IT architecture. Notice that the IT Architecture type is defined as being solely composed of IT Components, which in turn can be sites, platforms, operating systems and data storage units. In its original form, the model suggests that all architectural parts are equal w.r.t. the IT Architecture and, hence, that an IT architecture is simply a collection of IT components. If this is the intended semantics, a more suitable formalization would be to represent IT Architecture as a *collective* connected to its parts by a *memberOf* relation

Table 2. Characterization of the *HomoFunc* anti-pattern.

Acronym	Name
HomoFunc	Homogeneous Functional Complex

Description

A functional complex connected to a single part through a «componentOf» relation suggests that all instances of the part play the same role w.r.t. their whole, a homogeneous structure that does not characterize a functional complex.

Pattern Roles

#	Name	Allowed Sterotypes
1	whole	«kind», «subkind», «phase», «role», «category», «roleMixin», «mixin»
1	part	«kind», «subkind», «phase», «role», «category», «roleMixin», «mixin»
1	partOf	«componentOf»

Additional Constraints

1. Both *Whole* and *Part* should be: (i) stereotyped as kind; (ii) stereotyped as subkind, role or phase and be a direct or indirect subtype of another type stereotyped as kind; or (iii) be stereotyped as mixin, category or roleMixin and have all its direct or indirect children meeting conditions (i) or (ii)

2. The lower bound multiplicity of *partOf*'s association end connected to *Part* is greater than one

3. *Whole* is not a subtype of another class who has other types of part

Generic Example

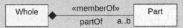

Refactoring Plans

1. **Change to membership:** change the nature of *Whole* from a functional complex to a collection and change the stereotype of *partOf* from «componentOf» to «memberOf»

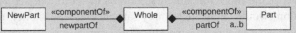

2. **Add functional parts:** specify additional functional parts for *Whole*

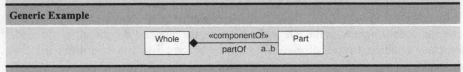

3. **Add subtypes for *Part*:** specify subtypes for *Part* and connect them to *Whole* through additional «componentOf» relations. If the original *partOf* relation is kept in the model, the added relations must subset, redefine or specialize it (see discussion in [24]).

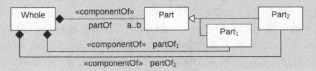

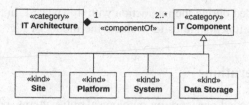

Fig. 3. HomoFunc occurrence in a fragment of an IT infrastructure model.

(as described by the first refactoring plan on Table 2). Conversely, if this would not the intended semantics, the model would be more accurate if the specific parthood relations between IT Architecture and its different functional components were made explicit (following a the third refactoring plan proposed in Table 2).

5 Tool Support

The Menthor Editor, formerly known as OntoUML Lightweight Editor (OLED), is an open-source ontology-driven conceptual modeling environment. A full support for anti-pattern management has been implemented in this editor. Following the strategy adopted in [9], these anti-pattern management functionalities include anti-pattern detection, analysis (via a wizard-based feature) and elimination (using the rectification plans proposed here). In other words, by employing the explicitly defined MOF metamodel on which this editor is based, we have: firstly, implemented algorithms to automatically detect anti-pattern occurrences, accessible through a detection dialog window (see example in the left part of Fig. 3); in sequence, based on our pre-defined solutions (rectification plans), we implemented wizards to interact with users to support anti-pattern analysis (the right part of Fig. 3 depicts a wizard for the HetColl anti-pattern); finally, we implemented algorithms to automatically rectify the model

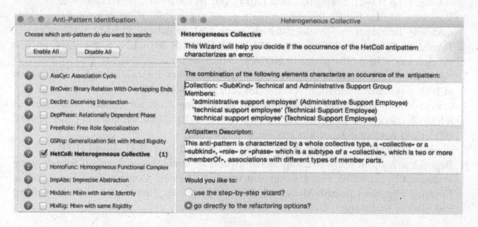

Fig. 4. Tool support for anti-pattern management.

using the input provided during the interaction with this wizard. In the Fig. 4, we have used these automated functionalities to evaluate the model of Fig. 2.

6 Final Considerations

In this paper, we extended our work on ontological anti-patterns, proposing three new error-prone structures in combination with pre-defined rectification solutions. In particular, we focused on anti-patterns related to the modeling of parthood (collectives and functional complexes) in conceptual modeling. Parthood is of fundamental importance in conceptual modeling, in general, and in areas such as Enterprise Modeling, Economy and Finance, Manufacturing, Life Sciences, among others, in particular. For this reason, the identification of these anti-patterns and their associated rectification plans as well as their automation in a model-based computational tool constitutes important contributions to the theory and practice of these disciplines.

We emphasize that it is not among our goals in this paper to defend particular modeling choices for specific concepts such as fleet, IT architecture or Work Group. In other words, we have no stand here in whether, in general, concepts such as these are better represented as *functional complexes* or *collectives*. The adequacy of a representation choice over another (or even both choices used simultaneously, following a modeling pattern such as the one of Fig. 1) depends exclusively on the rationale of a particular model and its underlying conceptualization. Nonetheless, the anti-patterns proposed in this paper are able to identify situations in which modelers recurrently make mistaken representation choices in that respect, i.e., choosing to use a collection to represent what is a functional complex in the domain, or vice-versa.

References

1. Guizzardi, G.: Ontological patterns, anti-patterns and pattern languages for next-generation conceptual modeling. In: Yu, E., Dobbie, G., Jarke, M., Purao, S. (eds.) ER 2014. LNCS, vol. 8824, pp. 13–27. Springer, Cham (2014). doi:10.1007/978-3-319-12206-9_2
2. Verdonck, M., Gailly, F.: Insights on the use and application of ontology and conceptual modeling languages in ontology-driven conceptual modeling. In: Comyn-Wattiau, I., Tanaka, K., Song, I.-Y., Yamamoto, S., Saeki, M. (eds.) ER 2016. LNCS, vol. 9974, pp. 83–97. Springer, Cham (2016). doi:10.1007/978-3-319-46397-1_7
3. Nardi, J.C., Falbo, R.A., Almeida, J.P.A.: Foundational ontologies for semantic integration in EAI: a systematic literature. Review. I3E(2013), 238–249 (2014)
4. Guizzardi, G.: Theoretical foundations and engineering tools for building ontologies as reference conceptual models. Semant. Web J. **1**, 3–10 (2010). Editors-in-Chief: Pascal Hitzler and Krzysztof Janowicz, IOS Press, Amsterdam
5. Koenig, A.: Patterns and antipatterns. J. Object-Oriented Prog. **8**(1), 46–48 (1995)
6. Benevides, A.B., et al.: Validating modal aspects of OntoUML conceptual models using automatically generated visual world structures. J. Univers. Comput. Sci. Special Issue Evolving Theories Concept. Model. **16**, 2904–2933 (2010). Editors: Klaus-Dieter Schewe and Markus Kirchberg

7. Guizzardi, G.: Ontological Foundations for Structural Conceptual Modeling. Telematics Institute Fundamental Research Series, Enschede, The Netherlands (2005)
8. Guizzardi, G., et al.: Towards ontological foundation for conceptual modeling: the unified foundational ontology (UFO) story. Appl. Ontol. **10** (2015). IOS Press
9. Sales, T.P., Guizzardi, G.: Ontological anti-patterns: Empirically uncovered error-prone structures in ontology-driven conceptual models. DKE **99**, 72–104 (2015)
10. Sales, T.P., Guizzardi, G.: Anti-patterns in ontology-driven conceptual modeling: the case of role modeling in OntoUML. In: Gangemi, A., Hizler, P., Janowicz, K., Krisnadhi, A., Presutti, V. (eds.) Ontology Engineering with Ontology Design Patterns: Foundations and Applications. IOS Press, The Netherlands (2016)
11. Pribbenow, S.: Meronymic Relationships: From Classical Mereology to Complex Part-Whole Relations, The Semantics of Relationships. Kluwer Academic Publishers, Dordrecht (2002)
12. Gerstl, P., Pribbenow, S.: Midwinters, end games, and bodyparts: A classification of part-whole relations. Int. J. Hum.-Comput. Stud. **43**, 865–889 (1995)
13. Varzi, A.C.: Parts, wholes, and part-whole relations: the prospects of mereotopology. J. Data Knowl. Eng. **20**, 259–286 (1996)
14. Simons, P.M.: Parts: An Essay in Ontology. Clarendon Press, Oxford (1987)
15. Guizzardi, G.: Ontological foundations for conceptual part-whole relations: the case of collectives and their parts. In: 23rd International Conference on Advanced Information System Engineering (CAiSE 2011), London, UK (2011)
16. Guizzardi, G.: The problem of transitivity of part-whole relations in conceptual modeling revisited. In: 21st International Conference on Advanced Information Systems Engineering (CAISE 2009), Amsterdam, The Netherlands (2009)
17. Guizzardi, G.: On the representation of quantities and their parts in conceptual modeling. In: Proceedings of FOIS 2010. IOS Press, Toronto (2010)
18. Ravin, Y., Leacock, C.: Polysemy: Theoretical and Computational Approaches, p. 240. Oxford University Press, USA (2002)
19. Jackson, D.: Software Abstractions: Logic, Language, and Analysis. MIT press, Cambridge (2012)
20. Calhau, R.F., Falbo, R.A.: A configuration management task ontology for semantic integration. In: Proceedings of the 27th Symposium on Applied Computing, SAC 2012, pp. 348–353. ACM, New York (2012)
21. Bastos, C.A.M., et al.: Building up a model for management information and knowledge: the case-study for a Brazilian regulatory agency. In: International Workshop on Software Knowledge (SKY) (2011)
22. Nardi, J.C., et al.: Towards a commitment-based reference ontology for services. In: Proceedings of the 17th International Enterprise Distributed Object Computing Conference (EDOC 2013), pp. 175–184. IEEE (2013)
23. Ferrandis, A.M.M., et al.: Applying the principles of an ontology-based approach to a conceptual schema of human genome. In: Proceedings of ER 2013, Hong Kong
24. Costal, D., et al.: Formal semantics and ontological analysis for understanding subsetting, specialization and redefinition of associations in UML. In: 30th International Conference on Conceptual Modeling (ER 2011), Brussels, Belgium (2011)
25. e Silva, H.C., Cassia Cordeiro de Castro, R., Gomes, M.J.N., Garcia, A.S.: Well-founded IT architecture ontology: an approach from a service continuity perspective. In: Benlamri, R. (ed.) NDT 2012. CCIS, vol. 294, pp. 136–150. Springer, Heidelberg (2012). doi:10.1007/978-3-642-30567-2_12

Context-Aware Decision Information Packages: An Approach to Human-Centric Smart Factories

Eva Hoos[1,3(✉)], Pascal Hirmer[2], and Bernhard Mitschang[2,3]

[1] Daimler AG, 71034 Böblingen, Germany
Eva.Hoos@daimler.com
[2] Institute of Parallel and Distributed Systems, University of Stuttgart,
70569 Stuttgart, Germany
{Pascal.Hirmer,Bernhard.Mitschang}@ipvs.uni-stuttgart.de
[3] Graduate School of Excellence Advanced Manufacturing Engineering,
University of Stuttgart, 70569 Stuttgart, Germany

Abstract. Industry 4.0 enables the integration of new trends, such as data-intensive cyber physical systems, Internet of Things, or mobile applications, into production environments. Although it concentrates on highly data-intensive automated engineering and manufacturing processing, the human actor is still important for decision making in the product lifecycle process. To support correct and efficient decision making, human actors have to be provided with relevant data depending on the current context. This data needs to be retrieved from distributed sources like bill of material systems, product data management and manufacturing execution systems, holding product model and factory model. In this paper, we address this issue by introducing the concept of decision information packages, which enable to compose relevant engineering data for a specific context from distributed data sources. To determine relevant data, we specify a context-aware engineering data model and corresponding operators. To realize our approach, we provide an architecture and a prototypical implementation based on requirements of a real case scenario.

Keywords: Industry 4.0 · Context-awareness · Data provisioning · Smart factory

1 Introduction

Industry 4.0 is a new trend that drives the digitization of production environments. Especially data-driven cyber physical systems [14] and Internet of Things [27] enable new approaches such as self-organization of production processes. However, human interaction and decision making is still mandatory and beneficial in the smart factory [8,30]. This especially holds in the domain of pre-production plants that manufacture the first prototypes of a product. Hence, lots of failures may occur that have to be quickly resolved by human workers. An efficient decision making process to find appropriate solutions to these failures has to consider different kinds of data. Examples are 3D-geometry data or a bill of material (BOM) representing the product structure, simulation data, process data, or measurement data of the exact dimensions of the product. These data are distributed onto

© Springer International Publishing AG 2017
M. Kirikova et al. (Eds.): ADBIS 2017, LNCS 10509, pp. 42–56, 2017.
DOI: 10.1007/978-3-319-66917-5_4

heterogeneous IT systems that are not integrated yet [15, 23, 26]. At the moment, workers have to manually collect all data relevant for decision making. We call this set of relevant data *decision information packages* (DIP). The composition of relevant data is a cumbersome task, since workers are trained to solve domain-specific problems, and not to find relevant data that requires comprehensive IT skills. To address this issue, we provide an approach to automatically compose DIPs in a compact manner in order to relief the decision maker from browsing complex IT systems and to redirect his focus on domain-specific problem-solving. We augment and go beyond existing approaches in providing relevant data via context-aware filtering [2, 4, 6] and data integration in engineering [15, 26] in order to fulfill domain-specific requirements. Key requirements are generation of DIPs without comprehensive IT knowledge and integration of context-aware filtering into the existing engineering IT landscape. This paper is the realization of the vision introduced in [12]. Detailed contributions of the paper are:

(1) Definition of DIP Structure: We performed a case study in the pre-production plant to identify requirements regarding DIPs. The IT landscape is composed of heterogeneous IT systems. Each of them represents a specific view of product data. To combine these views and assign the appropriate data to it, we design a generic schema for DIPs reflecting the product structure as well. The focus on the product structure is important so that workers can easily interpret relevant data. The generic schema serves as basis for the virtual integration facilitating decision making processes by integrating data from different sources.

(2) Context-Aware Composition of DIPs: Since there is a large amount of engineering data, it is important for an efficient decision making process that the amount of data is reduced and only relevant data is composed. Context data can be used to filter meaningful data. In the pre-production plant, for example, the context of shop floor workers is dependent on their current task, location and of the state of the environment, which often can, e.g., be captured by sensors [11]. Hence, context data has to be linked to engineering data to derive meaningful data for decision making. This should be possible without comprehensive IT knowledge. To address these issues, we develop a context-aware engineering data model and operators to process the data. They abstract from technical details of IT systems and context data, which facilitates the linking by domain experts.

(3) Architectural Realization for the Engineering Domain: The architecture to automatically compose and acquire DIPs needs to be integrated into the existing engineering IT landscape which is composed of legacy systems and dynamically appearing IT systems. Therefore, we define a system architecture that realizes our approach. The architecture serves as basis for our prototypical implementation, which is evaluated based on our real-world case scenario.

This paper is structured as follows: Sect. 2 introduces the use case scenario and defines the structure of DIPs. Section 3 contains the main contribution of our paper: an approach to compose DIPs. Section 4 introduces a system architecture, which serves as basis for the evaluation of our approach conducted in Sect. 5. Finally, Sect. 6 covers related work and Sect. 7 summarizes the paper.

2   Decision Information Packages in Engineering

This section introduces our case study at a German car manufacturer, which emphasizes the demand and utility of DIPs. Based on this, we derive a generic schema for DIPs in the engineering domain.

2.1   Case Study: Pre-production Plant

The case study is part of the manufacturing of car prototypes, also known as *pre-production test*. During the production in a preproduction plant, the cars pass several assembly stations of a production line. At each station, a shop floor worker assembles multiple parts. Since the product and the process are not as well defined as in series production, plenty of failures may occur. For example, parts cannot be assembled because the tolerances do not match. In the following, we describe a case scenario, in which the part "console" does not fit into the apparatus of the *front-end assembly station*, that assembles the front section of the car shell: The worker at the front-end assembly station recognizes the problem and needs to resolve it. There could be plenty of causes for this problem. In the following and due to space reason, we only look into three representative kinds of information to identify an error:

(1) Basic information about the part is required to identify name, version and material. These are stored in the *bill of material* (BOM) system, which contains information about all parts necessary to manufacture the product [9].
(2) The visualization of the 3D-Geometry model is necessary to check the correctness of the part's geometry. The worker has to access the *product data management system* (PDM) to find the appropriate 3D geometry file with respect to different versions and variants [22].
(3) Measurement reports are necessary to check tolerances. Measurement reports are stored in a file system in PDF format and contain the exact dimensions of an assembly.

The corresponding DIP is shown in Fig. 1 and discussed in the next subsection.

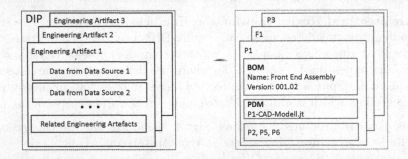

Fig. 1. left: Generic Schema of the DIP, right: DIP of the Case Study

2.2   Decision Information Packages

DIPs provide required information for decision making. They compose data from multiple, heterogeneous data sources with respect to the context of the problem. Thereby, different views of the data should be combined. To enhance the comprehensibility of the worker, DIPs should reflect the product structure. The product structure defines the relation between parts. Accordingly, we use engineering artifacts as structuring elements. Engineering artifacts virtually represent all parts and components required to build a product, also including manufacturing parts. They have a unique identifier already used in product model and factory model. Figure 1 shows an abstract model of a DIP. A DIP encapsulates data of multiple *engineering artifacts*. For each engineering artifact, information from different data sources are composed. The data format of the data sources varies from relational data and XML data to file-based data. Also, related engineering artifacts are provided. For example, related engineering artifacts of a part can be subcomponents or machines at which the part is manufactured.

In the depicted example, on the right of Fig. 1, the DIP contains information about the engineering artifacts *P1*, *F1* and the *P3*. These are the unique identifiers. The name of the part P1 and the current version number are acquired from the BOM. From the PDM system, the file containing the 3D model is included, called P1-CAD-Modell.jt, which is necessary to visualize the part. *P1* is related to *P2*, *P5*, and *P6*, which are subcomponents.

3   Context-Aware Composition of Decision Information Packages

This section presents (i) the context-aware engineering data model (CAEM) to link engineering data and context data, and (ii) context-aware operators to compose the DIPs based on this model.

3.1   Context-Aware Engineering Data Model

In order to link engineering data and context data in the CAEM, we introduce a context model which structures the context data and enables to define situations of users. We use the definition of context models as introduced in [13]. The context model is based on *context entities* and *context attributes*, characterizing the entities. Furthermore, it contains relations between entities. Figure 2 shows a simple context model for the engineering domain. This simple model is sufficient since the focus of the paper is not to provide a comprehensive context model but rather to link context with engineering data. Context entities are *station*, *actor*, and *project*. Actor is characterized by its context attributes name, role, and task. Each new product, e.g., a new car model, is defined as *development project*, which is divided into various phases such as pre-production test batches. The values of the attributes define the state of an entity. Situations derived from this context model can describe at which station an actors works, and also in

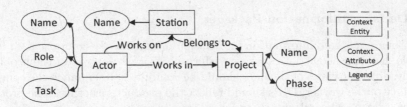

Fig. 2. Our context model

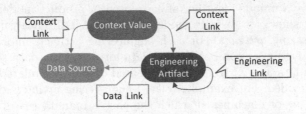

Fig. 3. Context-aware engineering model

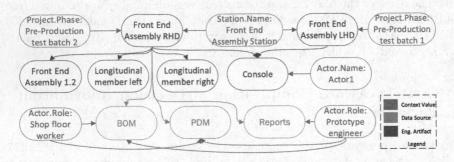

Fig. 4. Excerpt of a CAEM instance for the use case scenario

what project. An exemplary situation for our case scenario is: (i) the actor has the role *prototype engineer*, (ii) the phase is the *pre-production test batch 2*, and (iii) the station is the *front-end assembly station*.

The **context-aware engineering data model** is shown in Fig. 3. It models context values, engineering artifacts and data sources as entities as well as links between these entities. Note that the context value is the value of the context attribute defined in Fig. 2, such as that "front-end-assembly station" is the value of the context attribute "station name". *Data sources* represent an abstraction of data sources, in which engineering data are stored, e.g., the bill of material. Engineering links describe the semantic relation between engineering artifacts, such as "is part of" or "is manufactured by". Data links describe in which data sources information about the engineering artifact is stored. Context links describe in which context the engineering artifact is relevant. They can be established between context elements and data sources.

Figure 4 shows an excerpt of a CAEM instance according to our use case scenario. The semantics of the excerpt is as follows: (i) at the station *front-end assembly*, the part *front-end* is manufactured, (ii) the prototype engineer is interested in information from the data sources: BOM, PDM and reports, whereas the shop floor worker is only interested in BOM and PDM data. (iii) The front-end assembly station manufactures the right-hand-drive and the left-hand drive. (iv) At the pre-production test batch 2 (second test batch) only the right-hand-drive variants are manufactured.

3.2   Operators for Decision Information Package Composition

The composition of DIPs is supported by four operators which interact with the CAEM. In order to define operators, we interpret the CAEM as set of items and define set-operators for it. According to our context-aware engineering model shown in Fig. 3, we define three sets:

$$EngineeringArtifacts : EA = \{ea_1, ea_2, ..., ea_n\}$$
$$ContextValues : CV = \{cv_1, cv_2, ..., cv_m\}$$
$$DataSources : DS = \{ds_1, ds_2, ..., ds_p\}$$

We model the links of the CAEM as relations between target and source artifacts.

$$ContextEALinks : CEA \subseteq \{(cv_i, ea_j)|cv_i \in CV, ea_j \in EA\}$$
$$ContextDSLinks : CDS \subseteq \{(cv_i, ds_j)|cv_i \in CV, ds_j \in DS\}$$
$$DataLinks : DL \subseteq \{(ea_i, ds_j)|ea_i \in EA \land ds_j \in DS\}$$
$$EngineeringLinks : EL \subseteq \{(ea_i, ea_j)|i \neq j \land ea_i, ea_j \in EA\}$$

We define a particular situation of a user X as $SIT_x = \{cv_{x1}, cv_{x2}, ...\}$.

Engineering Artifact Selection: The operator *SelectEA* selects all engineering artifacts which are relevant for the given situation SIT_x:

$$SelectEA(SIT_x) = \bigcap_{cv_{xj} \in SIT_x} \{ea_i|(cv_{xj}, ea_i) \in CEA\} \tag{1}$$

Selection of relevant data sources: The operator *SelectDS* filters the relevant data sources for a particular situation SIT_x:

$$SelectDS(SIT_x) = \{ds_i|(cv_{xj}, ds_i) \in CDS \land cv_{xj} \in SIT_x\} \tag{2}$$

Discovery of data sources: The operator *DiscoverDS* identifies all data sources which provide information about the relevant engineering artifact *ea*:

$$DiscoverDS(ea) = \{ds_i|(ea, ds_i) \in DL\} \tag{3}$$

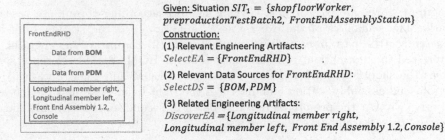

Given: Situation $SIT_1 = \{shopfloorWorker,$
$preproductionTestBatch2, FrontEndAssemblyStation\}$

Construction:

(1) Relevant Engineering Artifacts:
$SelectEA = \{FrontEndRHD\}$

(2) Relevant Data Sources for $FrontEndRHD$:
$SelectDS = \{BOM, PDM\}$

(3) Related Engineering Artifacts:
$DiscoverEA = \{Longitudinal\ member\ right,$
$Longitudinal\ member\ left,\ Front\ End\ Assembly\ 1.2, Console\}$

Fig. 5. Construction of a DIP using the operators with respect to the use case scenario

Discovery of related engineering artifacts: The operator $DiscoverEA$ discovers engineering artifacts that are related to a particular other engineering artifact:

$$DiscoverEA(ea_j) = \{ea_i|(ea_i, ea_j) \in EL \vee (ea_j, ea_i) \in EL\} \qquad (4)$$

To compose a DIP for a given situation SIT_x, $SelectEA$ creates the list of engineering artifacts relevant for the situation. This operator uses an intersection to be more restrictive and reduces the number of engineering artifacts more than using union. Afterwards, the data sources are determined using $SelectDS$ or $DiscoverDS$. If $SelectDS$ has an empty set as result, which means that no context value is assigned to a data source, $DiscoverDS$ explore the data sources. Finally, $DiscoverEA$ finds the related engineering artifacts for each relevant engineering artifact defined by $SelectEA$. Figure 5 visualizes a DIP and how to construct it using the operators. Furthermore, it shows the result for applying the operators to the context model.

Note that all these operators are set-oriented, because the result of an operation execution may end in a set of selected or discovered EA and DS. Furthermore, the operators hide from e.g., connectivity, data source types and access mechanisms to the data sources. Hence, the realization approach of the operators has to cope with this abstract operator level and provide a transformation to the underlying IT environment, e.g., to the sources of the systems. All these issues are covered in the next chapter.

4 Architecture to Provide Decision Information Packages

In order to realize our approach, we introduce the system architecture depicted in Fig. 6. The architecture enables to integrate the concept into the engineering environment. The environment is characterized by the users and their applications and the engineering context, which comprises all data in the engineering domain such as data stored in IT systems and context data such as sensor data, machine data, and user data. The architecture consists of three components, namely *Context-Aware Provisioning*, *Resource Access Platform* and *Context Management*.

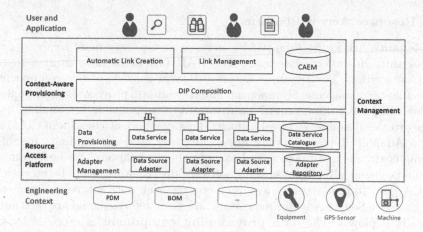

Fig. 6. System architecture to provision DIPs

Environmental context is gathered by the **Context Management**, which processes the low-level engineering context into higher-level context, also known as situations. For example, it derives given GPS coordinates of a user into the situation that the user "is in the manufacturing plant". Note that we do not focus on the Context Management and assume that there is an appropriate implementation available such as the ones introduced in [11,29].

4.1 Context-Aware Provisioning

The context-aware provisioning layer consists of a database managing context-aware engineering data model (CAEM) and of the subcomponents *DIP Composition*, *Link Management* and *Automatic Link Creation*.

In order to realize the **CAEM**, we use an entity-relation model to implement it into a relational database. The **link management** provides an API to create and delete links as well as to store the engineering artifacts, context elements, and data sources in the CAEM. This simplifies the definition of context links and engineering links for domain experts. The **automatic link creation** facilitates the creation of context links, which is a high effort task when conducted manually. It enables to create a set-based definition of links, for example, if the engineering artifacts have common attribute values. Second, it allows to import relations from other systems. For example, sometimes the information which part is manufactured, and on which station, is stored in another IT system. Finally, it is possible to integrate learning mechanisms so that the domain expert gets proposals for context links if relations with similar engineering artifacts exist. The **DIP composition** component gathers all the information required to compose a DIP as shown in Fig. 5. It receives the situation from the context management. According to the situation determined by the context values, the DIPs are constructed using the operators.

4.2   Resource Access Platform

The Resource Access Platform (RAP) serves as single entry point to access data sources through uniform interfaces. The RAP provisions data sources to the DIP Composition. The Resource Access Platform consists of two components: (i) the adapter management component, and (ii) the data provisioning component.

The **adapter management component** is responsible for binding data sources to the Resource Access Platform. This component consists of the Data Source Adapter Repository that stores adapters used for binding, and of a runtime environment to deploy and execute them. Adapters can be automatically and dynamically deployed using software provisioning technologies such as TOSCA [19]. To bind data sources, adapters for each data source are extracted from the Data Source Adapter Repository, are parameterized, and are then automatically deployed. The **data provisioning component** is accessed by the DIP Composition and contains the *Data Service Catalog* that provides meta-information about all available data sources. Furthermore, it contains *Data Services* that offer access to the actual data. First, the RAP is accessed by the DIP Composition in order to search for relevant data sources found in the CAEM in the Data Service Catalog. The RAP then provides references to corresponding Data Services that encapsulate access to underlying data using proper adapters.

The Resource Access Platform is based on the Resource Management Platform as introduced by Hirmer et al. [10]. We extend this platform to support specific needs of the engineering domain. By doing so, we support data sources of the engineering domain, e.g., CAD models, sensor data, or simulation data. In addition, we automate the lookup in the Data Service Catalog, which is previously done manually. The interface to applications accessing the RMP (e.g., the DIP Composition) remains the same.

5   Evaluation

We evaluate our approach according to the goal of DIPs which is to improve decision making on the shopfloor. With our approach, we enable to automatically compose and provide DIPs in the engineering domain. To evaluate this, we created a proof of concept implementation. Furthermore, we aim to reduce the information according to the context of the users in order to provide compact DIPs. This is important in order to relief the worker from meaningless information. Therefore, we applied our approach on a real data set of the introduced case scenario and investigate the reduction of information.

5.1   Proof-of-Concept Implementation

We implemented a prototype as proof-of-concept for our approach. The architecture depicted in Fig. 6 serves as basis for this implementation. As user interface, we implemented a mobile app using HTML5, CSS, and JavaScript and offers functionality to provide DIPS to the user. The implementation of the

context-aware provisioning is as follows: The **CAEM** itself is implemented as a relation model used as schema for an SQLite[1] database, which can be accessed through a Java-based interface connecting to the database with the corresponding SQLite driver. The Java interface is hosted on an Apache Tomcat application server. For the implementation of the **Context Management**, we used the existing tool SitOPT [11,29], which provides an interface to register on situations that are derived based on context data. As mentioned above, the **Resource Access Platform** consists of two components. For the implementation of the adapter layer, we use MongoDB[2] as adapter repository as well as an Apache Tomcat Java runtime environment for the adapters. Hence, all adapters are implemented in Java, exclusively. For the deployment of adapters, we use the TOSCA runtime OpenTOSCA [3]. To implement the data provisioning layer, we use Java REST services as well as a Java-based implementation of the Data Service Catalog, which stores its data into a MySQL database.

5.2 Case-Oriented Evaluation of Information Reduction

In order to evaluate how well context can be used to filter relevant data, we apply our approach on a real data set and determine the information reduction. The data set originates from our case study introduced in Sect. 2. We look into data to assemble the bottom of the car shell. Table 1 shows the number of items in the CAEM. Engineering artifacts are the parts of the bottom of the car shell. The context values are belonging to the entities station, actor, and project. The data sources are BOM, PDM, and measurement reports. Engineering links describe the *part-of* relation of the parts. In order to evaluate how well context can be used to filter data, we investigate the compactness of the DIPs. The compactness of DIPs is dependent on the *number of engineering artifacts* and *the data size of the information packages* of engineering artifacts. We assume, the smaller the DIPs are, the more effective the context-aware filtering.

Reduction of Number of Engineering Artifacts. We analyze the reduction of engineering artifacts per DIPs using different kinds of situations. A situation consists of values from the names of actors and stations as well as from the phase

Table 1. CAEM data

Data type	Number
EngineeringArtifacts	785
ContextAttributes values	88
ContextLinks	5261
EngineeringLinks	964
DataLink	2355

---

[1] https://sqlite.org/.
[2] http://mongodb.com/.

Table 2. Reduction of engineering artifacts

Station	Phase	Actor	Number of DIPs	Average number of EA per DIP
Yes	No	No	34	3,44
No	No	Yes	41	9,41
No	Yes	No	13	366
Yes	No	Yes	46	2,3
Yes	Yes	No	383	1,17
No	Yes	Yes	453	6,5
Yes	No	Yes	46	2,3
Yes	Yes	Yes	332	1,17

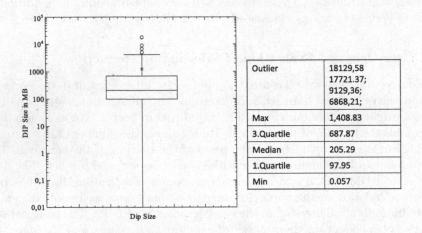

Fig. 7. Boxplot of DIP size

of projects. We have calculated 1348 DIPs and count the number of engineering artifacts per DIP. The results are shown in Table 2. The first three columns indicate which context attributes are used to define a situation. We build every valid situation according to the context attributes. Column *number of DIPs* reflects only DIPs for valid situations. The next column reports the *average number of EAs per DIP*. Without our approach, in the worse case a human worker would have to examine all 785 engineering artifacts. The results show that with two different kinds of context attributes, we get comparatively small DIPs, which contain between 1 and 6 engineering artifacts in average. We conclude that using at least two different kinds of context values reduces the number of engineering artifacts by a factor of 100.

Reduction of the DIP size. We also investigate the DIP sizes resulting from the considered situations. Therefore, we calculate the DIP size for each possible situation based on two or three different kinds of context attributes. In our case, the critical file size is the one of the geometry model, which is required to visualize the 3D-Model of the car. We neglect data from the BOM since they are only key-value pairs and the measurement reports, since their size is constant. Figure 7 shows the results via a box-plot. Despite few huge outliers, such as a DIP with size of 18.12 GB, over half of the DIPs range between 97 MB and 687 MB as well as the size of DIPs.

The evaluation of the reduction shows that the size of DIPs can be reduced significantly by filtering it via the context. This enables a more efficient decision making process because the worker does not have to browse meaningless data. Station, actor and phase are appropriate types of context, since they reduce the number of EAs in DIPs drastically and, thus, influence the size of DIPs.

6 Related Work

The DIP approach can be seen as virtual integration. Yet, in contrast to federated database systems [21] and recent ontology-based integration approaches [18, 28], DIPs do not require a complex mediated schema nor schema matching process, which is required to integrate domain experts without IT knowledge into the integration process. With respect to related work, we differentiate four groups according to automatic context-aware provisioning of DIPs. The first group reviews approaches to abstract data sources for automatic data provisioning. The second group comprises approaches in the engineering domain that acquire engineering data in the client without filtering relevant data. The third and fourth group focus on filtering relevant data either on data level or on application level.

A lot of middleware platforms have been developed that abstract from accessing data sources [1,7,16], belonging to the first group. However, none of the existing approaches provide the functionality needed to provision DIPs. First of all, dynamic environments of the engineering domain need to be handled with frequently (dis-)appearing data sources. In this paper, we introduce the RAP to bind these data sources dynamically through adapters that are deployed using TOSCA. Furthermore, the RAP is generic, i.e., it supports all kinds of sources assuming that a corresponding adapter exists for them. Consequently, the RAP fits our approach very well and was integrated instead of other approaches.

Approaches belonging to the second group, provide data acquisition from multiple engineering IT systems. Katzenbach et al. introduce a common engineering client, where data are provisioned by an engineering service bus [15]. Similarly, the authors of [26] suggest a system-level integration using standards and harmonized human interfaces. However, none of them consider filtering relevant data.

The third group integrates context into data sources. Martinenghi and Torlone [17] and Roussous et al. [20] extend the relational data model with context and define an appropriate query language. Stavrakas et al. [24] integrate context into XML by developing a multidimensional semi-structured data model and a query language to process them [25]. Since all approaches integrate context into an existing data model, there is no generally defined context model involved. Furthermore, the databases have to be adapted to integrate the context in contrast to our approach which integrate context virtually. The fourth group addresses linking context and data on the application level. Bobillo et al. [4] develop a model to manage context-relevant knowledge in ontologies. This is based on the domain ontology of the knowledge-based system and on a context ontology. Barkat et al. [2] define a context ontology to integrate context into the semantic databases, called OntoDB. Hence, their approach is restricted to applications using exactly this database. Bolchini et al. [5] introduce a context ontology based on a self-developed context model to define the portion of the ontology which are relevant. Similar to this, they provide a method to define context-aware views for relational databases [6]. Hence, many related approaches try to achieve similar goals regarding context-aware filtering of relevant data using ontology models. In our approach, we decided to omit the use of ontologies to reduce the complexity. Most advantages of ontologies come with reasoning and linking to other ontologies. For our approach, a simple meta-model is sufficient.

7   Summary

In this paper, we introduce an approach to compose and provide *decision information packages* (DIPs) to support problem resolving in the smart factory. We introduce DIPs on the basis of a real use case at a German car manufacturer. DIPs are constructed using a meta-model and corresponding operators. The meta-model links engineering data to context data. The operators process this model to find data relevant for specific situations, e.g., of a shop floor worker being faced with issues assembling a specific part. To realize this approach and to integrate it into the engineering domain, we introduce a system architecture. The approach is evaluated through a prototypical implementation to demonstrate its feasibility and a case-oriented evaluation using real data to highlight the compactness of DIPs. Our approach leads to a significant information reduction.

References

1. Barbosa, A.C.P., Porto, F.A.A., Melo, R.N.: Configurable data integration middleware system. J. Braz. Comput. Soc. **8**, 12–19 (2002)
2. Barkat, O., Bellatreche, L.: Linking context to ontologies. In: 2015 11th International Conference on Semantics, Knowledge and Grids (SKG) (2015)
3. Binz, T., Breitenbücher, U., Haupt, F., Kopp, O., Leymann, F., Nowak, A., Wagner, S.: OpenTOSCA – a runtime for TOSCA-based cloud applications. In: Basu, S., Pautasso, C., Zhang, L., Fu, X. (eds.) ICSOC 2013. LNCS, vol. 8274, pp. 692–695. Springer, Heidelberg (2013). doi:10.1007/978-3-642-45005-1_62
4. Bobillo, F., Delgado, M., Gómez-Romero, J.: Representation of context-dependant knowledge in ontologies: a model and an application. Expert Syst. Appl. **35**, 1899–1908 (2008)
5. Bolchini, C., Curino, C., Schreiber, F.A., Tanca, L.: Context integration for mobile data tailoring. In: 7th International Conference on Mobile Data Management (MDM 2006) (2006)

6. Bolchini, C., Quintarelli, E., Schreiber, F.A., Baldassarre, M.T.: Context-aware knowledge querying in a networked enterprise. In: Anastasi, G., Bellini, E., Nitto, E., Ghezzi, C., Tanca, L., Zimeo, E. (eds.) Methodologies and Technologies for Networked Enterprises. LNCS, vol. 7200, pp. 237–258. Springer, Heidelberg (2012). doi:10.1007/978-3-642-31739-2_12

7. Grant, A., Antonioletti, M., Hume, A.C., et al.: OGSA-DAI: middleware for data integration: selected applications. In: 2008 IEEE Fourth International Conference on eScience, eScience 2008 (2008)

8. Gröger, C., Kassner, L., Hoos, E., Königsberger, J., Kiefer, C., Silcher, S., Mitschang, B.: The data-driven factory - leveraging big industrial data for agile, learning and human-centric manufacturing. In: ICEIS 2016 (2016)

9. Hegge, H., Wortmann, J.C.: Generic bill-of-material: a new product model. Int. J. Prod. Econ. **23**, 117–128 (1991)

10. Hirmer, P., Wieland, M., Breitenbücher, U., Mitschang, B.: Automated sensor registration, binding and sensor data provisioning. In: Proceedings of the CAiSE 2016 Forum (2016)

11. Hirmer, P., Wieland, M., Schwarz, H., Mitschang, B., Breitenbücher, U., Sáez, S.G., Leymann, F.: Situation recognition and handling based on executing situation templates and situation-aware workflows. Computing **99**, 163–181 (2016)

12. Hoos, E., Hirmer, P., Mitschang, B.: Improving problem resolving on the shop floor by context-aware decision information packages. In: Proceedings of the CAiSE 2017 Forum (2017)

13. Hoos, E., Wieland, M., Mitschang, B.: Analysis method for conceptual context modeling applied in production environments. In: Abramowicz, W. (ed.) BIS 2017. LNBIP, vol. 288, pp. 313–325. Springer, Cham (2017). doi:10.1007/978-3-319-59336-4_22

14. Jazdi, N.: Cyber physical systems in the context of Industry 4.0. In: 2014 IEEE International Conference on Automation, Quality and Testing, Robotics (2014)

15. Katzenbach, A.: Automotive. In: Stjepandić, J., Wognum, N., Verhagen W.J.C. (eds.) Concurrent Engineering in the 21st Century, pp. 607–638. Springer, Cham (2015). doi:10.1007/978-3-319-13776-6_21

16. Langegger, A., Wöß, W., Blöchl, M.: A semantic web middleware for virtual data integration on the web. In: Bechhofer, S., Hauswirth, M., Hoffmann, J., Koubarakis, M. (eds.) ESWC 2008. LNCS, vol. 5021, pp. 493–507. Springer, Heidelberg (2008). doi:10.1007/978-3-540-68234-9_37

17. Martinenghi, D., Torlone, R.: Querying context-aware databases. In: Andreasen, T., Yager, R.R., Bulskov, H., Christiansen, H., Larsen, H.L. (eds.) FQAS 2009. LNCS (LNAI), vol. 5822, pp. 76–87. Springer, Heidelberg (2009). doi:10.1007/978-3-642-04957-6_7

18. Noy, N.F.: Semantic integration. ACM. SIGMOD Rec. **33**, 65–70 (2004)

19. OASIS: Topology and orchestration specification for cloud applications

20. Roussos, Y., Stavrakas, Y., Pavlaki, V.: Towards a context-aware relational model. In: International Workshop on Context Representation and Reasoning (2005)

21. Sheth, A.P., Larson, J.A.: Federated database systems for managing distributed, heterogeneous, and autonomous databases. ACM Comput. Surv. **22**, 183–236 (1990)

22. Stark, J.: Product lifecycle management. In: Product Lifecycle Management (Volume 2). Decision Engineering, pp. 1–35. Springer, Cham (2015). doi:10.1007/978-3-319-24436-5_1

23. Stark, R., Hayka, H., Israel, J.H., Kim, M., Müller, P., Völlinger, U.: Virtuelle Produktentstehung in der Automobilindustrie. Informatik-Spektrum (2011)

24. Stavrakas, Y., Gergatsoulis, M.: Multidimensional semistructured data: representing context-dependent information on the web. In: Pidduck, A.B., Ozsu, M.T., Mylopoulos, J., Woo, C.C. (eds.) CAiSE 2002. LNCS, vol. 2348, pp. 183–199. Springer, Heidelberg (2002). doi:10.1007/3-540-47961-9_15

25. Stavrakas, Y., Pristouris, K., Efandis, A., Sellis, T.: Implementing a query language for context-dependent semistructured data. In: Benczúr, A., Demetrovics, J., Gottlob, G. (eds.) ADBIS 2004. LNCS, vol. 3255, pp. 173–188. Springer, Heidelberg (2004). doi:10.1007/978-3-540-30204-9_12

26. Trippner, D., Rude, S., Schreiber, A.: Challenges to digital product and process development systems at BMW. In: Stjepandić J., Wognum N., Verhagen W.J.C. (eds.) Concurrent Engineering in the 21st Century. Springer, Cham, pp. 555–569 (2015). doi:10.1007/978-3-319-13776-6_19

27. Vermesan, O., Friess, P.: Internet of Things: Converging Technologies for Smart Environments and Integrated Ecosystems. River Publishers, Alsbjergvej (2013)

28. Wauer, M., Meinecke, J., Schuster, D., Konzag, A., Aleksy, M., Riedel, T.: Semantic federation of product information from structured and unstructured sources. In: Web-Based Multimedia Advancements in Data Communications and Networking Technologies. IGI Global (2013)

29. Wieland, M., Schwarz, H., Breitenbücher, U., Leymann, F.: Towards Situation-Aware Adaptive Workflows. In: PerCom (2015)

30. Zuehlke, D.: SmartFactory-towards a factory-of-things. Annu. Rev. Control **34**, 129–138 (2010)

Subsequence Matching
and Streaming Data

Fast Subsequence Matching in Motion Capture Data

Jan Sedmidubsky[1](✉), Pavel Zezula[1], and Jan Svec[2]

[1] Masaryk University, Brno, Czech Republic
xsedmid@fi.muni.cz
[2] University of West Bohemia, Pilsen, Czech Republic

Abstract. Motion capture data digitally represent human movements by sequences of body configurations in time. Subsequence matching in such spatio-temporal data is difficult as query-relevant motions can vary in lengths and occur arbitrarily in a very long motion. To deal with these problems, we propose a new subsequence matching approach which (1) partitions both short query and long data motion into fixed-size segments that overlap only partly, (2) uses an effective similarity measure to efficiently retrieve data segments that are the most similar to query segments, and (3) localizes the most query-relevant subsequences within extended and merged retrieved segments in a four-step postprocessing phase. The whole retrieval process is effective and fast in comparison with related work. A real-life 68-minute data motion can be searched in about 1 s with the average precision of 87.98% for 5-NN queries.

1 Introduction

Motion capturing technologies can accurately record human movements at high spatial and temporal resolutions. The recorded motion sequence (or simply *motion*) is represented as an ordered sequence of *poses* that describe skeleton configurations in corresponding video frames. Note that terms "frame" and "pose" are sometimes used interchangeably when referring to a given time moment. Each pose is represented by a set of 3D coordinates determining positions of captured body *joints* in space. An example of a single pose with 31 joints is visualized in Fig. 1a.

Recorded spatio-temporal motions are analyzed in a variety of applications, e.g., in security to monitor and detect suspicious events from surveillance cameras or in computer animation to search large databases of human motions for production of realistically looking games or movies. These applications require efficient subsequence matching: Given a short *query motion* and a long *data motion*, search the data motion and locate its subsequences that are the most similar to the query motion. For example, find occurrences of acrobatic elements within a 5-minute dancing performance. Locating such query-relevant subsequences constitutes a hard task since their exact lengths and starting/ending positions are not known in advance. Even when textual annotations are available, they cannot be employed because the query need not correspond to any semantic action. To deal with these problems, a general subsequence matching approach for spatio-temporal motion data is needed.

© Springer International Publishing AG 2017
M. Kirikova et al. (Eds.): ADBIS 2017, LNCS 10509, pp. 59–72, 2017.
DOI: 10.1007/978-3-319-66917-5_5

2 Related Work

Subsequence matching methods for motion capture data generally require a (1) segmentation technique to partition a data motion into meaningfully-long data segments, (2) similarity measure to compare query and data segments, and (3) matching algorithm to efficiently localize query-similar subsequences by grouping the most relevant data segments.

Segmentation. A segmentation technique partitions the data motion into short segments that are directly comparable with segments of the query motion. The segmentation can be done by determining non-overlapping segments that correspond to repetitive movements [17] or predefined actions (e.g., walking, kicking and jumping) [10]. However, these techniques are not suitable for general subsequence retrieval because they do not cope well with queries that do not belong to any predefined action class. Moreover, query-relevant subsequences that occur on the boundaries of non-overlapping segments are difficult to be identified. To guarantee findability of any data subsequence, both query and data motions are partitioned into short segments of a fixed-size. The query can be partitioned into overlapping segments using the sliding-window principle and the data motion into disjoint (i.e., non-overlapping) segments to reduce the data replication, or vice versa [5]. In both cases, query segments need to be matched with data segments respecting their temporal order.

Similarity Measure. Feature extraction is applied to spatio-temporal motion data to discover key motion characteristics in a low dimensional feature space. Features can be either carefully chosen [10] or learned automatically on the domain of motion data using Neural Networks [3,18] or Support Vector Machines [6]. Although machine-learning approaches are generally very effective in classification, they can hardly be employed to calculate similarity in the task of subsequence matching. There are some exceptions [13,18] that additionally extract discriminative features for similarity comparison.

A variety of measures exists to determine pair-wise similarity of motion features, such as joint-angle rotations [15] or distances between joints [8]. Motion features are usually compared by temporal alignment techniques, e.g., by the Dynamic Time Warping and its variants [1,10,12]. Although such techniques deal well with temporal discrepancies, such as faster and slower movements of otherwise same actions, they have quadratic time complexity. To decrease complexity, fixed-size features are extracted and efficiently compared, e.g., 4,096D motion features by the Euclidean distance [13], 160-bit features by the Hamming distance [18], and 282D feature vectors by a weighted metric function [4].

Matching Algorithm. Having the data motion partitioned, a retrieval algorithm is used to locate data segments that are similar to query segments. To speed-up similarity searching within a large amount of data segments, scalable index structures can be utilized. For example, the iSAX2+ index is employed to search large collections of time series [2], a trie-based structure is traversed to identify query-relevant parts in the data motion [7], or the M-Index is used to efficiently search for similar key poses that determine relevant data segments [15]. To decrease search costs, the query motion is not partitioned and directly compared against data segments that are organized in multiple levels [13]. Although this approach does not require postprocessing of retrieved segments in temporal order, it restricts the query motion in its size.

Our Contributions. In this paper, we propose a new subsequence matching approach for searching for subsequences that are similar to an arbitrary query, i.e., the query need not correspond to any predefined action. Even if the data-motion segmentation is inspired by traditional techniques [5], it generates much lower number of segments, compared to [5,13]. Both query and data segments are efficiently compared on the basis of the Euclidean distance, in contrast to the expensive Dynamic Time Warping used in [1,10]. We do not additionally limit the query in its size, in contrast to [13]. The suitability of the proposed approach is also demonstrated by comparing the results with related work from both accuracy and performance points of view.

3   Subsequence Matching

We propose a subsequence matching approach that inspects a long data motion and finds its query-relevant subsequences. In particular, we fix the segment size and partition the short query motion into a small number of disjoint (consecutive non-overlapping) segments, while the long data motion into a moderate number of segments that overlap only partly. Starting positions of adjacent data segments can differ by a large margin (e.g., about 30–50% with respect to the segment size) due to the ability of a neural-network similarity measure to effectively compare shifted motions. This measure uses the Euclidean distance on $4,096$D feature vectors that are extracted from segments using a deep convolutional neural network. Since the Euclidean distance can be indexed, query-relevant data segments are efficiently retrieved. In an advanced postprocessing phase, the retrieved segments are extended and merged to determine relevant data parts. Such parts are then inspected by another joint-coordinate similarity measure to identify starting and ending positions of the most similar candidate motions.

3.1   Problem Definition

Motion sequence (or simply motion) $M = (P_1, \ldots, P_n)$ is defined as a sequence of poses P_i ($i \in [1, n]$), where $n = |M|$ equals to the motion length (i.e., the number of captured poses/frames). Each pose $P_i = (C_1, \ldots, C_{31})$ consists of 3D real-world coordinates $C_i \in \mathbb{R}^3$ ($i \in [1, 31]$) that correspond to specific body joints visualized in Fig. 1a. Subsequence $M[i^s : i^e]$ is again a motion starting at the i^s-th frame (inclusive) and ending at the i^e-th frame (exclusive) within motion M, i.e., $|M[i^s : i^e]| = i^e - i^s$.

The subsequence matching problem is to inspect long *data motion* D and find its $k \in \mathbb{N}$ subsequences $D_1[i_1^s : i_1^e], \ldots, D_k[i_k^s : i_k^e]$ that are the most similar

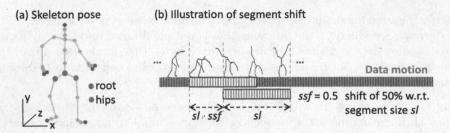

Fig. 1. (a) Pose captured at a given frame. (b) A relative shift of 50% of two consecutive data segments having the fixed size of sl frames.

Table 1. Table of symbols.

Symbol	Description
M	Motion $M = (P_1, \dots, P_n)$ defined as a sequence of poses P_i ($i \in [1, n]$)
$|M|$	Motion length corresponding to the number of poses (frames)
$M[i_s, i_e]$	Subsequence of motion M starting at the i_s-th pose (inclusive) and ending at the i_e-th pose (exclusive), i.e., $|M[i_s : i_e]| = i_e - i_s$
$\overrightarrow{F^M}$	4,096-dimensional feature vector that compactly describes short motion M
D	Long data motion (e.g., taking several hours)
Q	Short query motion (e.g., taking several seconds)
n^D/n^Q	Number of generated data/query segments
sl	Segment length in number of poses (user-defined parameter)
ssf	Segment shift factor ($ssf \in (0, 1]$) that determines a relative shift between consecutive data segments (user-defined parameter)
k/k'	Number of the most similar subsequences retrieved for a k-NN query/number of the most similar data segments retrieved for each query segment

to typically-short *query motion* Q. Since there can be a number of similar subsequences that heavily overlap, only the most similar and non-overlapping ones are returned as the query result. The notation along with user-defined parameters is summarized in Table 1.

3.2 Similarity of Motions

We calculate similarity between relatively short motions taking approximately several seconds, e.g., between query and data segments. Similarity is determined on the basis of differences between spatio-temporal joint trajectories of compared motions. Inspired by related papers [13, 16], we preprocess original 3D joint coordinates by normalizing the actor's position, orientation and body size. In particular, original 3D joint locations are translated to the body coordinate system with its origin on the "root" joint (red joint in Fig. 1a), followed by a 3D rotation to fix the x-axis parallel to the 3D line from "right hip" to "left hip". In the last step of normalization, all the 3D coordinates are scaled so that the normalized skeleton has the same proportions – sizes of bones – as a standardized (average) actor. Such normalization is supposed to be suitable for searching for general actions (e.g., jumping and walking) which we evaluate in the experiments. We define the two following distance measures to calculate similarity of normalized motions.

1. **Joint-Coordinate Distance Measure** $sim^{\mathbf{JC}}$. This measure evaluates differences between normalized positions of corresponding joints in each pose. The differences are calculated by the Euclidean distance and summed together to determine similarity of the same-length motions M_1 and M_2 (i.e., $|M_1| = |M_2|$) as:

$$sim^{\mathbf{JC}}(M_1, M_2) = \sum_{i=1}^{|M_1|} \sum_{j=1}^{31} \left\| C_j^{P_i^{M_1}} - C_j^{P_i^{M_2}} \right\|, \tag{1}$$

where $C_j^{P_i^{M_1}} / C_j^{P_i^{M_2}}$ is 3D position of j-th joint in i-th pose of motion M_1/M_2.

2. **Neural-Network Distance Measure** $sim^{\mathbf{NN}}$. Firstly, normalized motion M is transformed into a visual image representation [14]. The image has the size of $31 \times |M|$ pixels, where rows correspond to individual joints while columns to poses. Color of image pixel $[x, y]$ ($x \in [1, 31], y \in [1, |M|]$) approximates the normalized 3D position of the x-th joint in the y-th frame within a discrete RGB color space of 256^3 bins. The generated image is then processed by a reference model[1] of a deep convolutional neural network to discover inherent visual patterns in diversely colorful motion images. The output of the last hidden layer of the network is a $4,096$-dimensional feature vector $\overrightarrow{F^M}$ that describes motion M compactly. More details about feature extraction can be found in [14]. The feature vector has the fixed size even for motions of a variable length. Feature vectors $\overrightarrow{F^{M_1}}$ and $\overrightarrow{F^{M_2}}$ extracted for motions M_1 and M_2 are compared by the Euclidean distance to determine their similarity as:

$$sim^{\mathbf{NN}}(\overrightarrow{F^{M_1}}, \overrightarrow{F^{M_2}}) = \sqrt{\sum_{i=1}^{4,096} \left(\overrightarrow{F^{M_1}}[i] - \overrightarrow{F^{M_2}}[i] \right)^2}. \tag{2}$$

Utilization of Distance Measures. Although both distance measures calculate motion similarity based on normalized joint coordinates, the neural-network measure is more effective due to its ability to learn diverse motion patterns. Moreover, it is still effective even when a certain amount of noise is added to compared motions. This property is utilized to significantly decrease the number of generated data segments by shifting them about a much longer distance in comparison with 1-frame shift. On the other hand, there is a need to train a neural-network model based on training motion data and mainly to extract a feature vector for each motion, which is quite an expensive operation taking approximately 25 ms using a single GPU card. Although we spend some time to extract features of all data segments in a preprocessing phase, during query

---

[1] https://github.com/BVLC/caffe/tree/master/models/bvlc_reference_caffenet.

processing the number of feature extractions is limited only to a small number of query segments and identified query-relevant motion candidates.

The joint-coordinate measure does not require an expensive feature-extraction process because it calculates similarity directly on normalized motion data. Although it is not so effective as the sim^{NN} measure, it is sufficient to be utilized in the postprocessing step where starting/ending positions of query-relevant candidate motions are needed to be refined in real time. This includes calculating similarity between a query and motion parts that are not known in advance.

3.3 Offline Segmentation and Indexing of Data Motion

The neural-network distance measure is proposed to calculate similarity of mean-ingfully long motions, taking approximately several seconds. Since a searched data motion can be very long (e.g., taking several hours), it must be partitioned into short segments that are comparable by this distance measure. In this paper, we use a fixed length of segments which is quantified by *segment length* para-meter $sl \in \mathbb{N}^+$. An appropriate setting of this parameter primarily depends on a target application and future queries.

Inspired by Faloutsos et al. [5], we partition a query motion into *disjoint* (i.e., consecutive non-overlapping) segments, while the data motion into *overlapping* segments. Due to the "noise tolerance" property of the neural-network distance measure, data segments can be shifted by a large margin. We quantify a rel-ative shift of consecutive data segments by *segment shift factor* $ssf \in (0,1]$. The absolute shift (in number of poses) between two consecutive data segments is calculated with respect to segment length sl as $ssf \cdot sl$. In the experiments, we show that relative shifts of 30–50% contribute to high search effectiveness, while generating orders of magnitude fewer data segments compared to the tra-ditional 1-pose shift. The smaller number of data segments is generated, the faster retrieval is achieved. Figure 1b illustrates a shift of 50% (i.e., $ssf = 0.5$) which results in involving each data-motion pose within two data segments – it corresponds only to a twofold data replication.

Based on the absolute shift of $sl \cdot ssf$ poses between consecutive data seg-ments, data motion D is segmented into the total number of n^D segments:

$$n^D = \left\lceil \frac{|D| - sl}{sl \cdot ssf} \right\rceil. \tag{3}$$

Individual overlapping data segments $D_1, \ldots, D_{n^D}$ are cut from the data motion, where the position of j-th segment D_j ($j \in [1, n^D]$) is defined as:

$$D_j \left[i_s^j : \min \left\{ i_s^j + sl, |D| + 1 \right\} \right] \qquad i_s^j = (j-1) \cdot sl \cdot ssf + 1. \tag{4}$$

All the data segments have a fixed length of sl poses. The only exception is the last segment D_{n^D} that can be shortened maximally by $ssf \cdot 100\%$ with respect to the segment length.

Each data segment D_j is then processed by the reference model of the deep convolutional neural network to extract its $4,096$-dimensional feature vector.

Indexing Data Segments. The extracted feature vectors of data segments can be optionally indexed to speedup the retrieval process. As the feature vectors are compared by the Euclidean distance, any high-dimensional- or metric-based index structure can be utilized. We confront the naive sequential scan with the usage of index structure in the experiments.

3.4   Query Processing

The retrieval process evaluates a *k-nearest neighbor* (*k*-NN) query that inspects long data motion D and localizes its k subsequences that are the most similar to short query motion Q. In particular, the query motion is partitioned into disjoint (consecutive non-overlapping) segments which are independently used to retrieve similar data segments by the effective neural-network distance measure. The retrieved segments are extended and merged together to determine potentially relevant parts within the long data motion. Each relevant part is then explored to localize its best subsequence match, so-called candidate, by the efficient joint-coordinate distance measure. The localized candidates are finally ranked by the neural-network distance and the k most relevant ones are returned as the query result. The whole search process is schematically illustrated in Fig. 2 and described in the following steps in more detail.

Online Query Segmentation. Query Q is partitioned into n^Q ($n^Q = \lceil |Q|/sl \rceil$) disjoint segments $Q_1, \ldots, Q_{n^Q}$, each of them having the length of sl poses. The exception is the last n^Q-th segment that can be shorter – it has $|Q| - (n^Q - 1) \cdot sl$ poses. The position of the j-th segment Q_j ($j \in [1, n^Q]$) is defined as:

$$Q_j[i_s^j : \min\{i_s^j + sl, |Q| + 1\}] \qquad i_s^j = (j - 1) \cdot sl + 1. \tag{5}$$

All the query segments are then processed by the neural network to extract their 4,096-dimensional feature vectors $\overrightarrow{F^{Q_1}}, \ldots, \overrightarrow{F^{Q_{n^Q}}}$.

Retrieval. Feature vector $\overrightarrow{F^{Q_j}}$ of each query segment Q_j is used as a query object of k'-NN($\overrightarrow{F^{Q_j}}$) *sub-query* that independently searches by the sim^{NN} measure for the $k' \in \mathbb{N}$ most similar data segments among all features $\overrightarrow{F^{D_1}}, \ldots, \overrightarrow{F^{D_{n^D}}}$. Constant k' is a user-defined parameter and its appropriate setting is evaluated within the experiments. Similar data segments can be retrieved efficiently if an index structure on feature vectors of data segments is constructed.

Postprocessing. In total, $n^Q \cdot k'$ data segments are retrieved by all k'-NN sub-queries. The retrieved segments are postprocessed in the following steps.

(a) *Extension of data segments* – retrieved data segments are extended to the same length as the query motion by considering their surrounding poses in

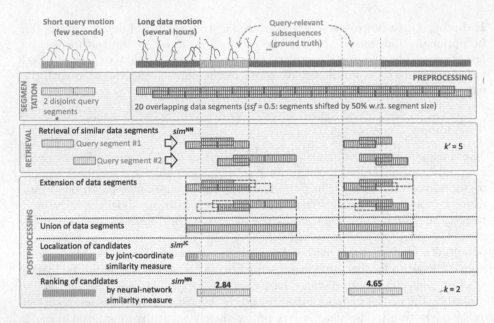

Fig. 2. Illustration of the retrieval process that inspects a data motion of 20 overlapping segments to obtain the two most similar subsequences (i.e., two ranked candidates) with respect to a query consisting of 2 disjoint segments. Although the retrieved candidates are not perfectly aligned with the actual query-relevant subsequences, they overlap with them in the majority of frames.

the data motion. The same number of poses surrounding a given query segment on its left/right side within the query motion is considered to extend the left/right side of data segments retrieved to that query segment. Formally, the data segment which is retrieved as similar to query segment Q_j ($j \in [1, n^Q]$), is extended by adding $(j-1) \cdot sl$ of its preceding poses and $(|Q| - j \cdot sl)$ of its succeeding poses, i.e., data segment $D[i_s, i_e]$ is extended to $D[i_s - (j-1) \cdot sl, i_e + |Q| - j \cdot sl]$. Such extensions can better mark relevant parts within the data motion.

(b) *Union of data segments* – extended data segments that mutually overlap are merged together to identify non-overlapping *relevant data parts*. Two segments $D[i_s, i_e]$ and $D'[i'_s, i'_e]$ overlap if and only if $\min\{i_e, i'_e\} \leq \max\{i_s, i'_s\}$.

(c) *Localization of candidates* – each relevant data part is processed to identify its best subsequence match, so-called *candidate*, having the same length while being the most similar with respect to the query motion. Merged data part $D[i_s, i_e]$ (longer than the query motion) is scanned by a sliding window of the same length as the query motion to determine the best candidate match. Specifically, all possible windows $D[i'_s : i'_s + |Q|]$ ($i'_s \in [i_s, i_e - |Q|]$) are compared by the sim^{JC} measure against the query motion and the most similar one is considered as the candidate. Since there can be dozens of windows, we use the fast joint-coordinate distance measure that does not involve an expensive feature-extraction step.

(d) *Ranking of candidates* – identified candidates are finally ranked by calculating the sim^{NN} distance between the $4,096$-dimensional feature vector of the query motion and the feature vectors extracted for these candidates. The k most ranked candidates are finally returned as the result of k-NN query Q.

Main bottlenecks of query processing are (1) retrieval of the most similar data segments to query segments and (2) the extraction of feature vectors for query segments and identified candidates. Since the extraction of $4,096$D features is expensive, we limit the number of extractions by considering a reasonable number of candidates – extraction times take several hundreds of milliseconds on average for each query, as demonstrated in the experiments.

4   Experimental Evaluation

We experimentally evaluate both effectiveness and efficiency of the proposed subsequence matching approach and compare the results against existing solutions. We also make a study on a large-scale scenario to enable subsequence matching potentially in an 87-day motion in couple of seconds.

4.1   Dataset

Both effectiveness and efficiency are evaluated on the motion capture dataset HDM05 [9]. This dataset contains 324 motions performed by 5 different actors (with sampling frequency of 120 Hz). Similarly as in [10,13,15], we use a subset of 102 motions (68 min in total) for which a ground truth is provided. This ground truth annotates $1,464$ subsequences within the 102 motions. The annotated subsequences are divided into 15 non-uniformly populated categories that correspond to exercise actions, such as moving, jumping or punching. The average action takes 2.36 s (283 frames) while the shortest and longest ones have 0.34 s (41 frames) and 17.2 s ($2,063$ frames), respectively. Without loss of generality, we concatenate the 102 motions into a single 68-minute data motion and keep a track of motion boundaries in a supplemental data structure.

4.2   Methodology

Based on settings of parameters of segment shift ssf and segment length sl, the 68-minute data motion is preprocessed to determine a particular number of data segments to be generated. These segments are then processed by the reference model of neural network (see Sect. 3.2) to extract their $4,096$-dimensional feature vectors. The network model is trained by fine-tuning the original reference model with another subset of the HDM05 dataset that classifies $2,328$ short motions into 122 categories (see [14] for more details).

The proposed approach is analyzed by evaluating k-nearest-neighbor (k-NN) queries. As the sparsest ground-truth category contains only 6 motion instances,

k is fixed to 5. A 5-NN query is constructed for each of $1,464$ ground-truth samples, that are used as query objects. The retrieved subsequences that overlap with the query-object sample (i.e., exact matches) are excluded. Due to unionizing data segments within the query postprocessing step, only the non-overlapping subsequences are returned as the query result.

Search effectiveness is measured for each query by the *precision* as the fraction of true-positive retrieved subsequences. A subsequence is true positive if it overlaps with some ground-truth sample that is labeled with the same category as the query object. The global precision is then averaged over all $1,464$ queries.

4.3   Analysis of Preprocessing Phase

The preprocessing phase primarily involves the extraction of $4,096$D feature vectors for all data segments. The total number of generated data segments depends on the setting of parameters of segment length sl and shift factor ssf. A specific setting is always a trade-off between search performance and accuracy. Lower values of ssf constitute higher overlaps between consecutive data segments, which increases a chance of locating query-relevant data parts but, on the other hand, increases the amount of data replication. An appropriate setting of both parameters depends on estimated lengths of future queries and the ability of the $sim^{\mathbf{NN}}$ distance measure to retrieve relevant data segments whose content is not properly aligned (i.e., the content is shifted) with respect to the content of query segments.

As parameters of segment length and shift factor are set, a data motion is partitioned into n^D segments (see Eq. 3). The most expensive operation is the extraction of feature vectors taking $25\,$ms on average for a single segment. However, we can still extract the feature vectors in real time using a single GPU card even in the worst considered scenario when $ssf = 0.1$ and $sl = 90$. With this setting, the 68-minute data motion is partitioned into about $54\,$k segments and it takes approximately $23\,$min to extract $54\,$k feature vectors.

Figure 3a demonstrates search accuracies in 60 different combinations of settings of segment length and shift factor (by fixing parameter $k' = 50$). In particular, for each of six different segment lengths ranging from 90 to 240 poses, ten different settings of segment shift factor ($ssf \in \{0.1, 0.2, \ldots, 1.0\}$) are evaluated. As expected, the search accuracy generally decreases with an increasing value of shift factor ssf, i.e., with smaller overlaps between data segments. For the used dataset of $1,464$ queries that significantly vary in lengths, the segment length of 150 poses along with the shift factor of 0.3 are the most appropriate setting, achieving the precision of 87.98%. Note that queries being shorter than a single segment (i.e., sl poses) are processed in the same way as longer segments because a $4,096$D feature vector can be principally extracted for any motion size.

4.4   Analysis of Retrieval Effectiveness

We fix the most appropriate setting of segment-length and shift-factor parameters as $sl = 150$ and $ssf = 0.3$ and measure how the search precision changes

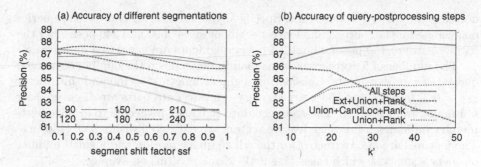

Fig. 3. (a) Search accuracy of the retrieval algorithm for different segmentation settings – segment lengths $sl \in \{90, 120, 150, 180, 210, 240\}$ and shift factor $ssf \in [0.1, 1.0]$. (b) Influence of excluding query-postprocessing steps "Extension of data segments" (*Ext*) and "Localization of candidates" (*CandLoc*) on the precision. Steps "Union of data segments" (*Union*) and "Ranking of candidates" (*Rank*) are included in all variants.

with an increasing number of retrieved data segments, which is quantified by parameter k'. Since each k'-NN sub-query retrieves exactly k' segments, the total number of retrieved data segments for each query equals to $k' \cdot n^Q$. At the same time, we analyze a significance of individual query-postprocessing steps on the search precision. In Fig. 3b, we can see that the precision generally increases with an increasing number k' of retrieved segments. The only exception is the variant with the excluded candidate-localization step (*Ext+Union+Rank*). Such variant unionizes extended data segments into longer and longer parts with increasing k'. Although these parts contain a query-relevant subsequence, they are too long to be ranked as globally similar to the query motion. When both "Localization of candidates" and "Extension of data segments" steps are excluded from the query evaluation (*Union+Rank*), the search precision is quite low – 84.58% for $k' = 50$. It increases to 86.16% when relevant data parts are additionally inspected to localize best-match candidates (*Union+CandLoc+Rank*). In addition, when the best-match candidates are localized within extended (longer) relevant data parts, the precision achieves up to 87.98% (*All steps*).

4.5   Analysis of Retrieval Efficiency

Search costs are primarily influenced by (1) evaluating k'-NN sub-queries and (2) extracting 4,096D feature vectors for query segments and best-match candidates localized within relevant data parts. Using a single CPU thread (i7 960 at 3.2 GHz) and main-memory sequential scan, we are able to evaluate a single k'-NN sub-query in 8 ms and 216 ms with respect to the best and worst scenario with 2 k ($sl = 240$, $ssf = 1.0$) and 54 k ($sl = 90$, $ssf = 0.1$) data segments, respectively. The scenario achieving the highest effectiveness ($sl = 150$, $ssf = 0.3$) generates about 11 k data segments that can be searched in 44 ms by a single sub-query. Although there are n^Q independent sub-queries to be evaluated (as the number of query segments), they can be processed in parallel to decrease the query response time. Specifically, there are 1.6 and 3.6 sub-queries on average for the shortest and longest segment sizes of 90 and 240 poses. For the segment size of 150 poses, about 2.4 query segments are generated on average.

The number of feature extractions mainly depends on the number of merged query-relevant data parts. Considering segment length $sl = 150$ and shift factor $ssf = 0.3$, from 14.3 to 55.9 merged query-relevant data parts are discovered for k' ranging from 10 to 50. These merged data parts are inspected to localize best-match candidates from which feature vectors are extracted within 350–$1,400$ ms. The features are also extracted for the whole query motion and a small number of query segments, which takes $(1 + 2.4) \cdot 25\,\text{ms} = 85\,\text{ms}$ on average.

Without any parallel processing, we need about 1.35 s to search the 68-minute motion and localize its query-relevant subsequences. In particular, we need (1) 106 ms to retrieve similar data segments with respect to 2.4 query segments on average using by a simple sequential scan in main memory, (2) 85 ms to extract feature vectors of the query motion and query segments, and (3) $1,163$ ms to extract features of 47 localized candidates on average. These times are measured for the setting achieving the highest effectiveness of 87.98%.

Efficiency Study on a Disk-Oriented Index Structure. Our objective is to efficiently search very long data motions having lengths in order of days or even months. As such large motion dataset is not publicly available, we simulate search efficiency by adopting data from the experimental evaluation presented by Novak et al. [11]. They introduce a disk-oriented approximate index structure, called the PPP-Codes, and evaluate it on the same type of features. In particular, they use the same reference model of neural network to extract $4,096$D feature vectors for common photographs. Even if the domain of photographs is different from our motion images, the extracted feature vectors exhibit similar characteristics.

The experiment in [11] indexes 20 million image features that are stored on an SSD disk. In the retrieval phase, a single 40-NN query is evaluated in 770 ms on average, while achieving a recall $\geq 85\%$ (it is the percentage of the same vectors retrieved by the PPP-Codes with respect to the sequential scan). Such recall value is reached by accessing only about $10,000$ vectors out of 20 M. The level of PPP-Codes search approximation can be additionally controlled by a user to find an appropriate trade-off between search effectiveness and efficiency. More detailed information about the experiment and the PPP-Codes structure is available in [11].

We could possibly employ the PPP-Codes to index all data segments and then retrieve query-similar ones much more efficiently in comparison with the sequential scan. By considering the most effective setting (i.e., $sl = 150, ssf = 0.3, k' = 40$), 20 M data segments would correspond to about an **87-day** motion. The extraction of 20 M feature vectors would take about 6 days using a single graphics card, while indexing by the PPP-Codes roughly 12 h only. Such very long motion could be then searched practically in real time (i.e., 1.85 s for retrieving query-similar segments and 1.25 s for feature extraction, without any parallelism) to obtain the five most similar subsequences with respect to a short query motion having a length of about 3 s on average.

4.6   Comparison with State-of-the-Art Approaches

There are few approaches that focus on efficient subsequence matching in motion capture data. Unfortunately, some of them [7,18] do not evaluate a search accuracy on any standardized ground truth. We compare the results against the most recent approach in [13] that beats earlier papers [4,15]. The competitive approach [13] achieves the precision of 82.98% for 5-NN queries on the same HDM05 subset of 15 categories. We outperform this result by achieving 87.98% using the same segment shift ($ssf = 0.3$). From the efficiency point of view, query response times are comparable on very long motions. Note that individual sub-queries can be processed in parallel and the postprocessing phase requires a constant time, disregarding the data motion length. More importantly, our approach does not limit the size of query and has a much smaller space complexity. The number of data segments in our approach increases linearly with respect to an increasing segment shift ssf, which is not true for the competitive approach in [13] having an exponential increase of data segments. For example, by fixing $ssf = 0.1$, our approach generates 32,842 data segments while the competitive approach 631,846 segments for the same 68-minute data motion, which is nearly 20-times more data.

Some other papers do not consider search efficiency, e.g., the approach in [1] needs about 36 s to search 100 k motion samples. By using a sequential scan without any index support, we need only 0.4 s to search such dataset.

5   Conclusions

We propose a new subsequence matching approach that employs two distance measures. Due to the "noise ability" of the neural-network distance measure, adjacent data segments can be shifted by a large margin (e.g., about 30–50% w.r.t. the segment size), which results in partitioning a long data motion into a moderate number of segments. The neural-network measure is also effective and indexable which enables accurate and fast retrieval of query-similar data segments. The retrieved segments are then extended and merged into relevant parts that are inspected on the basis of another joint-coordinate distance measure. This measure is very fast to localize starting and ending positions of the best-match candidates within the merged relevant parts. The candidates are finally ranked by the effective neural-network distance measure. The whole approach achieves high effectiveness, e.g., reaching 87.98% for 5-NN queries evaluated on the 68-minute motion. Such precision clearly outperforms those approaches reported in related work and can even be achieved by searching the motion using the sequential scan in real time. The disk-based PPP-codes index could additionally increase the searchable length to 87 days with query response time around 3 s.

Acknowledgements. This research was supported by GBP103/12/G084.

References

1. Beecks, C., Hassani, M., Obeloer, F., Seidl, T.: Efficient query processing in 3D motion capture databases via lower bound approximation of the gesture matching distance. In: ISM 2015, pp. 148–153 (2015)
2. Camerra, A., Shieh, J., Palpanas, T., Rakthanmanon, T., Keogh, E.: Beyond one billion time series: indexing and mining very large time series collections with isax2+. Knowl. Inf. Syst. **39**(1), 123–151 (2014)
3. Du, Y., Wang, W., Wang, L.: Hierarchical recurrent neural network for skeleton based action recognition. CVPR **2015**, 1110–1118 (2015)
4. Elias, P., Sedmidubsky, J., Zezula, P.: Motion images: an effective representation of motion capture data for similarity search. In: Amato, G., Connor, R., Falchi, F., Gennaro, C. (eds.) SISAP 2015. LNCS, vol. 9371, pp. 250–255. Springer, Cham (2015). doi:10.1007/978-3-319-25087-8_24
5. Faloutsos, C., Ranganathan, M., Manolopoulos, Y.: Fast subsequence matching in time-series databases. SIGMOD Rec. **23**(2), 419–429 (1994)
6. Kadu, H., Kuo, C.C.: Automatic human mocap data classification. IEEE Trans. Multimedia **16**(8), 2191–2202 (2014)
7. Kapadia, M., Chiang, I.K., Thomas, T., Badler, N.I., Kider Jr., J.T.: Efficient motion retrieval in large motion databases. In: ACM SIGGRAPH Symposium on Interactive 3D Graphics and Games (I3D 2013), pp. 19–28. ACM (2013)
8. Liu, X., He, G.F., Peng, S.J., Cheung, Y., Tang, Y.Y.: Efficient human motion retrieval via temporal adjacent bag of words and discriminative neighborhood preserving dictionary learning. IEEE Trans. Hum.-Mach. Syst. **PP**(99), 1–14 (2017)
9. Müller, M., Röder, T., Clausen, M., Eberhardt, B., Krüger, B., Weber, A.: Documentation Mocap Database HDM05. Technical report, CG-2007-2, Universität Bonn (2007)
10. Müller, M., Baak, A., Seidel, H.P.: Efficient and robust annotation of motion capture data. In: SCA 2009, p. 10. ACM Press (2009)
11. Novak, D., Cech, J., Zezula, P.: Efficient image search with neural net features. In: Amato, G., Connor, R., Falchi, F., Gennaro, C. (eds.) SISAP 2015. LNCS, vol. 9371, pp. 237–243. Springer, Cham (2015). doi:10.1007/978-3-319-25087-8_22
12. Papapetrou, P., Athitsos, V., Potamias, M., Kollios, G., Gunopulos, D.: Embedding-based subsequence matching in time-series databases. ACM Trans. Database Syst. **36**(3), 17:1–17:39 (2011)
13. Sedmidubsky, J., Elias, P., Zezula, P.: Similarity searching in long sequences of motion capture data. In: Amsaleg, L., Houle, M.E., Schubert, E. (eds.) SISAP 2016. LNCS, vol. 9939, pp. 271–285. Springer, Cham (2016). doi:10.1007/978-3-319-46759-7_21
14. Sedmidubsky, J., Elias, P., Zezula, P.: Effective and efficient similarity searching in motion capture data. Multimed. Tools Appl. 1–22 (2017)
15. Sedmidubsky, J., Valcik, J., Zezula, P.: A key-pose similarity algorithm for motion data retrieval. In: Blanc-Talon, J., Kasinski, A., Philips, W., Popescu, D., Scheunders, P. (eds.) ACIVS 2013. LNCS, vol. 8192, pp. 669–681. Springer, Cham (2013). doi:10.1007/978-3-319-02895-8_60
16. Shahroudy, A., Liu, J., Ng, T.T., Wang, G.: NTU RGB+D: a large scale dataset for 3D human activity analysis. In: CVPR 2016 (2016)
17. Vögele, A., Krüger, B., Klein, R.: Efficient unsupervised temporal segmentation of human motion. In: ACM Symposium on Computer Animation (2014)
18. Wang, Y., Neff, M.: Deep signatures for indexing and retrieval in large motion databases. In: 8th ACM Conference on Motion in Games, pp. 37–45. ACM (2015)

Interactive Time Series Subsequence Matching

Danila Piatov, Sven Helmer[✉], and Johann Gamper

Free University of Bozen-Bolzano, Bolzano, Italy
{danila.piatov,sven.helmer,johann.gamper}@unibz.it

Abstract. We develop a highly efficient access method, called Delta-Top-Index, to answer top-k subsequence matching queries over a time series data set. Compared to a naïve implementation, our index has a storage cost that is up to two orders of magnitude smaller, while providing answers within microseconds. We demonstrate the efficiency and effectiveness of our technique in an experimental evaluation with real-world data.

1 Introduction

Similarity search in time series plays an important role for monitoring dynamic environments in areas such as meteorology, road traffic, financial markets, sensor and computer networks. In line with the growing amount of available data and the need to put it to use, there has been a lot of work on storing, querying, and analyzing such data. In the past decade, researchers have investigated novel methods and techniques for representing, indexing, classifying, clustering, searching, and approximating time series data (see [2,3,5,13] for a non-exhaustive list).

We focus on efficient mechanisms for searching for subsequences or patterns in time series, doing so in the context of an agricultural setting in South Tyrol. The South Tyrolean association for consulting fruit and wine growers, "Südtiroler Beratungsring für Obst- und Weinbau" (SBR), collects weather data from 115 weather stations located in orchards and vineyards throughout the province. The measurements include quantities such as temperature (at different elevation levels above ground), wind speed, rainfall, and humidity and they are recorded at a sampling rate of five minutes.[1] SBR uses this data to advise the local growers on their decisions, in which weather plays a crucial role, as wrong ones may have a major impact and result in the loss of valuable crops, e.g. caused by severe hail- or thunderstorms.

A typical query consists of a time series for a given location and time period in the data set and searches for the most similar periods in the data set. The idea is to see which decisions were made during similar circumstances and how the situation unfolded. Figure 1 shows the formulation of an example query using a prototype of our system by choosing a weather station and a date range. Here we select a station in Nals on 11 August 2011, which recorded a sharp rise in temperature during the early morning hours on that day. In the lower half of the screenshot the ten most similar time series are displayed. Our goal was to build

---

[1] Altogether there are around twenty different parameters.

© Springer International Publishing AG 2017
M. Kirikova et al. (Eds.): ADBIS 2017, LNCS 10509, pp. 73–87, 2017.
DOI: 10.1007/978-3-319-66917-5_6

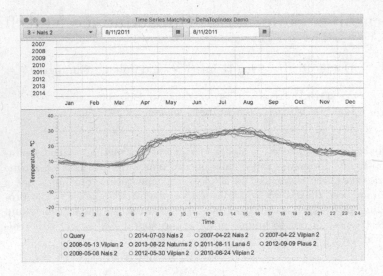

Fig. 1. Example for a temperature time series

a system that is able to deliver results instantaneously, providing an interactive, real-time interface. Moving the indicator, which is the vertical blue bar in the upper half of the figure, will immediately show new results in the lower half. In summary, we make the following contributions:

- We generalize prefix-sum Euclidean distance matrices by introducing segmentation to speed up the computation of the distances between subsequences of long time series.
- Taking this further, we develop Top-Index, an access method that can produce answers to top-k subsequence matching queries ultrafast.
- Our Delta-Top-Index exploits the redundancy found in a Top-Index to store the same information in a hundredth of the space.
- In an experimental evaluation with real-world data sets we demonstrate that a Delta-Top-Index can decompress the answer to a query in a matter of microseconds.

2 Related Work

Early work in the area of time series matching is based on ε-matching, which means we are looking for time series that have a Euclidean distance smaller than ε to each other. Agrawal et al. worked on indexing whole sequences via the Discrete Fourier Transform (DFT) [1]. DFT is applied to each time series, but only the first few coefficients are taken and indexed using an R*-tree. This access method acts as a filter, creating a candidate set that may include false positives, which have to be removed in a second step. Faloutsos et al. extended this to matching subsequences instead of whole sequences [3]. A sliding window of width w is placed at every possible position in the time series and the resulting subsequences are indexed similarly to [1]. In order to reduce the number of data points in the index, a heuristic groups close points into minimum-bounding rectangles (MBRs) and then indexes the MBRs instead. However, a query time series longer than w has to be split into disjoint sequences of length w and each one is used to query the index (with an adjusted value for ε). Moon et al. developed a variation of this by splitting the time series into disjoint sequences and the query sequence is scanned via a sliding window [8]. This has the advantage that it needs to index a much smaller number of points (so MBRs are not needed anymore), but at the price of a larger number of queries (every sliding window over the query becomes one index access).

Further improvements were suggested: Kim et al. proposed using a spatial join to search the index for multiple query points simultaneously [6] and Moon et al. developed a generalized version of [3,8], allowing some trade-offs between the size of the index and the number of queries. One drawback of all these approaches is the need to calibrate the parameter w. Lim et al. note that the performance of these approaches deteriorates when the query sequence is much longer than the window length w and propose building multiple indexes for different values of w, accelerating queries at the price of a larger storage overhead [7].

In 2000, Yi and Faloutsos [12] proposed the idea of segmented means as a basis for fast indexing of time series data. A similar idea has been proposed a short time later by Keogh et al. [5], called piecewise aggregate approximation. Both approaches segment the data in equal parts along the time dimension and calculate the average value for each segment, achieving a dimensionality reduction along the time dimension. Shieh and Keogh then moved on to a symbolic representation called SAX, which was refined and extended to an indexable version of SAX, called iSAX [11]. Zoumpatianos et al. [13] present an adaptive version of the iSAX index. Ding et al. [2] provide an overview of different state-of-the-art techniques by conducting a series of experiments for eight different representation methods and nine different similarity measures. There are also index solutions for other distance measures, such as dynamic time warping (DTW) [10] and longest common subsequence (LCS) [4].

Closest to us is the work by Mueen et al., who use the same principle as our basic prefix-sum Euclidean distance matrix to compute so-called shapelets, but they do not optimize this matrix in any way [9]. As we will see, the non-optimized version of the matrix is only usable for short time series. With the help of our optimized matrix, we can handle much larger time series.

3   Problem Formalization

3.1   Basic Terminology

A *time series* x is a sequence of n measurements in chronological order with a constant time interval between the measurements:

$$x := x_1, x_2, \ldots, x_n = \{x_i : 1 \leqslant i \leqslant n\}, \quad n = |x|.$$

For instance, the records for the air temperature[2] at a specific weather station ws_i taken every five minutes can be represented as a time series:

$$\text{temp}_{ws_i} = 15, 15, 14, 13, 13, 12, \ldots$$

There are different ways to measure the similarity or distance between two time series. The one we use, the *Euclidean distance*, is one of the most common ones. The distance between the time series x and y is computed in the following way:

$$ED(\boldsymbol{x}, \boldsymbol{y}) := \sqrt{\sum_{i=1}^{n}(x_i - y_i)^2}, \quad n = |\boldsymbol{x}| = |\boldsymbol{y}|.$$

We define a *subsequence* of a time series by the index of its first element s and its length l (the resulting time series elements are re-indexed starting from 1):

$$\boldsymbol{x}[s, l] := \{x_i : s \leqslant i < s + l\}.$$

For example, $\text{temp}_{ws_i}[2, 3]$ is equal to 15, 14, 13.

3.2   Computing Similarities

On account of the needs of SBR, we are interested in three particular ways of querying the time series data, which we describe below.

All-Pairs Subsequence Matching. The first one, *all-pairs subsequence matching* is defined as follows: given two time series x and y, compute the Euclidean distances between all possible pairs of subsequences of x and y of the same length (x and y can be the same time series). More formally,

$$\forall l = 1, \ldots, \min\{|\boldsymbol{x}|, |\boldsymbol{y}|\}, \quad \forall s_1 = 1, \ldots, |\boldsymbol{x}| - l + 1, \quad \forall s_2 = 1, \ldots, |\boldsymbol{y}| - l + 1$$
$$\text{compute } ED(\boldsymbol{x}[s_1, l], \boldsymbol{y}[s_2, l])$$

Top-k Subsequence Matching. Next, given two time series x and y (not necessarily distinct) and a query subsequence $\boldsymbol{x}[s_1, l]$ of x, we want to find the *top-k subsequences* $\boldsymbol{y}[s_2, l]$ of y closest to the query subsequence in terms of Euclidean distance.

Full Top-k Subsequence Matching. Finally, in the *full top-k subsequence matching*, given a set of time series $x_1, \ldots, x_m$ and the query $\langle t, s_1, l \rangle$, defined by the subsequence $\boldsymbol{x}_t[s_1, l]$, we want to find the top-k closest subsequences within all m time series. This includes $\boldsymbol{x}_t$, but excludes the query subsequence itself.

---

[2] We use integers to simplify the matter, the actual temperatures are represented by real numbers.

4 Prefix-Sum Euclidean Distance Matrix

Assuming for a moment that two time series x and y have the same length, the complexity of computing the distances between all pairs of subsequences of x and y naïvely is $O(n^4)$, which is prohibitively expensive. We now turn towards computing the Euclidean distances more efficiently.

4.1 Basic Distance Matrix

The *Euclidean distance matrix* A^{xy} for two time series x and y is defined as:

$$A^{xy} \in \mathbb{R}^{|x| \times |y|}, \quad A^{xy}[i, j] := (x_i - y_j)^2.$$

It allows us to express the Euclidean distance between any pair of subsequences of x and y of the same length l as:

$$ED(x[s_1, l], y[s_2, l]) = \sqrt{\sum_{k=0}^{l-1} A^{xy}[s_1 + k, s_2 + k]}.$$

We observe that we always compute the sum of matrix elements along its diagonals (parallel to the main diagonal). Thus, we can replace the elements of each diagonal with their prefix sum sequence, obtaining the *prefix-sum Euclidean distance matrix* P^{xy}:

$$P^{xy} \in \mathbb{R}^{|x| \times |y|}, \quad P^{xy}[i, j] := A^{xy}[i, j] + P^{xy}[i - 1, j - 1],$$

where $P^{xy}[i, 0] := 0$ and $P^{xy}[0, j] := 0$ for any i and j. This gives us

$$\sum_{k=0}^{l-1} A^{xy}[s_1 + k, s_2 + k] := P^{xy}[s_1 + l - 1, s_2 + l - 1] - P^{xy}[s_1 - 1, s_2 - 1].$$

Therefore, computing the Euclidean distance between any pair of subsequences of x and y of the same length l boils down to:

$$ED(x[s_1, l], y[s_2, l]) = \sqrt{P^{xy}[s_1 + l - 1, s_2 + l - 1] - P^{xy}[s_1 - 1, s_2 - 1]}.$$

As an example, let us assume that $x = 1, 2, 3$ and $y = 2, 3, 4$. Then we have

$$A^{xy} = \begin{pmatrix} 1 & 4 & 9 \\ 0 & 1 & 4 \\ 1 & 0 & 1 \end{pmatrix}, \quad P^{xy} = \begin{pmatrix} 1 & 4 & 9 \\ 0 & 2 & 8 \\ 1 & 0 & 3 \end{pmatrix}.$$

and computing $ED(x[1, 2], y[2, 2])$ is equal to $\sqrt{P^{xy}[2, 3] - P^{xy}[0, 1]} = \sqrt{8}$.[3]

In the special case of $x = y$, we have square matrices with $A^{xx}[i, i] = 0$, $P^{xx}[i, i] = 0$, $A^{xx}[i, j] = A^{xx}[j, i]$, and $P^{xx}[i, j] = P^{xx}[j, i]$. Consequently, we only have to compute and store the elements below (or above) the main diagonal of A or P, saving storage space.

---

[3] Remember that $P^{xy}[i, 0] := 0$ and $P^{xy}[0, j] := 0$.

4.2   Generalized Prefix-Sum Euclidean Distance Matrix

Building and storing a full matrix is not always an option, as it will contain n^2 elements, n being the length of the time series we want to match. For instance, for our weather station data set, which spans a period of eight years with recordings for every five minutes, this would mean that each individual time series (e.g., temperature readings for weather station ws_i) contains $\sim$840,000 points, resulting in a matrix with $\sim$700 billion elements.

However, we do not always need a very fine granularity for the subsequence offsets and lengths. For several of the queries on the weather station data set we only need a precision on the level of a day. Typical queries are "what are the seven days most similar to last week?" or "what are the 30 days most similar to the month a week ago?".

We propose the *generalized prefix-sum Euclidean distance matrix* to get the size of the matrix down to manageable levels. The idea is to define a segment size p and allow only subsequences that start and end at segment boundaries. We call such subsequences *segment subsequences*:

$$\boldsymbol{x}[s, l, p] := \boldsymbol{x}[ps - p + 1, pl].$$

The generalized Euclidean distance matrix $A^{\boldsymbol{x}\boldsymbol{y}p}$ then looks like this:

$$A^{\boldsymbol{x}\boldsymbol{y}p} \in \mathbb{R}^{\lfloor |\boldsymbol{x}|/p \rfloor \times \lfloor |\boldsymbol{y}|/p \rfloor}, \quad A^{\boldsymbol{x}\boldsymbol{y}p}[i, j] := (ED(\boldsymbol{x}[i, 1, p], \boldsymbol{y}[j, 1, p]))^2.$$

The generalized prefix-sum Euclidean distance matrix $P^{\boldsymbol{x}\boldsymbol{y}p}$ is defined as before, by replacing each diagonal of $A^{\boldsymbol{x}\boldsymbol{y}p}$ with its prefix-sum sequence.

Also, the Euclidean distance between segment subsequences can be computed in exactly the same manner in constant time as before:

$$ED(\boldsymbol{x}[s_1, l, p], \boldsymbol{y}[s_2, l, p]) =$$
$$\sqrt{P^{\boldsymbol{x}\boldsymbol{y}p}[s_1 + l - 1, s_2 + l - 1] - P^{\boldsymbol{x}\boldsymbol{y}p}[s_1 - 1, s_2 - 1]}.$$

5   Speeding up Similarity Computations

The prefix-sum Euclidean distance matrix reduces the complexity of computing the Euclidean distance from linear to constant, takes $O(n^2)$ time to build, and requires $O(n^2)$ space to store, helping us to compute the queries defined in Sect. 3.2 more efficiently.

5.1   All-Pairs Subsequence Matching

As mentioned above, the complexity of the naïve computation is $O(n^4)$. Using the prefix-sum Euclidean distance matrix we can bring this down to $O(n^3)$. This follows from the fact that we need $O(n^2)$ time to build the matrix and $O(n^3)$ time to query it for all possible combinations of parameters: length l, subsequence offset s_1, and subsequence offset s_2.

5.2   Top-k Subsequence Matching

The complexity of a naïve computation of the top-k subsequence matching is $O(n^2)$, as we have to compute the Euclidean distance for every target subsequence offset. We can reduce this to $O(n)$ if we use a precomputed matrix that is stored in an index. Although the size of the matrix is $O(n^2)$, we only need to access a small part of it to compute a top-k subsequence match.

5.3   Full Top-k Subsequence Matching

Building generalized prefix-sum Euclidean distance matrices for all pairs of the time series and querying them would still not scale for full top-k subsequence matching. We develop another index structure, called *Top-Index*, which is based on the idea of precomputing and storing the answers to all possible queries. Implementing a Top-Index in a naïve way would still not scale. Nevertheless, here we explain the basic concepts of the Top-Index without any optimizations (to illustrate how it works) and then move on to the compressed Delta-Top-Index in Sect. 6.

Top-Index. Assume we have m time series $x_1, x_2, \ldots, x_m$ with (for simplicity and without loss of generality) the same length of n segments of length p. We also fix an upper bound for k, meaning that we are able to answer all top-k queries up to this upper bound.

A query is defined by the triplet $\langle t, s, l \rangle$, denoting the segment subsequence $x_t[s, l, p]$. The query result are the top-k subsequences from all m time series and $n - l + 1$ offsets, excluding the query subsequence itself. The basic unit of the Top-Index is the *top list*, that represents the result of a single query:

$$toplist^k(t, s, l) := \langle d_1, t'_1, s'_1 \rangle, \langle d_2, t'_2, s'_2 \rangle, \ldots, \langle d_k, t'_k, s'_k \rangle.$$

Each tuple in the top list denotes the address of a matching subsequence (time series index t'_i, offset s'_i, length l) and the Euclidean distance from the query subsequence to the matching subsequence (d_i).

The Top-Index is then defined as the set of top lists for all possible queries:

$$TopIndex^k := \{toplist^k(t, s, l) : t = 1..m, \ l = 1..n, \ s = 1, \ldots, n - l + 1\}.$$

With this Top-Index in place, it becomes very easy to find the top-k subsequences closest to $x_t[s, l, p]$. We simply retrieve $toplist^k(t, s, l)$, which is the complete solution.

The pseudocode for building a Top-Index is shown in Algorithm 1. The basic idea is to create an empty index, look up the Euclidean distances between all possible combinations of subsequences in the precomputed prefix-sum matrix, and insert them into the index, keeping only the best matches and discarding the rest. When adding tuple $\langle d, t, s \rangle$ to a top list, we first check whether d is smaller than the biggest d in the list. If not, we discard the tuple. If yes, we remove the element with the biggest d from the list and insert the new tuple to the appropriate position (so the list is again ordered by d).

Algorithm 1. Build the Top-Index

---

    input   : Segment size p, number of matches k, sequence of m time series
                $x_1, \dots, x_m$ of length n segments
    output : $TopIndex^k$ for the input data

1 Allocate $TopIndex^k$, initialize all fields d to $+\infty$
2 **foreach** $t_1 = 1$ **to** m **do**
3   **foreach** $t_2 = 1$ **to** t_1 **do**
4   Compute the generalized prefix-sum matrix for x_{t_1} and x_{t_2}
5   **foreach** $l = 1$ **to** n **do**
6   **foreach** $s_1 = 1$ **to** $n - l + 1$ **do**
7   **foreach** $s_2 = 1$ **to** (if $t_1 = t_2$ **then** $s_1 - 1$ **else** $n - l + 1$) **do**
8                    $d \leftarrow ED(x_{t_1}[s_1, l, p], x_{t_2}[s_2, l, p])$
9   Add $\langle d, t_1, s_1 \rangle$ to $toplist^k(t_2, s_2, l)$ of the $TopIndex^k$
10   Add $\langle d, t_2, s_2 \rangle$ to $toplist^k(t_1, s_1, l)$ of the $TopIndex^k$

11   Discard the prefix-sum matrix

12 **return** $TopIndex^k$

Updating a Top-Index. Even though a Top-Index is expensive to build, updating and maintaining it is not very expensive. The reason for this is the way that time series are updated: we either append new data items to them or we create completely new time series (which are, in turn, updated by appending items). Similar to Algorithm 1, we add new top lists for new segment offsets and lengths. We also need to compute distances of the newly added segments to other subsequences and insert them into the Top-Index.

Further Notes. Although the Top-Index can be used for any distance measure, we optimize it for the Euclidean distance using the generalized prefix-sum Euclidean distance matrices. We do not have to keep all matrices in memory, as the algorithm builds a matrix for a pair of time series, uses it, and then discards it before building a new matrix.

Even using generalized prefix-sum Euclidean distance matrices, Algorithm 1 is rather inefficient, as it contains five nested loops. Nevertheless, it is embarrassingly parallelizable: in our implementation we can (and actually did) execute lines 4–11 in parallel.[4] Moreover, in the following section we illustrate how to optimize full top-k subsequence matching further.

6 Delta-Top-Index

The storage overhead of a Top-Index is high, but in turns out that it contains a lot of redundant data. In the following we illustrate how to build a compressed

---

[4] We avoid race conditions by protecting top list modifications with a critical section.

version of the Top-Index, called Delta-Top-Index, which achieves a compression rate of more than 100.

6.1   Basic Idea

First of all, we investigate redundancy in the index by taking every top list

$$toplist^k(t, s, l) := \langle d_1, t'_1, s'_1 \rangle, \langle d_2, t'_2, s'_2 \rangle, \ldots, \langle d_k, t'_k, s'_k \rangle$$

and stripping out the distance information and element order, treating them as a set of references to top-k matched subsequences:

$$strippedtoplist^k(t, s, l) := \{\langle t'_1, s'_1 \rangle, \langle t'_2, s'_2 \rangle, \ldots, \langle t'_k, s'_k \rangle\}.$$

Let us compare an arbitrary $strippedtoplist^k(t, s, l-1)$ to its slightly longer counterpart $strippedtoplist^k(t, s, l)$, and assume for the moment that these two lists have an overlap (we will show later that this is often the case):

$$strippedtoplist^k(t, s, l-1) = \{u_1, \ldots, u_{k-r}, \ d_1, \ldots, d_r\},$$
$$strippedtoplist^k(t, s, l) = \{u_1, \ldots, u_{k-r}, \ i_1, \ldots, i_r\}.$$

Some elements can be found in both sets (unchanged), while the rest will have been replaced by deleting old elements from the first set and inserting new elements into the second set. We denote the number of replaced elements by r.

If we look at a whole sequence of stripped top lists, which we call a *top list row*,

$$strippedtoplist^k(t, s, 1), strippedtoplist^k(t, s, 2), \ldots,$$
$$strippedtoplist^k(t, s, n-s+1).$$

we only store a complete list of elements for the first list, $strippedtoplist^k(t, s, 1)$, the other lists can be represented by the deltas to their preceding lists. A simple way to implement this delta is to store the set of deleted and the set of inserted elements.[5] With this information, given $strippedtoplist^k(t, s, i)$, we can reconstruct $strippedtoplist^k(t, s, i+1)$. Clearly, this only saves storage space if $r < \frac{k}{2}$.

So, an important questions remains: how large is r typically? We have found that for real-world data sets r tends to be very small when computing the delta between two stripped top lists. Very often it is zero and it is rarely greater than one. Figure 2 shows a typical distribution of values for r (for a fixed query time series index and different offsets s and lengths l). We observe that this is a very sparse matrix, all the empty cells contain a zero. Secondly, the larger l, the more empty cells we have. This can be explained by the fact that adding one more segment to an already long subsequence will not make a large difference in terms of similarity.

---

[5] We will look at a more sophisticated implementation in the following section.

Subsequence length l, segments

s	2	3	4	5	6	7	8	9	10	11	12	13	14	15	16	17	18	19	20	21	22	23	24	25	26	27	28	29	30	31	32	33	34	35	36	37	38	39	40	41	42	43	44	45	46
1	6	2	2	2		2		2	1										1										1					1	1		1							1	1
2	8	5	3		1	1	1		1										1							1										1					1	1		
3	9	7	1																1													1	1											1	
4	5	1														1															1	1									1				
5	4	1	2	1				1	1							1													1			1								1					
6	5	3	1	1			1	1	1					1															1		1											1			
7	6	2	1	1	1	1	1					1											1				1	1	1	1		1	1				1				2				
8	8	1		2	1	1	1		1	1											1			1							1			1		1					2				
9	3		3	1	1						1									1				1							1				1				2						
10	3	3	3	2	1		1		1										1			1							1			1			2										

Subsequence offset s, segments

Fig. 2. Number of changes (r) in stripped top lists when switching from subsequence length $l - 1$ to l. Zero values are shown as empty cells.

6.2   Efficient Implementation of Deltas

A straightforward implementation of the delta as described in the previous section has two major issues, both caused by representing the elements of a stripped top list as sets. First, we have to include $2r$ elements in the delta: the set of deleted elements and the set of newly inserted elements. Computing a set difference, which we have to do to apply a delta, is also not very efficient. The options for implementing the set difference range from hash-maps or sort-merge algorithms all the way to nested loops. As we have to apply multiple consecutive deltas, one on top of the other, the performance of this step is crucial.

Encoding the Delta. To eliminate the need for the computation of a set difference and the explicit storage of deleted elements, we impose an order on the elements found in stripped top lists. Instead of storing the deleted items, we can then use their index in the previous list. Moreover, since the indices are integers in the range from 0 to $k - 1$, we can store them in a bitmask b, further reducing the storage overhead. For example, encoding the deletion of the elements at positions 0, 2, and 5 for $k = 8$ results in $2^0 + 2^2 + 2^5 = 0010\,0101_2$ for the bitmask. Consequently, a delta uses the following encoding:

$$\langle b, (i_1, i_2, \ldots, i_r) \rangle,$$

with b standing for the bitmask of deleted items and i_j for an inserted element.

Applying the Delta. Contemporary CPUs have a constant-time instruction for looking up the index of the least significant bit set in a bitmask, which is also supported by compilers. We use this instruction to obtain from the bitmask the index of the first deleted item and replace this element with the first inserted item i_1. We then shift the bitmask, find the index of the second deleted element, replace it with the second inserted element i_2 and so on, until the bitmask is empty. This procedure is fast, in-place, and does not use additional memory, therefore all the data is kept and processed in the L1 CPU cache.

6.3 Handling a Top List Row

Encoding. Given a single top list row, we first write to the output stream all k elements of the first top list (with $l = 1$). We then skip to the next top list with a non-empty delta ($r > 0$) and write the following to the output stream: the l of the top list, the bit-mask b of the deleted elements, and the sequence of the inserted elements $i_1, \ldots, i_r$. We do not have to write r itself, as it can be derived from b. We keep skipping to the next list with a non-empty delta and processing it, until no more top lists are left.

Querying. Given an encoded top list row, we now want to extract the original top list for query length l_q. For that we load the first k elements from the encoded row stream to an array of k elements. We then sequentially read and apply the deltas from the stream. We stop when we reach the end of the stream or a length l greater than our query length l_q. The array now contains the information we are looking for.

6.4 Putting It All Together

A Delta-Top-Index is a disk- or memory-based file with a dictionary and a set of encoded top list rows. The dictionary maps a query time series index t and the subsequence offset s to the corresponding top list row.

Processing a query works as follows: we look up the series index t and the subsequence offset s in the dictionary, getting back a delta-compressed top list row. We then query this encoded row for the query length l_q as described in Sect. 6.3. Overall one Delta-Top-Index query requires two random lookups and one sequential scan of an encoded top list row.

7 Experimental Evaluation

7.1 Environment

All algorithms and structures were implemented in-memory in C++ and compiled with GCC 4.9.2 using -O3 optimization flag to 64-bit binaries. The execution was performed on a machine with two Intel Xeon E5-2667 v3 processors under Linux.

7.2 Test Time Series

Synthetic Time Series. Synthetic time series give us more control over investigating the impact of time series length on the performance of the algorithms. We used a random walk time series, where the value of each element is the value of the previous element (or 0 if it is the first) plus a normally-distributed random value with a mean of 0 and a variance of 1. If several time series were used in an experiments, all were generated using a different random generator seed.

Real-World Time Series. For real-world data we took the air temperature time series of the SBR data (described in Sect. 1). The set consists of 115 time series, each having ~840,000 points (five-minute sampling rate for a period of eight years).

7.3 Experiments and Results

Prefix-Sum Euclidean Distance Matrix Operations. We begin with the empirical evaluation of the prefix-sum Euclidean distance matrix. For this experiment we used two synthetic time series and varied their length. The query was: for a given subsequence length l, find a pair of subsequences with the smallest Euclidean distance between them. The results are shown in Fig. 3. We measured the time needed to build the matrix, the time needed to answer the query using a horizontal scan of the matrix, the time needed to answer the query using a vertical scan of the matrix, and a baseline (answering the query by directly computing the Euclidean distances).

First of all, we observe that the performance of the naïve approach deteriorates very quickly as the query subsequence length grows (note the double-logarithmic scales). We also observe that the order in which we query the elements of the Euclidean distance matrix is important—doing so by row is several times faster than by column, and the difference grows with increasing subsequence length l and increasing time series length.

When the query length is roughly comparable to the time series length, there are fewer offsets to be considered by the query. Thus, for short time series lengths the algorithms are faster for query length $l = 1000$ compared to $l = 10$ in Fig. 3.

Top-Index Creation. In this experiment we investigate the performance of creating a Top-Index. We use 20 synthetic time series of length $n = 1000$, and vary the number of results k, the segment size p, and the thread count. The results are shown in Fig. 4.

We observe that by increasing the segment size p we can achieve a significant performance boost (using multiple threads also helps) and that the algorithm

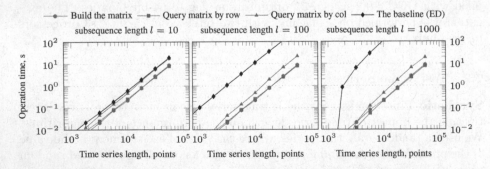

Fig. 3. Prefix-sum Euclidean distance matrix operation performance

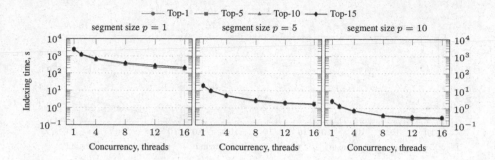

Fig. 4. Top-Index indexing time for 20 synthetic time series of length 1000

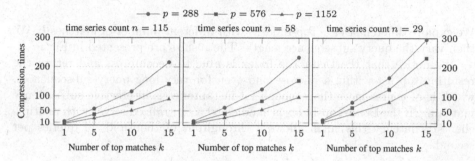

Fig. 5. Delta-Top-Index compression rate

scales very well (note the logarithmic y-axis). Finally, we see that the number of top matches, k, does not have a big impact on the performance.

Delta-Top-Index Compression Rate. The compression rate of the Delta-Top-Index is the ratio of the size of the Top-Index to the size of the corresponding Delta-Top-Index. To study it, we used the full set of 115 real-world time series and 50% and 25% subsets. Again, we varied the segment size p and the number of top elements k maintained by the index. The results are shown in Fig. 5.

A Delta-Top-Index scales much better with an increasing k and an increasing precision (i.e., decreasing p) compared to the Euclidean distance matrix and the Top-Index, starting at a factor of around ten and going beyond 200. Decreasing p means that a Top-Index has a finer granularity with more redundancy that can be exploited by the Delta-Top-Index. In absolute numbers, the size of a Top-Index is 55 GB compared to the 260 MB of the corresponding Delta-Top-Index for $n = 115$, $p = 288$, and $k = 15$. For $n = 115$, $p = 1152$, and $k = 1$ the numbers are 234 MB and 38 MB, respectively.

Delta-Top-Index Query Performance. In this experiment we explore the query latency of the Delta-Top-Index. Again, we use the real-world dataset.

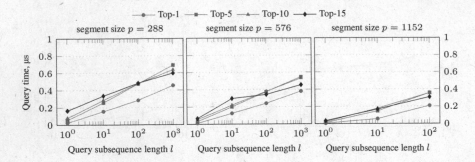

Fig. 6. Delta-Top-Index query performance

We explore different Delta-Top-Indexes with different number of matches k. We then vary the query subsequence length. The results are presented in Fig. 6.

We notice that the Delta-Top-Index is able to decompress and return the resulting top list within a microsecond even for very long query subsequences, which is of course more than enough for instantaneous interactive subsequence matching. If the Delta-Top-Index is handled by a centralized server with multiple clients, a latency this small allows a throughput of thousands of queries per second.

8   Conclusions

We developed a prototype for the efficient analysis of time series data, with the help of which queries can be answered instantaneously, i.e., within a time frame of microseconds. In order to do so, we generalized a prefix-sum Euclidean distance matrix, in which a lot of precomputed data we need to answer queries is stored. We go a step further by proposing an access method, called Top-Index, with which queries can be answered even more quickly then with the distance matrix alone. While it is not cheap to build such an index for a high-resolution segmentation of the data, updating and maintaining the index via append-only operations is efficient. The creation of the index can also be accelerated by parallelizing the task. Using a clever compression scheme, this index is implemented in a way that reduces the size of the naïve Top-Index by up to two orders of magnitude, making it applicable to longer time series as well. In summary, our Delta-Top-Index is suitable for interactive applications where low-latency responses have a high priority.

References

1. Agrawal, R., Faloutsos, C., Swami, A.: Efficient similarity search in sequence databases. In: Lomet, D.B. (ed.) FODO 1993. LNCS, vol. 730, pp. 69–84. Springer, Heidelberg (1993). doi:10.1007/3-540-57301-1_5

2. Ding, H., Trajcevski, G., Scheuermann, P., Wang, X., Keogh, E.J.: Querying and mining of time series data: experimental comparison of representations and distance measures. PVLDB **1**(2), 1542–1552 (2008)
3. Faloutsos, C., Ranganathan, M., Manolopoulos, Y.: Fast subsequence matching in time-series databases. In: SIGMOD, pp. 419–429 (1994)
4. Han, T.S., Ko, S.-K., Kang, J.: Efficient subsequence matching using the longest common subsequence with a dual match index. In: Perner, P. (ed.) MLDM 2007. LNCS (LNAI), vol. 4571, pp. 585–600. Springer, Heidelberg (2007). doi:10.1007/978-3-540-73499-4_44
5. Keogh, E.J., Chakrabarti, K., Pazzani, M.J., Mehrotra, S.: Dimensionality reduction for fast similarity search in large time series databases. Knowl. Inf. Syst. **3**(3), 263–286 (2001)
6. Kim, S.W., Park, D.H., Lee, H.G.: Efficient processing of subsequence matching with the Euclidean metric in time-series databases. IPL **90**(5), 253–260 (2004)
7. Lim, S.H., Park, H., Kim, S.W.: Using multiple indexes for efficient subsequence matching in time-series databases. Inf. Sci. **177**(24), 5691–5706 (2007)
8. Moon, Y.S., Whang, K.Y., Loh, W.K.: Duality-based subsequence matching in time-series databases. In: ICDE, pp. 263–272 (2001)
9. Mueen, A., Keogh, E., Young, N.: Logical-shapelets: an expressive primitive for time series classification. In: Proceedings of the 17th ACM SIGKDD International Conference on Knowledge Discovery and Data Mining, KDD 2011, pp. 1154–1162. ACM (2011)
10. Papapetrou, P., Athitsos, V., Potamias, M., Kollios, G., Gunopulos, D.: Embedding-based Subsequence matching in Time-series Databases. ACM Trans. Database Syst. **36**(3), 17:1–17:39 (2011)
11. Shieh, J., Keogh, E.J.: iSAX: indexing and mining terabyte sized time series. In: SIGKDD, pp. 623–631 (2008)
12. Yi, B., Faloutsos, C.: Fast time sequence indexing for arbitrary Lp norms. In: VLDB, pp. 385–394 (2000)
13. Zoumpatianos, K., Idreos, S., Palpanas, T.: ADS: the adaptive data series index. VLDB J. **25**(6), 843–866 (2016)

Generating Fixed-Size Training Sets for Large and Streaming Datasets

Stefanos Ougiaroglou[1,2]($\boxtimes$), Georgios Arampatzis[1], Dimitris A. Dervos[1], and Georgios Evangelidis[2]

[1] Department of Information Technology, Alexander TEI of Thessaloniki,
57400 Sindos, Greece
stoug@uom.edu.gr, arampatzisgi@gmail.com, dad@it.teithe.gr
[2] Department of Applied Informatics, School of Information Sciences,
University of Macedonia, 54636 Thessaloniki, Greece
gevan@uom.gr

Abstract. The k Nearest Neighbor is a popular and versatile classifier but requires a relatively small training set in order to perform adequately, a prerequisite not satisfiable with the large volumes of training data that are nowadays available from streaming environments. Conventional Data Reduction Techniques that select or generate training prototypes are also inappropriate in such environments. Dynamic RHC (dRHC) is a prototype generation algorithm that can update its condensing set when new training data arrives. However, after repetitive updates, the size of the condensing set may become unpredictably large. This paper proposes dRHC2, a new variation of dRHC, which remedies the aforementioned drawback. dRHC2 keeps the size of the condensing set in a convenient, manageable by the classifier, level by ranking the prototypes and removing the least important ones. dRHC2 is tested on several datasets and the experimental results reveal that it is more efficient and noise tolerant than dRHC and is comparable to dRHC in terms of accuracy.

Keywords: k-NN classification · Data reduction · Prototype generation · Data streams · Clustering

1 Introduction

The problem of handling fast data streams [1] or large datasets that cannot reside in main memory has attracted the attention of the Data Mining and Machine Learning research communities. Moreover, researchers focus on how to run data mining algorithms on devices with limited memory (e.g., sensor devices) avoiding data transferring costs to powerful processing servers. Classification is a typical data mining task that has many applications on all aforementioned environments.

Classification algorithms (or classifiers) try to assign unclassified items to a set of predefined classes, on the basis of the available training dataset, i.e., a set of already classified items. Classifiers can be either model-based (eager) or instance-based (lazy). Both aim at accurate class prediction but they differ on

© Springer International Publishing AG 2017
M. Kirikova et al. (Eds.): ADBIS 2017, LNCS 10509, pp. 88–102, 2017.
DOI: 10.1007/978-3-319-66917-5_7

how they work. An eager classifier pre-processes the training set and builds a model that is then used to classify unclassified items. In contrast, lazy classifiers do not build any model. They classify a new item by scanning the whole training set. The size and the quality of the training set is vital for both types of classifiers. These factors determine the effectiveness and the efficiency of the classifier. However, if the size of the training set is very large, the usage of any classifier is prohibitive due to the high computational cost involved.

k-Nearest Neighbors (k-NN) is a well-known and extensively used lazy classifier [5]. When an unclassified item arrives, the algorithm scans the available training data and retrieves the k nearest items or neighbors to it according to a distance metric (e.g., Euclidean distance). Then, the unclassified item is assigned to the most common class among the classes of the k nearest neighbors.

The k-NN classifier is an effective classifier especially when it is used on small training sets. For larger training sets, its performance degrades because all distances between the new item and the training data items must be computed. In addition, contrary to the eager classifiers that can discard the training data after the construction of the classification model, the k-NN classifier has higher storage requirements since it must have the training set always available. In addition, k-NN classifier is not noise tolerant. Noise misleads the classifier and downgrades classification accuracy. The application of a Data Reduction Technique (DRT) that builds a small representative (condensing) set of the initial training data as a preprocessing step can deal with these weaknesses.

Since the k-NN classifier does not build any classification model, it can be easily adapted in streaming environments [1]. However, it can be used only on a portion of a training stream (e.g., a data window thresholded by the user). Another approach could be the maintenance of a condensing set built from the data stream.

Dynamic RHC (dRHC) [12] is a DRT that incrementally builds its condensing set. A new training data segment can be used to update an existing condensing set without applying the prototype generation procedure from scratch over the complete training data (new and old training data). Therefore, dRHC is appropriate for dynamic environments where new training data becomes gradually available and for very large datasets that can not fit in main memory.

The experimental results presented in [12] demonstrate that dRHC is a fast DRT, achieves high reduction rates and does not degrade the classification accuracy achieved by the k-NN classifier on the original training set. However, after repetitive updates of the condensing set, its size may exceed the size of the available memory. This observation is behind the motivation of the present work. The contribution is the development of a new version of dRHC that remedies this drawback by keeping the size of the condensing set fixed. In particular, the paper proposes dRHC2 that ranks the prototypes and removes the weakest ones when the size of the condensing set exceeds a predefined threshold value. The experimental study shows that dRHC2 is faster than dRHC while keeping accuracy at high levels.

The rest of this paper is structured as follows: Sect. 2 discusses the background knowledge on DRTs and their limitations. Section 3 reviews the dRHC algorithm. Section 4 considers in detail the proposed dRHC2 algorithm. In Sect. 5, both algorithms are experimentally compared to each other on fourteen datasets. The experimental results are validated by the Wilcoxon signed rank test. Section 6 concludes the paper and proposes directions for future work.

2   Background Knowledge

Data Reduction Techniques can be grouped into two main categories: (i) Prototype Selection (PS) algorithms that collect representative items (or prototypes) from the initial training set [8], and (ii) Prototype Generation (PG) algorithms that generate prototypes by summarizing on similar items [15]. PS algorithms can be either condensing or editing. The latter aim for improved classification accuracy by removing noise from the training data and by "cleaning" the decision boundaries between the discrete classes. On the other hand, PS-condensing and PG algorithms aim for data condensation, i.e., the construction of a condensing set of the initial training data.

Most PS-condensing and PG algorithms are based on a simple observation: the items that do not define the decision boundaries between classes can be removed without loss of accuracy. Consequently, PS-condensing algorithms try to collect only the items that are close to the decision boundaries (border items). On the other hand, PG algorithms generate many prototypes for the close-border areas and few for the "internal" data areas. It is worth mentioning that most of the PG and PS-condensing algorithms are sensitive to noise. Hence, for training sets with noise, an editing algorithm must be applied beforehand.

A great number of DRTs have been proposed in the literature. PS and PG algorithms are reviewed, categorized and compared to each other in [15] and [8], respectively. Although there are some exceptions (e.g., IBL algorithms [2,4]), DRTs are usually memory-based. This implies that the whole training set must reside in main memory. This property renders DRTs inappropriate for very large training sets that cannot fit into the device's memory or for devices with limited memory (e.g., sensor devices) without transferring data to a server over the network for processing.

Furthermore, these DRTs cannot consider new training items after the construction of the condensing sets, i.e., they cannot update their condensing set in a dynamic manner. Suppose that a DRT is applied over a training set TS and builds a condensing set. Moreover, suppose that new training items D become available. For the construction of an updated condensing set, the DRT must run from scratch over the complete training set $TS \cup D$. Therefore, all training items must always be available. Hence, DRTs are inappropriate for dynamic/streaming environments where new training items become gradually available. The Dynamic RHC is a PG algorithm that can be used in dynamic/streaming environments [12].

3   The dRHC Algorithm

Dynamic RHC (dRHC) is a descendant of the Reduction through Homogeneous Clusters (RHC) algorithm [11,12]. The latter is based on the concept of cluster homogeneity. RHC utilizes k-means clustering. Initially, it considers the training set with D classes as a non-homogeneous cluster C. The algorithm computes a class-mean for each class in C by averaging on the corresponding items. Then, k-means clustering is applied on C by using these class-means as initial seeds and D clusters are formed. Each item in C is assigned to one of the D clusters. Then, RHC examines the D clusters. If a cluster is homogeneous (i.e., has items of only one class), its cluster-mean constitutes a prototype and is placed in the condensing set. If a cluster is non-homogeneous, the aforementioned procedure is applied recursively on it. RHC terminates when there are no non-homogeneous clusters left. In other words, each homogeneous cluster contributes a prototype.

RHC builds many prototypes for close-border areas and fewer for the "internal" areas. By using the class-means as initial seeds for k-means clustering, quick discovery of large homogeneous clusters is feasible (the larger clusters discovered, the higher reduction rates achieved). The main disadvantage of RHC is that it is memory-based. Moreover, it cannot cope with training sets that cannot fit in memory or streaming data.

The dRHC algorithm retains all the properties of RHC. In addition, it can also manage large or streaming datasets by considering the available data in the form of data segments. The application of dRHC involves two phases (see Fig. 1): (i) *initial condensing set construction* and (ii) *condensing set update*. The *initial condensing set construction* phase is executed only once, when the first data segment arrives. The following data segments are processed by the *condensing set update* phase. The *initial condensing set construction* phase is almost identical to RHC. The only difference is that each prototype of the Condensing Set (CS) stores a weight attribute that counts how many items are represented by the specific prototype. The *condensing set update* phase uses the prototypes of the current condensing set and the items of a new data segment in order to build a set of initial clusters and then it proceeds similarly to RHC.

The *condensing set update* phase can be easily understood by considering Algorithm 1. The algorithm has two input parameters: an already constructed (old) condensing set (OCS) and a data segment (Seg) with new training items. The output is an updated (new) condensing set (NCS). The *condensing set update* phase utilizes a queue (Q) data structure to hold the unprocessed clusters (lines 2–3). Initially, the *condensing set update* phase initializes as many clusters, as the number of prototypes in OCS (lines 3–6). Each cluster contains only the prototype. Then, for each item x in Seg, the algorithm identifies the nearest prototype and assigns x to the corresponding cluster (lines 7–11). All items in Seg have $weight = 1$. Hence, each cluster contains an old prototype with $weight >= 1$ and items (from Seg) with $weight = 1$. The clusters are enqueued to Q for further processing (lines 12–14).

The *condensing set update* phase generates the updated condensing set NCS considering the weight values. For each homogeneous cluster, it computes the

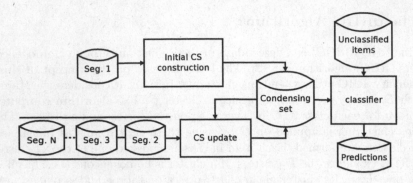

Fig. 1. Classification using dRHC

weighted mean of the cluster (lines 19–22). For each non-homogeneous cluster C, a weighted mean is estimated for each class in C (lines 24–29). Like RHC, dRHC utilizes the weighted class means as initial seeds for k-Means clustering (line 30). Of course, dRHC uses the version of k-means that computes the cluster means by considering the corresponding weights. Thus, a vector attribute a_j, $j = 1, 2, \ldots, n$ of a class or cluster mean m_C (lines 20, 26) is computed by the following formula:

$$m_C.a_j = \frac{\sum_{x_i \in C} x_i.a_j \times x_i.weight}{\sum_{x_i \in C} x_i.weight}$$

k-means clustering builds as many clusters as the number of different classes in C. They are enqueued to Q for further processing. The repeat-until loop (lines 17,35) ends when all clusters become homogeneous, i.e., when Q is empty. Note that old prototypes usually have weights greater than one and have higher influence in the computation of a new weighted class or cluster mean than any item of a new data segment, whose weight is one.

An example of the execution of the *condensing set update* phase is depicted in Fig. 2. More specifically, Fig. 2(a) presents an existing condensing set created by either the *initial CS construction* phase or by a previously executed *condensing set update* phase. The condensing set has three prototypes with the corresponding weights. Suppose that a segment with seven new training items is available and is about to be processed (Fig. 2(b)). Each new item has a weight equal to one. The first step is the assignment of each new item to the cluster of the nearest prototype (Fig. 2(c)). The new items assigned to cluster A have the same class as the class of the prototype in A. Therefore, the prototype "moves" towards the new items (Fig. 2(d)). This is achieved by computing the weighted mean in A. The latter constitutes the new prototype and is placed in the condensing set along with its new weight. On the other hand, there is no item that has been assigned to cluster B. Hence, the corresponding prototype remains unchanged. Cluster C becomes non-homogeneous. For each class in C, a weighted class mean is computed, k-means is executed and two homogeneous clusters are built (see Fig. 2(d,e)).

Algorithm 1. dRHC: condensing set update phase

Input: OCS, Seg
Output: NCS

1: {**Queue Initialization**}
2: $Q \leftarrow \varnothing$
3: $CLi \leftarrow \varnothing$ {empty list of clusters}
4: **for** each prototype $m \in OCS$ **do**
5:     add new cluster $C = \{m\}$ in CLi
6: **end for**
7: **for** each item $x \in Seg$ **do**
8:     $x.weight = 1$
9:     find the Cluster $C_x \in CLi$ with the nearest to x prototype
10:     $C_x \leftarrow C_x \cup \{x\}$ {The mean of C_x is not recomputed}
11: **end for**
12: **for** each cluster C in CLi **do**
13:     Enqueue(Q, C)
14: **end for**
15: {**Construction of NCS**}
16: $NCS \leftarrow \varnothing$
17: **repeat**
18:     $C \leftarrow$ Dequeue(Q)
19:     **if** C is homogeneous **then**
20:         $m \leftarrow$ weighted mean of C
21:         $m.weight \leftarrow \sum_{x_i \in C} x_i.weight$
22:         $NCS \leftarrow NCS \cup \{m\}$
23:     **else**
24:         $M \leftarrow \varnothing$ {M is the set of weighted class means}
25:         **for** each class L in C **do**
26:             $m_L \leftarrow$ weighted mean of L
27:             $m_L.weight \leftarrow \sum_{x_i \in L} x_i.weight$
28:             $M \leftarrow M \cup \{m_L\}$
29:         **end for**
30:         $NewClusters \leftarrow k$-means(C, M)
31:         **for** each cluster $NC \in NewClusters$ **do**
32:             Enqueue(Q, NC)
33:         **end for**
34:     **end if**
35: **until** IsEmpty(Q)
36: **return** NCS

Finally, a weighted cluster mean for each cluster is computed and its weight is estimated. They constitute new prototypes and they are placed in the condensing set (Fig. 2(f)).

In the experimental study presented in [12], RHC and dRHC were compared to each other and against state-of-the-art PS [2,9,10,17] and PG [14] algorithms. It turns out that dRHC is the fastest algorithm (with the lowest preprocessing cost) and builds the smallest condensing set without loss of classification accuracy.

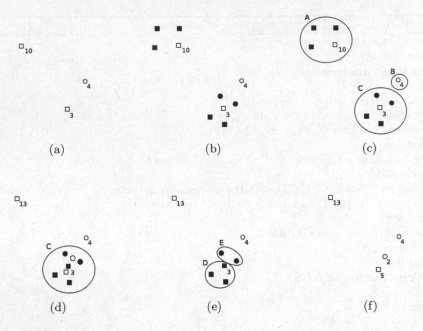

Fig. 2. Example of execution of the *condensing set update* phase of dRHC

4   The Proposed dRHC2 Algorithm

Although dRHC seems to be appropriate for large training datasets or streaming environments, it has a major drawback: after repetitive executions of the
condensing set update phase, the condensing set may become very large. Therefore, there is a need to keep the size of the condensing set fixed. The dRHC2
algorithm is a new version of dRHC that copes with this drawback.

dRHC2 accepts an input parameter that is the desirable maximum size of
the condensing set. In effect, dRHC2 executes similarly to dRHC, but when
the size of the condensing set exceeds the maximum size (T), the algorithm
removes prototypes to save space. In other words, dRHC maintains the size of
the condensing set equal to T by removing the least important prototypes from
the updated condensing set. Thus, dRHC2 introduces a post-processing step in
the *condensing set update phase* presented in Sect. 3. T is adjusted by the user
taking into account the trade-off between accuracy and computational cost, the
level of noise in the data and of course the system limitations.

The dRHC2 algorithm includes a mechanism that ranks the prototypes
according to their importance. Next, it retains only the T first prototypes and
removes the rest. An essential role to the determination of the prototypes' "importance" plays the concept of prototype weight. One may claim that a prototype
with high weight value is more important than a prototype with lower weight.
This is not absolutely true. If a prototype was generated in a recent *condensing
set update* phase, it probably has a lower weight value than an older prototype.

Therefore, it is doomed to be discarded. On the other hand, an old prototype that has survived many executions of the *condensing set update* phase, probably has a high weight value. Thus, the latter tends to be favored against the new prototypes and will survive in the condensing set. Consequently, if dRHC2 ranks the prototypes according to their weight, it may discard most of the prototypes arriving in later stages.

In order to more appropriately rank the prototypes, dRHC2 keeps a counter of the data segments that have arrived and adds an extra attribute to each prototype that denotes the number (r) of the data segment on which the prototype was generated. After each execution of the *condensing set update* phase, dRHC2 ranks the prototypes according to a measure called AnA, which stands for Average number of Arrivals and takes into account the weight and the age of each prototype. In effect, AnA estimates the prototype weight per data segment. AnA is computed as follows:

$$AnA = \frac{w}{ds - r + 1}$$

where

- w is the prototype weight that is the number of items summarized together and represented by the specific prototype
- r is the number of the data segment on which the prototype was generated
- ds is the number of the current data segment

By using AnA, no prototypes are favored in the ranking. The AnA of the most recent prototypes, i.e., the ones created by the last *condensing set update* phase, is equal to their w. On the other hand, the AnA of older prototypes is practically their w divided by their age.

Algorithm 4 summarizes the post-processing step introduced by dRHC2. This step is the only difference between dRHC and dRHC2. The algorithm accepts the set maximum condensing set size T and the condensing set produced by either the *initial condensing set construction* phase or a *condensing set update* phase (NCS) as input parameters. When $|NCS| > T$, the algorithm returns a new condensing set NCS with $|NCS| = T$.

Algorithm 2. dRHC2: Post-processing step

Input: NCS, T
Output: NCS
1: **if** $|NCS| > T$ **then**
2:     $NCS \leftarrow$ keep the T prototypes with the highest AnA
3: **end if**
4: **return** NCS

In case of data with noise, dRHC inevitably generates prototypes that represent noisy items. However, these prototypes usually have low weight and AnA values. If we adopt dRHC2 instead of dRHC, these prototypes are eventually removed.

Therefore, we expect dRHC2 to handle noisy training data and, in such cases, achieve higher accuracy than dRHC.

One can claim that dRHC2 maintains a condensing set that adapts according to the concept drift [16] that may exist in the data stream. This assumption may not be absolutely correct. The newly generated prototypes or the prototypes that are updated after a *condensing set update* are not favored against the older ones.

5   Performance Evaluation

5.1   Experimental Setup

The performance of dRHC2 was tested against dRHC using fourteen datasets distributed by the KEEL dataset repository[1] [3]. Table 1 summarizes on the datasets used. In the experimental study presented in [12], RHC and dRHC were evaluated against five state-of-the-art data reduction techniques [2,9,10,14,17] on the same fourteen datasets. Therefore, we "indirectly" compare dRHC2 to these techniques. The reader can execute dRHC and dRHC2 over the particular datasets using WebDR[2] [13].

The Euclidean distance was adopted as the distance metric. All algorithms were implemented in C. Except KDD, all the other datasets were used without normalization. We randomized the datasets that were distributed sorted on the class label. For each dataset and algorithm, three average measurements obtained via five-fold cross-validation were estimated: (i) Accuracy (ACC), (ii) Reduction Rate (RR), and, (iii) Preprocessing Cost (PC) in terms of distance computations. To be fair, we should notice that PC measurements do not include the small cost overhead introduced by the ranking of the prototypes. Accuracy was estimated by running k-NN classification with $k = 1$. Note that we did not use special evaluation methods for data streams (e.g., test-then-train) [7], since the used datasets do not exhibit concept drift.

For KDD, we removed the three nominal and the two fixed-value attributes that exist in the dataset. Moreover, KDD contains many duplicates that were also removed. The attribute ranges of KDD vary extremely. We normalized them to the $[0, 1]$ range.

The dRHC and dRHC2 algorithms consume data in the form of data segments. Thus, the training sets of the datasets were split into segments of a specific size. The data segment that we adopted for each dataset is presented by the last column in Table 1. The segment size corresponds to either the size of the buffer that accepts data from a stream or the size of the available memory (scenario of limited main memory and/or large datasets). The experimental study presented in [12] shows empirically that the size of the data segment does not affect the performance of dRHC. Therefore, we did not conduct experiments with different segment sizes. The performance measurements were estimated following the arrival of all the data segments.

---

[1] http://sci2s.ugr.es/keel/datasets.php.
[2] https://atropos.uom.gr/webdr.

Table 1. Dataset description

Dataset	Size	Attributes	Classes	Memory/Buffer size
Letter Image Recognition (LIR)	20,000	16	26	2,000
Magic G. Telescope (MGT)	19,020	10	2	1,902
Pen-Digits (PD)	10,992	16	10	1,000
Landsat Satellite (LS)	6,435	36	6	572
Shuttle (SH)	58,000	9	7	1,856
Texture (TXR)	5,500	40	11	440
Phoneme (PH)	5,404	5	2	500
Balance (BL)	625	4	3	100
Pima (PM)	768	8	2	100
Ecoli (ECL)	336	7	8	200
Yeast (YS)	1,484	8	10	396
Twonorm (TN)	7,400	20	2	592
MONK 2 (MN2)	432	6	2	115
KddCup (KDD)	141,481	36	23	4,000

The dRHC2 algorithm accepts as an input parameter the set maximum condensing set size (T). For each dataset, T was adjusted to a certain percentage of the size of the condensing set generated by dRHC. The T values chosen were the 85%, 70%, 55% and 40% of the size of the condensing sets constructed by dRHC. In other words, when dRHC builds a condensing set with 1000 prototypes, four experiments were conducted to evaluate the performance of dRHC2 with the following T values: $T = 850$, $T = 700$, $T = 550$, $T = 400$.

5.2 Comparisons

Table 2 presents the performance measurements of dRHC and dRHC2. The best measurements are highlighted in boldface. The preprocessing cost measurements are in million distance computations. Although, both dRHC and dRHC2 are adopted when the conventional k-NN classifier cannot be applied due to its limitations, for reference, we report the accuracy measurements achieved by applying k-NN on the original complete training set.

Obviously, the lower the T value used, the higher reduction rates achieved, the lower preprocessing cost needed and, of course, the faster the classifier is. Therefore, the results that concern RR and PC measurements are to be expected since dRHC2 adopts a ceiling value for the maximum size of its condensing set and maintains it throughout the whole execution. Figure 3 depicts this property of dRHC2 by presenting diagrams for two indicative datasets. The diagrams illustrate how the preprocessing cost and the size of the condensing set initially increases and remains constant when T is reached.

The ACC performance measurements for dRHC2 are most promising (see in Table 2). In cases of datasets with low level of noise, dRHC2 achieves accuracy

Table 2. Comparison of dRHC2 and dRHC in terms of Accuracy (ACC(%)), Reduction Rate (RR(%)) and Preprocessing Cost (PC (millions of distance computations))

Data	T %	ACC (%)			RR (%)		PC (M)	
		1NN	dRHC	dRHC2	dRHC	dRHC2	dRHC	dRHC2
LIR	85:	95.83	**93.92**	93.40	88.18	89.95	19.57	19.18
	70:			92.84		91.73		17.72
	55:			91.85		93.50		15.36
	40:			90.08		95.27		12.29
MGT	85:	78.14	72.97	74.19	74.62	78.42	26.03	25.85
	70:			74.64		82.24		24.41
	55:			75.11		86.04		21.71
	40:			**75.97**		89.95		17.73
PD	85:	99.35	98.49	98.60	97.23	97.65	1.44	1.41
	70:			**98.63**		98.06		1.32
	55:			98.34		98.48		1.16
	40:			97.73		98.90		0.93
LS	85:	90.60	88.50	**88.61**	88.35	90.09	1.53	1.51
	70:			88.53		91.84		1.42
	55:			88.58		93.59		1.27
	40:			87.89		95.34		1.03
SH	85:	99.82	**99.70**	99.69	99.50	99.58	7.98	7.61
	70:			99.61		99.65		6.91
	55:			99.56		99.72		5.95
	40:			99.37		99.80		4.73
TXR	85:	99.02	**97.60**	97.38	94.95	95.71	0.68	0.67
	70:			97.00		96.46		0.62
	55:			96.46		97.23		0.53
	40:			95.76		97.98		0.43
PH	85:	90.10	85.38	**86.14**	82.34	84.99	1.64	1.62
	70:			85.62		87.65		1.52
	55:			85.21		90.29		1.33
	40:			84.88		92.95		1.08
BL	85:	78.40	70.56	71.84	78.12	81.40	0.029	0.029
	70:			73.12		84.60		0.027
	55:			77.28		88.00		0.025
	40:			**81.60**		91.20		0.021
PM	85:	68.36	63.93	65.23	65.11	70.41	0.064	0.063
	70:			67.96		75.61		0.060
	55:			68.09		80.81		0.055
	40:			**68.23**		86.02		0.046

(*continued*)

Table 2. (*continued*)

Data	T %	ACC (%)			RR (%)		PC (M)	
		1NN	dRHC	dRHC2	dRHC	dRHC2	dRHC	dRHC2
ECL	85:	79.78	71.46	74.73	68.92	73.61	0.015	0.015
	70:			76.22		78.44		0.015
	55:			78.28		82.90		0.014
	40:			**79.75**		87.73		0.013
YS	85:	52.02	48.38	48.31	51.23	58.59	0.306	0.306
	70:			48.65		65.91		0.306
	55:			48.99		73.23		0.278
	40:			**52.83**		80.47		0.244
TN	85:	94.88	93.08	94.03	95.37	96.06	0.695	0.688
	70:			94.54		96.76		0.654
	55:			95.45		97.45		0.590
	40:			**95.93**		98.14		0.495
MN2	85:	90.51	**97.68**	96.28	96.88	97.63	0.0040	0.0039
	70:			96.29		97.80		0.0038
	55:			94.45		98.27		0.0038
	40:			93.52		98.84		0.0040
KDD	85:	99.71	99.42	99.47	99.22	99.34	54.70	53.56
	70:			**99.51**		99.46		49.81
	55:			99.50		99.58		43.48
	40:			99.48		99.69		34.60
AVG	85:	86.89	84.36	84.84	84.29	86.67	8.19	8.04
	70:			85.22		89.01		7.49
	55:			85.51		91.36		6.55
	40:			**85.93**		93.73		5.26

comparable to that of dRHC. On the other hand, in cases of datasets with a high level of noise, dRHC2 achieves higher accuracy than dRHC. This happens because some less important dRHC prototypes originate from noisy data. These prototypes probably have low AnA values, they are ranked low, and they get discarded, eventually. Hence, in cases of datasets with high level of noise, we observe that dRHC2 with the lowest T value achieves even higher accuracy than the conventional k-NN classifier.

One final comment concerning the average measurements depicted in the last row of Table 2: dRHC2 can achieve even better accuracy than dRHC by avoiding the arbitrary growth in size of the condensing set, and by reducing the preprocessing cost.

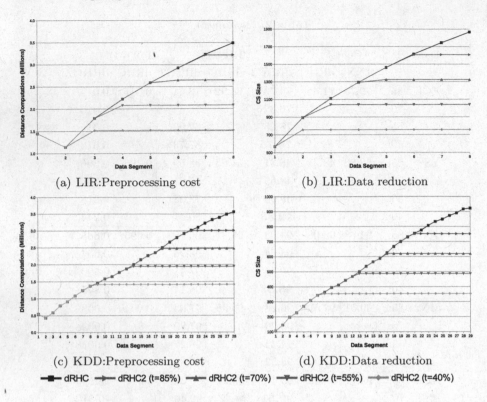

(a) LIR:Preprocessing cost   (b) LIR:Data reduction

(c) KDD:Preprocessing cost   (d) KDD:Data reduction

dRHC   dRHC2 (t=85%)   dRHC2 (t=70%)   dRHC2 (t=55%)   dRHC2 (t=40%)

Fig. 3. Processing per data segment

Table 3. Results of Wilcoxon signed rank test

Methods	Accuracy	
	w/l	Wilc.
dRHC vs dRHC2 ($T=85\%$)	5/9	0.158
dRHC vs dRHC2 ($T=70\%$)	4/10	0.103
dRHC vs dRHC2 ($T=55\%$)	6/8	0.363
dRHC vs dRHC2 ($T=40\%$)	7/7	0.397

5.3   Wilcoxon Signed Rank Test

The experimental study is complemented by providing the results of Wilcoxon signed rank test [6] to statistically validate the ACC measurements presented in Table 2. The test compares dRHC and dRHC2 in pairs considering the performance on each dataset. All four versions of dRHC2 (with different T values) were included in the test. Since dRHC2 is the dominant algorithm in terms of RR and PC, there is no need to include the corresponding measures in the test.

Table 3 presents the results of the Wilcoxon test. Columns labeled with "w/l" show the number of wins and losses, respectively. The Wilcoxon value

("Wilc." column) shows how significant the difference between the algorithms is. If it is lower than 0.05, one can claim that the difference is statistically significant. The results of the test reveal that dRHC and dRHC2 do not statistically differ in terms of accuracy. Thus, dRHC2 can be used instead of dRHC when there is need for a condensing set with a fixed size.

6 Conclusions and Future Work

The paper reports on dRHC2, a noise-tolerant PG algorithm that maintains a fixed size condensing set by monitoring a training data stream or by managing a large dataset which cannot reside in memory. The experimental study yields promising results. Even when the condensing set generated by dRHC2 is less than half the size of that generated by dRHC, there is no loss in accuracy and in many cases the accuracy achieved by dRHC2 is even higher. In addition, because the size of the condensing set is not arbitrary increased and remains constant, the preprocessing cost remains low and constant till the end of the execution.

Our plans for future work include the development of variations of dRHC2 that will be able to fully handle data streams with concept drift. This could be achieved by increasing the importance of newly generated prototypes.

References

1. Aggarwal, C.: Data Streams: Models and Algorithms. Advances in Database Systems Series. Springer, Heidelberg (2007)
2. Aha, D.W., Kibler, D., Albert, M.K.: Instance-based learning algorithms. Mach. Learn. **6**(1), 37–66 (1991). http://dx.doi.org/10.1023/A:1022689900470
3. Alcalá-Fdez, J., Fernández, A., Luengo, J., Derrac, J., García, S.: KEEL datamining software tool: data set repository, integration of algorithms and experimental analysis framework. Multi. Valued Logic Soft Comput. **17**(2–3), 255–287 (2011)
4. Beringer, J., Hüllermeier, E.: Efficient instance-based learning on data streams. Intell. Data Anal. **11**(6), 627–650 (2007). http://dl.acm.org/citation.cfm?id=1368018.1368022
5. Cover, T., Hart, P.: Nearest neighbor pattern classification. IEEE Trans. Inf. Theor. **13**(1), 21–27 (2006). http://dx.doi.org/10.1109/TIT.1967.1053964
6. Demšar, J.: Statistical comparisons of classifiers over multiple data sets. J. Mach. Learn. Res. **7**, 1–30 (2006). http://dl.acm.org/citation.cfm?id=1248547.1248548
7. Gama, J.A., Sebastião, R., Rodrigues, P.P.: Issues in evaluation of stream learning algorithms. In: Proceedings of the 15th ACM SIGKDD International Conference on Knowledge Discovery and Data Mining, pp. 329–338, KDD 2009. ACM, New York (2009). http://doi.acm.org/10.1145/1557019.1557060
8. Garcia, S., Derrac, J., Cano, J., Herrera, F.: Prototype selection for nearest neighbor classification: taxonomy and empirical study. IEEE Trans. Pattern Anal. Mach. Intell. **34**(3), 417–435 (2012). http://dx.doi.org/10.1109/TPAMI.2011.142
9. Hart, P.E.: The condensed nearest neighbor rule. IEEE Trans. Inf. Theory **14**(3), 515–516 (1968)

10. Olvera-Lopez, J.A., Carrasco-Ochoa, J.A., Trinidad, J.F.M.: A new fast prototype selection method based on clustering. Pattern Anal. Appl. **13**(2), 131–141 (2010)
11. Ougiaroglou, S., Evangelidis, G.: Efficient dataset size reduction by finding homogeneous clusters. In: Proceedings of the Fifth Balkan Conference in Informatics, pp. 168–173, BCI 2012. ACM, New York (2012). http://doi.acm.org/10.1145/2371316.2371349
12. Ougiaroglou, S., Evangelidis, G.: RHC: a non-parametric cluster-based data reduction for efficient k-NN classification. Pattern Anal. Appl. **19**(1), 93–109 (2014). http://dx.doi.org/10.1007/s10044-014-0393-7
13. Ougiaroglou, S., Evangelidis, G.: WebDR: a web workbench for data reduction. In: Calders, T., Esposito, F., Hüllermeier, E., Meo, R. (eds.) Machine Learning and Knowledge Discovery in Databases. LNCS, vol. 8726, pp. 464–467. Springer, Heidelberg (2014). http://dx.doi.org/10.1007/978-3-662-44845-8_36
14. Sánchez, J.S.: High training set size reduction by space partitioning and prototype abstraction. Pattern Recogn. **37**(7), 1561–1564 (2004)
15. Triguero, I., Derrac, J., Garcia, S., Herrera, F.: A taxonomy and experimental study on prototype generation for nearest neighbor classification. Trans. Sys. Man Cyber Part C **42**(1), 86–100 (2012). http://dx.doi.org/10.1109/TSMCC.2010.2103939
16. Tsymbal, A.: The problem of concept drift: definitions and related work. Technical report TCD-CS-2004-15, The University of Dublin, Trinity College, Department of Computer Science, Dublin, Ireland (2004)
17. Wilson, D.R., Martinez, T.R.: Reduction techniques for instance-based learning algorithms. Mach. Learn. **38**(3), 257–286 (2000). http://dx.doi.org/10.1023/A:1007626913721

OLAP

Detecting User Focus in OLAP Analyses

Mahfoud Djedaini[✉], Nicolas Labroche, Patrick Marcel, and Verónika Peralta

University of Tours, Tours, France
{mahfoud.djedaini,nicolas.labroche,patrick.marcel,
veronika.peralta}@univ-tours.fr

Abstract. In this paper, we propose an approach to automatically detect focused portions of data cube explorations by using different features of OLAP queries. While such a concept of focused interaction is relevant to many domains besides OLAP explorations, like web search or interactive database exploration, there is currently no precise formal, commonly agreed definition. This concept of focus phase is drawn from Exploratory Search, which is a paradigm that theorized search as a complex interaction between a user and a system. The interaction consists of two different phases: an exploratory phase where the user is progressively defining her information need, and a focused phase where she investigates in details precise facts, and learn from this investigation. Following this model, our work is, to the best of our knowledge, the first to propose a formal feature-based description of a focused query in the context of interactive data exploration. Our experiments show that we manage to identify focused queries in real navigations, and that our model is sufficiently robust and general to be applied to different OLAP navigations datasets.

1 Introduction

Interactive Data Exploration (IDE) is the task of efficiently extracting knowledge from data even if we do not know exactly what we are looking for [4]. Typically, an exploration includes several queries where the result of each query triggers the formulation of the next one. OLAP analysis of data cubes is a particular case of IDE, that takes advantage of simple primitives like drill-down or slice-and-dice for the navigation. For example, an analyst may explore several attributes and cross several dimensions, in order to find clues, causes or correlations to explain unexpected data values, until identifying the most relevant data subsets and deeply analyzing them. While OLAP has been around for more than 20 years, little is known about typical navigational behavior.

Exploratory Search, however, is a sub-domain of Information Retrieval that studies user behaviors during their explorations [11]. The basic model of exploration distinguishes two main phases. In a first phase, called exploratory browsing, users are likely to explore the space, as well as better defining and understanding their problem. At this stage, the problem is being limited, labeled, and a framework for the answer is defined. Over time, the problem becomes more clearly defined, and the user starts to conduct more targeted searches.

© Springer International Publishing AG 2017
M. Kirikova et al. (Eds.): ADBIS 2017, LNCS 10509, pp. 105–119, 2017.
DOI: 10.1007/978-3-319-66917-5_8

In this second phase, called focused phase, users (re)formulate query statements, examine search results, extract and synthesize relevant information.

Detecting focused phases in an exploration can be exploited in a variety of applications, for instance in the context of user exploration assistants. When focused, an analyst would expect more *precise* queries, related to what she is currently analyzing. On the contrary, when exploring the data, the analyst would prefer more *diverse* queries, for a better data space coverage. Focus detection could also be used in data visualization. In a focus phase, an analyst would prefer a highly focused interface, presenting to her in details what she is currently investigating. Oppositely, an analyst who is exploring the data would rather expect an interface presenting an overview of available data, highlighting the diversity of available dimensions of analysis.

In this paper, we propose an approach to automatically detect focused phases in OLAP explorations. While there exists no formal definition or consensual formula to decide whether an OLAP exploration or a query is focused or not, the concept of focus can be intuitively described by different characteristics that indicate a focused activity. Our hypothesis is indeed that a definition of focus is highly dependent of a fine characterization of the queries composing an exploration. For instance, the granularity level or the number of filters of a query, or the number of OLAP operations that separate two consecutive queries, are such characteristics. In our proposal, we identify a total of *19 characteristics* to finely describe different aspects of a query, either intrinsically, relatively to its predecessor or relatively to the whole exploration containing it. We show that it is possible to define a metric to quantify each of these characteristics. It is then possible to see the central question of defining a formal model of the focus based on the characteristics as a classification problem where the descriptive features are the metrics' scores and the output variable indicates if a query is focused or not. By choosing an appropriate classification approach and well specified metrics, our work demonstrates that it is possible to build an interpretable, yet efficient, model for the focus that is consistent with expert evaluation on real OLAP navigations and predefined behavioral patterns on simulated OLAP navigations that were defined agnostic of any focus definition.

The paper structure is as follows: Sect. 2 showcases an example for motivating our approach, which is introduced in Sect. 3. Section 4 describes the formal framework and Sect. 5 details the metrics used to characterize focus. Section 6 highlights experiments and discusses results. Finally, before concluding, Sect. 7 presents related works.

2   Motivating Example

Our example is taken from the Star Schema Benchmark (SSB) specification [8][1]. This benchmark defines a workload consisting of 4 flights of queries.

---

[1] We redirect the reader to the SSB specification for the logical schema and the exact SQL text of the queries. In our example, we consider the instance is generated with a scale factor of 1.

Table 1. Description of SSB query flight 3. The measure sum(lo_revenue) is the same in all queries.

Query flight 3	Q1	Q2	Q3	Q4
Where	Year in [1992,1997], c_region=ASIA, s_region =ASIA	Year in [1992,1997] c_nation = 'US', s_nation = 'US'	Year in [1992,1997] (c_city= 'UKI1'or 'UKI5'), (s_city= 'UKI1' or 'UKI5')	Yearmonth=Dec97, (c_city= 'UKI1'or 'UKI5'), (s_city='UKI1' or 'UKI5')
Group by	c_nation, s_nation, year	c_city, s_city, year	c_city, s_city, year	c_city, s_city, year
# Cells	150	596	24	3
# Tuples examined	200 k	8 k	329	5

Each flight can be seen as an exploration over a 5 dimensional cube whose schema corresponds to the relational star schema defined by the benchmark. We particularly pay attention to flight number 3, that consists of 4 queries Q1, Q2, Q3 and Q4, analyzing revenue volume (see Table 1). Q1 asks for revenue generated for region Asia (for both suppliers and customers) between 1992 and 1997, by nations. It examines over 200,000 tuples of the fact table and produces 150 cells of the cube. Q2 asks for revenue in the United States at the city level, for the same time period. It examines 8000 tuples and produces 596 cells. Q3 remains at the city level but asks for revenue in the United Kingdom, and only for two cities in the UK, examining 329 tuples and producing 24 cells. These two cities remain selected in Q4 that drills the time dimension at the month level, to just one month, examining only 5 tuples and producing 3 cells.

It can be seen that the beginning of the navigation is not focused, while the second half (Q3 and Q4) start to focus on a particular zone in the cube. The benchmark specification actually accounts for this, indicating that the last query was deliberately specified to be a "needle-in-haystack" query. We now review what differentiates the first half of the navigation, which is exploratory, from the second, focused, part.

The first two queries are only loosely related to each other. They move by relatively big jumps in the data space. They are coarse in terms of the filters and the granularity levels used, which results in big portions of the fact table being analyzed to produce relatively large answer sets. This may induce high execution time but surely also high consideration time (time taken by the user to analyze the answer set).

On the contrary, queries 3 and 4 are separated by one OLAP operation (specifically, a drill down operation), and the text of query 4 is obtained from that of query 3 with only a few modifications. The queries become finer, in the sense that more filters are accumulated on finer granularity levels, targeting a smaller portion of the fact table. The result sets are also much smaller. Content-wise, the "needle-in-haystack" effect indicates that this focus is justified by something surprising in the data.

As we will see in the next sections, our study of real navigation logs collected from users corroborate these intuitive considerations. Specifically, we have observed in these logs that longer navigations tend to incorporate quite long focused sequences, often at the middle or end of the navigation, corresponding to short jumps in the data space. In these focused sequences, queries are close to each other in terms of OLAP operations and text, and their answer sets often share cube cells. Being able to automatically detect such focus zones in navigations has many advantages, for example in experience sharing among users (the needle in the haystack, discovered in a former navigation, may be useful to other analysts) or suggesting recommendations in line with the user's immediate interests, and more generally to make user experience with big datasets less disorienting.

These observations led us to define 3 categories of features to characterize focused zones in navigations. The first category corresponds to intrinsic properties of the query taken independently of the other queries, like the filters, the answer set, etc. The second category positions a query respectively to its immediate predecessor in the navigation, for instance to detect OLAP operations. Finally, the last category positions the query relatively to the navigation itself, for instance to check if the query appears in a long chain of similar queries.

3   Characterizing and Detecting Focus Phases

Our approach aims at automatically detecting focus phases in user explorations. As mentioned above, there is yet no formula for deciding whether a query is focused or not. However, as illustrated by the previous motivation example, an expert is able to recognize a focus activity by looking at various characteristics of the queries and the exploration.

In order to quantify these intuitive characteristics, we define a set of metrics. As such, these metrics characterize different aspects of a query: the user intention (e.g., the desired granularity expressed through the aggregation level), the results (e.g., the number of cube cells retrieved), as well as its relationship to other queries (e.g., the differences between a query and its predecessor).

Then, the question of formally characterizing a focused query can be expressed as a classification problem in which all queries can be represented by scores issued from the metrics and the class output variables is binary, either "focused" or "not focused". These are the only two classes we are able to define regarding the fuzzy notion of focus. The main difficulty in this case relates to the building of a proper corpus of annotated queries by experts, to learn the model from it. Not all the classifiers meet this requirement. Indeed, as the ability to interpret what makes a query focused is a major objective in our work, we limit ourselves to linear models that learn a weight for each metric's score and then output a focus score that is computed as weighted sum over the metrics values for each query. In this context, we use an off-the-shelve SVM classifier whose separative hyperplane equation provides the expected relation to qualify the focus of a query based on our metrics scores and their associated weights. Moreover, this formalization allows to understand in a very intuitive way how each metric contributes to the detection of focus.

As detailed in Sect. 6, we used a set of real explorations over a real data cube to train and test the classifier. All the queries of all the explorations were labeled by a human expert, familiar with both the cube explored and the front end tool used for the explorations. Each query is then labeled either as focused or as exploratory. The labels we obtain are used as a ground truth for our classifier.

4   Formal Framework

This section introduces the formal framework underlying our approach, in which explorations are treated as first class citizens.

Exploration. An exploration is a triple $\langle e, lts, ets \rangle$, where $e = \langle q_1, \ldots, q_p \rangle$ is a sequence of p OLAP queries, lts is a function that gives for a query its launch time-stamp, and ets is a function that gives for a query its evaluation time-stamp. With a slight abuse of notation, we note $q \in e$ if a query q appears in the exploration e.

During their explorations, users inspect the elements of a cube instance (or simply, cube) retrieved by a query.

Cube model. Without loss of generality, the OLAP queries we consider are dimensional aggregate queries over a data cube [2]. We consider cubes under a ROLAP perspective, where dimensions consist of one or more hierarchies, each hierarchy consisting of one or more levels. Formally, a cube schema consists of: (i) n hierarchies, each hierarchy h_i being a set $Lev(h_i) = \{L_i^0, \ldots, L_i^d\}$ of levels together with a *roll-up* total order $\succeq_{h_i}$ of $Lev(h_i)$, (ii) a set of *measure* attributes M, each defined on a numerical domain. For a n-dimensional cube, a *group-by set* is an element of $Lev(h_1) \times \ldots \times Lev(h_n)$.

The elements of a cube are called cells.

Cells. Cells are tuples $\langle m_1, \ldots, m_n, meas \rangle$ where the m_i are taken in $Dom(L_i^{j_i})$, $L_i^{j_i} \in Lev(h_i)$, for all i, and $meas$ is a measure value.

Query model. A query $\langle G, P, M \rangle$ is defined by a group by set G (identifying the query granularity), a set of boolean predicates P, and a set of measures M. The answer to a query q, denoted $answer(q)$, is the set of non empty cells whose coordinates are defined by the query group by set and selection predicates.

Among the set of cells in an answer, we distinguish between base cells and aggregated cells. In base cells, m_i are taken in $Dom(L_i^0)$. In aggregate cells, there exists one m_i not in $Dom(L_i^0)$. An aggregate cell $\langle m_1, \ldots, m_n, meas \rangle$ can be defined as the result of a query $\langle \{L_1, \ldots, L_n\}, \{L_1 = m_1, \ldots, L_n = m_n\}, \{m\} \rangle$ for some measure m.

5   Metrics

This section details the metrics we identified to describe quantitatively the different aspects of focus for a query. We organize these metrics in three categories: (i) intrinsic to the query, i.e., only related to the query itself, (ii) delta metrics, i.e., dependent on the query's predecessor in the exploration, and (iii) contextual, i.e., dependent on the complete exploration, to provide more context to the metric. Each of these 3 categories can be then refined into 3 subcategories. Some of them relate to (T) the text of the query and cube schema, (R) the result of the query and (C) the chronology of the session, be it the time, or the sequentiality. Formal definitions are only given for non trivial metrics. In metric definitions, let q_k be a query in an exploration $e = \langle q_1, \ldots, q_p \rangle$.

Table 2. Overview of the considered metrics. *For convenience, we also put in this table coefficients (Coef) of the features, that relate to the experiments (see Sect. 6).*

Cat	Metric name	Description	Coef
Intrinsic metrics			
T	*Number of measures (NoM)*	Number of measures used in q_k	0,246
T	*Number of filters (NoF)*	Number of filters (or selections) used in q_k	0,553
T	*Number of aggregations (NoA)*	Number of aggregations (or projections) used in q_k	0,192
T	*Aggregation Depth (ADepth)*	Level of aggregation w.r.t. available cube levels	0,217
T	*Filter Depth (FDepth)*	Ratio of filtered data w.r.t. available cube data	0,147
R	*Number of cells (NoC)*	Number of non null cells in $answer(q_k)$	−0,395
R	*Relevant New Information (RNI)*	Amount of information in $answer(q_k)$	0,068
C	*Execution Time (ExecTime)*	Time taken for executing q_k	0,030
Delta metrics			
T	*Iterative Edit Distance (IED)*	Edition effort required to get q_k from q_{k-1}	−0,201
R	*Iterative Recall (IR)*	Recall of $answer(q_k)$ w.r.t $answer(q_{k-1})$	0,008
R	*Iterative Precision (IP)*	Precision of $answer(q_k)$ w.r.t $answer(q_{k-1})$	0,203
C	*Consideration Time (ConsTime)*	Time taken by the user to consider $answer(q_k)$	0,084
Contextual metrics			
T	*Click Per Query (CPQ)*	Number of subsequent queries at distance 1 from q_k	−0,100
T	*Click Depth (CD)*	Length of the query chain q_k belongs to	0,491
R	*Increase in View Area (IVA)*	Amount of new cells in q_k	−0,051
C	*Number of Queries (NoQ)*	Number of queries executed so far in the exploration	0,176
C	*Query relative position (QRP)*	Relative position of q_k within the exploration	−0,057
C	*Query Frequency (QF)*	Frequency at which the DB has been queried so far	0,019
C	*Elapsed Time (ElTime)*	Elapsed time since the beginning of the exploration	0,007

Intrinsic Metrics. This category concerns metrics that are exclusively related to a given query, independently of the exploration the query belongs to. Most metrics of subcategories (T) and (R) follow the intuition that the more focused a user, the more complex and detailed the queries she evaluates and the fewest the number of cells. In other words, if the user carefully chooses measures and filters, and sufficiently drills down, she has a precise idea of what she is looking for. These features can be computed straightforwardly from query text or query results and their definition is omitted due to lack of space. *Aggregation Depth (ADepth)* defines the aggregation depth of the query relatively to the levels of the cube. Consider a cube with l levels, $depth(l_i, h_i)$ being the depth of the level l_i in hierarchy h_i, and noting $l_i \in P$ if level l_i appears in the set P of predicates of query q, $ADepth(q) = \frac{\sum_{l_i \in P} depth(l_i, h_i)}{l}$. *Filter Depth (FDepth)* is computed similarly by considering for each filter its corresponding level. *Relevant New Information (RNI)* is a measure of entropy of the query result. For a query q_k, it evaluates the quantity of information contained in $answer(q_k)$. Formally, $RNI = 1 - (interest(answer(q_k)))$, where *interest* measures the interestingness degree as a simple normalized entropy, defined by: $interest(C) = \frac{(-\sum_{i=1}^{m} p(i) \log(p(i)))}{\log(m)}$, with $|C| = m$, $C(i)$ is the ith value of the set C and $p(i) = \frac{C(i)}{\sum_{i=1}^{m} C(i)}$ denotes the ith cell occurrence. *Execution Time (ExecTime)* is also related to query complexity, assuming all queries are executed in the same environment. It is computed as $ExecTime(q_k) = ets(q_k) - lts(q_k)$.

Delta Metrics. They characterize a query relatively to the previous query in the exploration. Here the intuition is that the closer two consecutive queries, the more they have in common, the more focused the user. *Iterative Edit Distance (IED)* represents the edition effort, for a user, to express the current query starting from the previous one. It is strongly related to OLAP primitives operations, and computed as the minimum number of atomic operations between queries, by considering the operations of adding/removing a measure, drilling up/down, and adding/removing a filter. The considered cost for each observed difference (adding/removing) is the same. *Iterative Recall (IR)* and *Iterative Precision (IP)* are computed as classical recall and precision by considering the current query result cells as the retrieved set, and the previous query result cells as the relevant set. Thus, the larger the intersection between queries in terms of accessed cells, the more focused we are. Formally, $IR(q_k, q_{k-1}) = \frac{(answer(q_k) \cap answer(q_{k-1}))}{answer(q_{k-1})}$ and $IP(q_k, q_{k-1}) = \frac{(answer(q_k) \cap answer(q_{k-1}))}{answer(q_k)}$. We give a default neutral score to the first queries of the exploration, by defining $IR(q_1) = IP(q_1) = 0.5$. We approximate *Consideration Time (CT)* by computing the time between the end of execution of the current query and the beginning of execution of the subsequent one: $ConsTime(q_k) = lts(q_{k+1}) - ets(q_k)$. We fixed $ConsTime$ to a neutral value (the average of all the previous queries) for the last query of the exploration. $ConsTime$ does not take into consideration the size of visualized data, as this is an independent feature (in NoC). The importance granted to the combination of $ConsTime$ and NoC features is delegated to the SVM.

Contextual Metrics. Contextual metrics characterize a query relatively to an exploration, and more specifically its position within it. In particular, a query occurring in different explorations, can get different scores for these metrics. The two contextual metrics in subcategory (T) adapt popular activity metrics used in Web Search. In this domain, *Clicks Per Query* is used to evaluate search engines through their Search Engine Results Pages (SERP). Given a SERP, CPQ represents the number of links in this page that have been clicked by the user. We adapt it by considering a click as obtaining a new query that differs in one operation from the current query. This model allows to represents typical user behaviors in front of OLAP systems. Formally, we count the number of queries occurring after q_k in the exploration, that are at edit distance one from q_k: $CPQ(q_k, e) = |\{q_p \in e \mid p > k, IED(q_k, q_p) = 1\}|$. In web search, *Click Depth* evaluates the number of pages that have been successively visited, by following hyper links, from one result in a SERP. For a given query q_k, we adapt it by calculating the length of the chain of queries starting from q_k that are distant of one OLAP operation from their immediate predecessor, without discontinuity. $CD(q_k, e) = |S_{kp}|$, where S_{kp} is the longest subsequence of exploration e starting at query q_k and ending at query q_p inclusive, such that $\forall q_i, q_{i+1} \in e, IED(q_i, q_{i+1}) \leq 1$. *Increase in View Area (IVA)* characterizes the increase in terms of new cells in $answer(q_k)$ compared to all the cells seen during the previous queries of the exploration. Formally, $IVA(q_k, e) = \frac{|answer(q_k)\backslash\bigcup_{i\in[1,k-1]} answer(q_i)|}{|\bigcup_{i\in[1,k]} answer(q_i)|}$. *Number of Queries (NoQ)* represents the absolute number of previous queries in the exploration. It is useful to capture the correlation between the tediousness of an exploration and the focus. *Query Relative Position (QRP)* allows to capture the influence of the position of the query in the exploration on the focus. We expect queries at the beginning of an exploration to be more exploratory, and the ones at the end to be more focused. It is computed as the rank of the query in the exploration, normalized by the size of the exploration: $QRP(q_k, e) = \frac{k}{|e|}$. *Query Frequency* captures the engagement of the user by measuring how many queries she submits per unit of time. $QF(q_k) = NoQ(q_k)/ElTime(q_k)$. Finally, *Elapsed Time* computes the time from the beginning of the exploration: $ElTime(q_k, e) = ets(q_k) - lts(q_1)$.

6   Experiments

This section presents the setup and outcomes of the experiments we conducted to evaluate our approach. We first discuss to which extent the coefficients learned by the model to weigh each descriptive metric (see Sect. 3) are consistent with the human expertise. Then, we show that our model, once learned on a dataset, can be generalized to other OLAP navigation datasets without any significant loss in prediction rate.

6.1   Experimental Setup

Data set. We worked with a real database instance, namely a cube called *MobPro*, built from open data on workers mobility. In *MobPro*, facts represent individuals moves between home and workplace, and dimensions allows to characterize a move depending on its frequency, the vehicle used, the traveled distance, etc. The cube is organized as a star schema with 19 dimensions, 68 levels in total, and 24 measures. 37,149 moves are recorded in the facts table.

User explorations. In this experiment, we asked 8 junior analysts (who are students in a master's degree specialized in BI), to analyze the cube using Saiku[2]. They were familiar with OLAP tools, but not necessarily with the data within *MobPro*. We gathered 22 explorations from the system logs. In total, these explorations represent 913 queries.

Query labeling. In order to learn our metric scores weights, we need to label the 913 queries, stating if they are focused or not. The 913 queries have been annotated by one expert, with the help of a web application specifically developed for this purpose. A second expert independently annotated 100 of those queries. Both experts are teachers in the masters degree in BI, and co-authors of this paper. On the 100 queries, a high agreement of 89% has been observed between the two experts, ensuring the representativeness of the 913 labels.

6.2   Model Training

Using our set of 913 queries, each described by the 19 features and the label, we trained a linear SVM classifier. The linear SVM outputs coefficients that traduce the relative importance of each feature. As the metrics are not normalized, the weights learned may be due to SVM compensating for initial low or high values of the metrics, and not only due to the relative intrinsic importance of the features. With a reasonable assumption of normal distribution of our metrics, we used a z-score for normalizing each metrics scores independently before training our model. Z-score ensures that for a given feature, each value is expressed relatively to this feature variance.

We used 76% of our data (700 queries) for training our model, while the other 24% (213 queries) constitute the test set. Both training and test sets are described in Table 3. We parameterized the SVM so that it performs a 10-fold cross validation while learning on the training set, and obtained an accuracy of 80%.

Table 3. Description of training and test sets for linear SVM

Description	Training set	Test set
# Queries	700	213
% Focus	46,7%	59,2%
% Non focus	53,3%	40,8%

Model Discussion. Feature coefficients we obtained are presented in Table 2. By observing them, weights can be easily classified into 4 categories, using 2 dimensions that we call polarity and intensity. The impact of a metric on the focus can be positive/negative in terms of polarity, and high/low in terms of intensity. Impact polarity depends on the sign of the coefficient of the metrics, whereas impact intensity depends on the absolute value of the coefficient. Here, we highlight trends and discuss in details some of the features.

A focused analyst has a relatively well defined information need in mind, which is clearly evidenced by the weights discovered. Indeed, among the metrics related to text (T) and results (R), we observe that all the metrics that restrict the perimeter of the analyzed data (like NoF, FDepth, NoA, ADepth, NoM) have a positive impact on focus. And as expected, metrics that relax the perimeter of analyzed data, like NoC and IVA, appear to have a negative impact on focus.

Metrics that characterize an important move within the data space have a negative impact on focus. IED and IVA are particularly concerned by this. Again, as expected, metrics that measure a closeness between two consecutive queries have a positive impact on the focus. IP is the best representative of that in the sense that its value decreases with the amount of new cells gathered compared to cells in the previous query.

Interestingly, most metrics relative to chronology (C) have little impact on the focus, with the notable exception of Number of Queries, which tends to confirm that focus phases indeed happen after rather long exploratory phases. Another rather surprising finding is that complex metrics in (R) like RNI do not show a significant impact on focus.

More generally, we observe that the importance of a feature is fairly related to the metrics categories and subcategories. No category or subcategory should be ignored, in the sense that all of them include metrics having high weights. A general trend is that metrics relative to the text of the query (T) have in general higher weights, which indicates that focus is highly correlated to the user intention expressed in the query syntax. Likewise, in general, intrinsic metrics tend to have a higher impact on the focus (as seen on NoC, NoF, NoM). But this is counterbalanced by the fact that CD, of category (C) has the second highest weight, meaning that the context of the exploration indeed provides semantics when assessing focus.

6.3   Model Performance

We previously described the meaningfulness of the features coefficients provided by the SVM. We also conducted different experiments, described below, to check the robustness of the predictive power of our classifier.

Testing on Artificial Explorations. The objective of this experiment is to validate our model on explorations whose focus is known. For that, we used Cube-Load [9] for generating realistic explorations. CubeLoad takes as input a cube schema and creates the desired number of sessions according to templates modeling various user exploration patterns. Patterns available in Cubeload simulate:

(*Goal Oriented*) users with limited OLAP skills pursuing a specific analysis goal, (*Slice And Drill* and *Slice All*) more advanced users navigating with a sequence of slice and/or drill operations, (*Exploratory*) users tracking unexpected results with exploratory sessions.

Following the definition of the patterns, we expect *Goal Oriented* explorations to be highly focused, while *Exploratory* are expected to be much less focused. For this experiment, we used the SSB schema [8] and generated a collection of 49 explorations (500 queries) over it. Table 4 presents, per exploration pattern, the ratio of (non) focused queries as predicted by our algorithm. The average ratio of focused queries per exploration pattern confirms our expectations. *Goal Oriented* explorations have a much higher ratio of focused queries compared to *Exploratory* ones. We also notice that *Exploratory* and *Slice All* have a similar ratio of focused queries. However, *Exploratory* explorations are much less focused than Slice All ones in terms of average focus intensity. *Slice and Drill* is correctly recognized as an intermediate behavior.

Table 4. Ratio of focus queries in terms of cubeload patterns

Patterns	Explorative	Goal oriented	Slice and Drill	Slice All
# of explorations	15	13	14	7
# of queries	171	123	126	80
% of focused queries	18,13	64,23	49,21	18,75
Avg focus	$-0,568$	$-0,070$	$-0,193$	0,738
Stdev focus	1,072	1,024	0,955	0,841

Testing on Real Explorations. We created a test set of 213 labeled queries from the 22 explorations we collected over the *MobPro* cube. We obtained an accuracy of 69, 9%, meaning that 69, 9% of the queries have been similarly classified by the expert and the model. Moreover, our model is pretty balanced, as we could retrieve focused (resp., non focused) queries with an accuracy of 71, 4% (resp., 67, 8%). These scores are good, especially given that a naive classifier that would always predict the focus class would reach an accuracy of 59, 2% on the test set.

6.4   Focus vs Analyst Skills

In general, a skilled analyst performs more focused explorations as she has a better knowledge of data. The objective of this experiment is to verify that our model is capable of retrieving this correlation.

We used a set of explorations, previously labeled by the same expert who labeled the queries for the focus. Our expert labeled each exploration with one of three labels: A, B, C. These labels categorize the user knowledge acquisition,

Table 5. Correlation between exploration focus and analyst skills

Analyst skills	A	B	C
Min Focus	$-0,558$	$-1,806$	$-2,366$
Max Focus	$1,657$	$1,126$	$-1,174$
Avg Focus	$0,241$	$-0,240$	$-1,767$
Stdev Focus	$0,861$	$0,895$	$0,513$

from A when the user conducts a in depth analysis and benefits from it in terms of knowledge, to C when the user conducts a very poor analysis and did not benefit from it. We used this as a ground truth. Besides, we used our model to predict the focus of the queries in these explorations. For each query, we predict not only its class (focus or not), but also the degree to which it belongs to the class, by computing the distance to the separation hyperplan found by the SVM. We compute the degree of focus of an exploration as the average of its queries focus degree. After matching explorations degree focus and skill label, we verify that our model is in accordance with intuition. Results are presented in Table 5. From this table, it is easy to identify that classes A and C are clearly distinguished by their degree of focus. Users who acquired knowledge conducted more focused explorations in average, with a minimum (resp., maximum) of focus relatively low (resp., high) compared to the others. This reasoning is inversely true for exploration in class C. Class B is an intermediate situation, quite ambiguous, where it cannot be stated clearly that the skill has been mastered or not.

6.5   Computation Efficiency

Besides the experiments that validate the robustness of our model, we evaluated the average computation time for each metric. Indeed, as we motivated focus detection as a way to improve user experience, we have to ensure that metrics computation runs in near real time. In average, it appears that for a given query, each metric computation does not require more than a few hundred of milliseconds. In average, the computation of all the metrics for a given query is 695 ms, which is negligible given that the average consideration time for a query result is 11200 ms. Having their metrics scores, the query classification is then instantaneous, which validates our approach.

7   Related Work

Analyzing user sessions has been studied for many years in web search. Recent works aim at characterizing the difficulty of search tasks and detecting variations in sessions. For instance, in [1], Athukorala et al. proposed a method for distinguishing between exploratory phases and lookup phases in the context of Information Retrieval. Their idea consists in experimentally discovering features that can be used for this distinction. They submitted several tasks to participants. Some of these tasks require exploration while other require more simple lookups. Based on objective measurements, they could identify that query length, completion time, and maximum scroll depth (in the browser), are the most distinctive indicators for distinguishing between exploratory/lookup tasks. Although they address a problem very similar to the one we tackle in this paper, they work at the exploration level, while our method gives finer results by characterizing each query of an exploration.

A recent trend in web search is to analyze web search sessions by means of machine learning, and more particularly with classifiers. In [3], the goal is to discover new intent and obtain content relevant to users' long-term interests. They develop a classifier to determine whether two search queries address the same information need. This is formalized as an agglomerative clustering problem for which a similarity measure is learned over a set of descriptive features (the stemmed query words, top 10 web results for the queries, the stemmed words in the titles of clicked URL, etc.). Perhaps closer to focus detection is the work of Odijk et al. [7] for characterizing user struggling during web searches. They propose a method for distinguishing between users exploring search results from user struggling for satisfying a given information need. They tackle this problem by using an approach similar to what we propose in terms of methodology. They use a bunch of keyword features, and using a large real set of explorations, they trained a machine learning algorithm for learning how to differentiate between the two aforementioned types of explorations.

In the databases domain, however, to the best of our knowledge, only a handful of works were interested in analyzing real database sessions. As noted by Jain et al. [5], there exists only two real world SQL workloads available to the research community: the SQLShare workload [5] and the Sloan Digital Sky Server workload [10]. In [5], authors present their enhanced SQL web based tool. They made their tool available online for several years, permitting users to upload their datasets and perform advanced analysis. From this, they could gather a rich workload of SQL queries, performed on different datasets, by different users. Based on different metrics computed on each single query (length of the query text, number of distinct operators, query runtime, ...), they propose to characterize each query according to its complexity. An interesting aspect of this work is the investigation of metrics for measuring the cognitive load of users that is indirectly translated in the complexity of the queries they issue. Moreover, at the workload scale, authors show that queries have a high diversity, considering a query similarity measure based on the query text. Authors also propose to classify user behaviors as exploratory and analytical, based on the number of queries per dataset. According to them, a user is analyzing when she submits a lot of queries to the same dataset. This simple distinction is rather coarse compared to the 19 dimensions we use in this paper for analyzing the same distinction.

Nguyen et al. [6] propose an approach for discovering the most accessed areas of a relational database to characterize user interests. Their notion of user interest relies on the set of tuples that are more frequently accessed, and is expressed as selection queries (mostly range queries). They use DBSCAN to cluster user interests in the Sky Server dataset. Their similarity metric relies on Jaccard coefficient of the accessed tables and on overlapping of predicates. In this work also, only one metrics (most frequent accessed tuples) is used. Additionally, we note that, being tailored for range queries, their metric is inappropriate for OLAP queries that are mostly dimensional (i.e., point based), due to the nature of the hierarchical dimensions used to select data.

With the growing interest around exploratory search in the context of interactive database exploration, we believe that our work constitute a first important contribution for understanding the different aspects of user navigations in structured data.

8 Conclusion

Exploratory search considers a search as a complex interaction between a user and a system, including exploratory phases and focus phases. In this paper, we highlight the usefulness of detecting which phases a user is currently in, in the context of OLAP exploration of data cubes. We propose an automatic method, based on a state of the art machine learning algorithm, modeling this detection as a classification problem. To our knowledge, our contribution is a pioneer of its kind. We successfully built a model, trained on a relatively large set of real explorations. We validated experimentally our model on a test set of real explorations, as well as on an artificially, driven, state-of-the-art exploration generator. On top of that, we checked the coherence of our model by using it to detect how skilled is a data analyst.

We plan to follow up our investigations in two main directions. First, we will study the use of focus as a score for evaluating the overall quality of an exploration. Second, we plan to generalize our approach to relational databases and more general types of explorations.

References

1. Athukorala, K., Glowacka, D., Jacucci, G., Oulasvirta, A., Vreeken, J.: Is exploratory search different? A comparison of information search behavior for exploratory and lookup tasks. JASIST **67**(11), 2635–2651 (2016)
2. Golfarelli, M., Rizzi, S., Design, D.W.: Modern Principles and Methodologies. McGraw-Hill, New York (2009)
3. Guha, R., Gupta, V., Raghunathan, V., Srikant, R.: User modeling for a personal assistant. In: WSDM, pp. 275–284 (2015)
4. Idreos, S., Papaemmanouil, O., Chaudhuri, S.: Overview of data exploration techniques. In: SIGMOD, pp. 277–281 (2015)
5. Jain, S., Moritz, D., Halperin, D., Howe, B., Lazowska, E.: Sqlshare: results from a multi-year SQL-as-a-service experiment. In: Proceedings of SIGMOD, pp. 281–293 (2016)

6. Nguyen, H.V., Böhm, K., Becker, F., Goldman, B., Hinkel, G., Müller, E.: Identifying user interests within the data space - a case study with skyserver. EDBT **2015**, 641–652 (2015)

7. Odijk, D., White, R.W., Awadallah, A.H., Dumais, S.T.: Struggling and success in web search. In: Proceedings of CIKM, pp. 1551–1560 (2015)

8. O'Neil, P., O'Neil, E., Chen, X., Revilak, S.: The star schema benchmark and augmented fact table indexing. In: Nambiar, R., Poess, M. (eds.) TPCTC 2009. LNCS, vol. 5895, pp. 237–252. Springer, Heidelberg (2009). doi:10.1007/978-3-642-10424-4_17

9. Rizzi, S., Gallinucci, E.: CubeLoad: a parametric generator of realistic OLAP workloads. In: Jarke, M., Mylopoulos, J., Quix, C., Rolland, C., Manolopoulos, Y., Mouratidis, H., Horkoff, J. (eds.) CAiSE 2014. LNCS, vol. 8484, pp. 610–624. Springer, Cham (2014). doi:10.1007/978-3-319-07881-6_41

10. Szalay, A.S., Gray, J., Thakar, A., Kunszt, P.Z., Malik, T., Raddick, J., Stoughton, C., vandenBerg, J.: The SDSS skyserver: public access to the sloan digital sky server data. In: Proceedings of SIGMOD, pp. 570–581 (2002)

11. White, R.W., Roth, R.A., Search, E.: Beyond the Query-Response Paradigm. Morgan & Claypool Publishers, San Rafael (2009)

Sparse Prefix Sums

Michael Shekelyan[(✉)], Anton Dignös, and Johann Gamper

Free University of Bozen-Bolzano, Bozen, Italy
{michael.shekelyan,dignoes,gamper}@inf.unibz.it

Abstract. The prefix sum approach is a powerful technique to answer range-sum queries over multi-dimensional arrays in constant time by requiring only a few look-ups in an array of precomputed prefix sums. In this paper, we propose the *sparse prefix sum* approach that is based on relative prefix sums and exploits sparsity in the data to vastly reduce the storage costs for the prefix sums. The proposed approach has desirable theoretical properties and works well in practice. It is the first approach achieving constant query time with sub-linear update costs and storage costs for range-sum queries over sparse low-dimensional arrays. Experiments on real-world data sets show that the approach reduces storage costs by an order of magnitude with only a small overhead in query time, thus preserving microsecond-fast query answering.

1 Introduction

Prefix sums are a widely known technique across many fields to enable range-sum queries over grids in constant time, i.e., independent of the size of the grid. Range-sum queries sum the cell values along rectangular shapes. The technique is used in many disciplines, e.g., in On-Line Analytic Processing (OLAP) to answer aggregate range queries [11,13], in computer vision to efficiently compute features for object recognition [18] and in data mining to efficiently find maximum rectangular subarrays [17].

Given a d-dimensional grid, the basic idea of the classical prefix sum technique [11] is to precompute and store for each grid cell a so-called *prefix sum*. The prefix sum of a cell is the sum over all grid cells in the range from the origin up to and including itself. At query time, the prefix sums allow to evaluate any range-sum query in constant time by looking up and combining 2^d prefix sums. Consider Fig. 1, which shows around 2.9 billion GPS coordinates collected for the OpenStreetMap project[1]. The range-sum over the query rectangle spanned by the points A and D can be derived from four range sums that are anchored at the origin O, i.e., $|AD| = |OD| - |OB| - |OC| + |OA|$, where $|XY|$ represents the value of the range sum determined by X and Y. The range-sums that are anchored at the origin O are precomputed as prefix sums, hence at query time only four look-ups are required to answer any range query. While the prefix sum technique enables constant query time, it leads to very high update

---

[1] www.openstreetmap.org.

© Springer International Publishing AG 2017
M. Kirikova et al. (Eds.): ADBIS 2017, LNCS 10509, pp. 120–135, 2017.
DOI: 10.1007/978-3-319-66917-5_9

costs, because updating an individual cell value requires updating all origin-anchored range-sums that contain that cell. To overcome this problem, relative prefix sums [10] were introduced that split a grid with N cells into $\sqrt{N}$ blocks with $\sqrt{N}$ cells each. The key behind this approach is to break the dependency between groups of cells within different blocks. Relative prefix sums preserve constant query time and vastly improve the update efficiency to $\mathcal{O}(\sqrt{N})$.

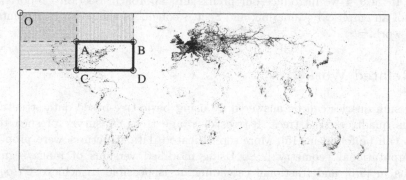

Fig. 1. Prefix sum: range-sum over query box $|AD| = |OD| - |OC| - |OB| + |OA|$ where $|OD|, |OC|, |OB|$ and $|OA|$ are the prefix sums of the query box's corners.

The key problem with the prefix sum and the relative prefix sum approaches is their high storage costs for sparse data and high-resolution grids. The prefix sums need to be precomputed for each grid cell, yielding a dense array of prefix sum values which need to be materialized. This is completely independent from the sparsity of the data.

To overcome this storage problem, in this paper we propose the *sparse prefix sum* technique. It is based on relative prefix sums, but additionally adopts a semi-sparse representation. Some cells in the grid always have to be materialized to ensure constant query time, but many cells can be stored implicitly using look-up tables, because there are many repetitions of values caused by empty cells in the data grid. The look-up tables cause almost no querying time overhead. As a result we get the first approach to achieve constant query time, sub-linear update costs, as well as sub-linear storage costs for sparse low-dimensional arrays.

The main technical contributions can be summarized as follows:

- We propose sparse prefix sums that are based on relative prefix sums, but adopt a semi-sparse representation for the prefix sums to achieve sub-linear storage costs for sparse low-dimensional arrays.
- We describe an algorithm to construct sparse prefix sums without materializing the whole grid that has sub-linear construction costs for sparse low-dimensional arrays.
- We propose an algorithm to answer range queries using sparse prefix sums in constant time.

– We empirically compare the sparse prefix sum approach with other constant-time approaches on real-world datasets. The results show that, for sparse low-dimensional arrays, our approach leads up to an order of magnitude reduction in storage costs with only a small overhead in query time, still keeping the query time in the range of microseconds.

Structure. Section 2 discusses related work. Section 3 recaps the prefix sum approach. In Sect. 4 we introduce our prefix sum approach. Section 5 reports the result of an empirical evaluation, and Sect. 6 concludes the paper and points to future work.

2   Related Work

Range-sum queries can be answered by using basic tree-based data structures, such as quadtrees, k-d trees, R-trees or range trees. A survey of such techniques can be found in [15]. More sophisticated data structures were proposed in computational geometry [2,8]. Using modified versions of range/segment trees as in [5,6] and fractional cascading, it is possible to achieve $\mathcal{O}(log^d S)$ update, $\mathcal{O}(log^{d-1} S)$ querying time, and $\mathcal{O}(S)$ storage, where S is the number of non-empty grid cells. Unlike the proposed sparse prefix sum method, all these approaches cannot offer constant query time.

The (conventional) prefix sum technique [11] precomputes cumulative sums to answer range-sum queries in constant time. However, the $\mathcal{O}(N)$ update and storage costs make it unsuitable for high-resolution grids with a very large number of cells N.

To overcome this problem, the relative prefix sum technique [10,16] splits the grid into blocks and stores conventional prefix sums only along the block borders. In the inner region of the blocks, local prefix sums are stored, which only sum over cells inside the block. By breaking dependencies between the cells in different blocks, the update cost is reduced to $\mathcal{O}(\sqrt{N})$. While the query time remains constant, the data sparsity cannot be exploited, yielding linear storage costs similar to the prefix sum approach.

Multiple techniques were proposed in the past to further reduce the update complexity [3,4,9,13]. However, unlike our sparse prefix sum approach, none of these techniques offers constant query time and is able to exploit sparsity in the data. One notable exception is the dynamic data cube technique [9], which can exploit data sparsity in cases when large contiguous blocks are empty.

Other techniques were proposed to exploit sparsely populated grids, but none of them offers comparable properties to our approach, in particular constant query time. The blocked array technique [11] creates prefix sums at a smaller resolution and stores the high-resolution grid without prefix sums as a sparse array, which reduces storage costs. The prefix cube pool technique [7] identifies dense subregions and constructs local prefix sums for each subregion. It can only reduce storage costs in case of large contiguous blocks of empty cells. The pCube technique [14] stores only sparse grid cells and uses data indices such as quadtrees or R-Trees to speed-up querying. The sub-cube compression technique [12] is a lossless compression technique that stores sparse prefix cubes by decomposing them into sub-cubes, where each sub-cube results from a different combination of sparse cells. Unlike our prefix sum approach, none of these approaches achieves constant query time and a significant reduction of storage costs.

3 Background

In this section, we provide the formal background for prefix sums [11] and relative prefix sums [10]. We use A to denote the d-dimensional grid over which range-sums are computed and $A_{c_1,\dots,c_d}$ to denote the grid cell value with coordinates $(c_1,\dots,c_d)$. The *prefix sum* of the cell with coordinates $(t_1,\dots,t_d)$ is defined as

$$P_{t_1,\dots,t_d} = \sum_{c_1=0}^{t_1} \cdots \sum_{c_d=0}^{t_d} A_{c_1,\dots,c_d}.$$

That is, a prefix sum $P_{t_1,\dots,t_d}$ is the sum of all cells in the range with lower corner $(0,\dots,0)$, i.e., the origin of the data space, and upper corner $(t_1,\dots,t_d)$. Consider the 2-dimensional grid A in Fig. 2a, which is used as a running example. The array of prefix sums P is shown in Fig. 2b. For instance, the prefix sum $P_{2,2}$ is computed as $P_{2,2} = A_{0,0} + A_{0,2} = 1 + 7 = 8$, whereas $P_{7,7}$ is the sum of all grid values.

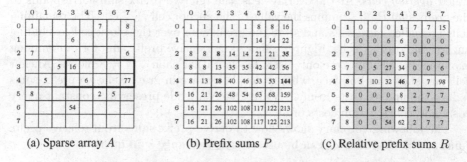

(a) Sparse array A

	0	1	2	3	4	5	6	7
0	1					7		8
1				6				
2	7							6
3			5	16				
4		5			6			77
5	8					2	5	
6				54				
7								

(b) Prefix sums P

	0	1	2	3	4	5	6	7
0	1	1	1	1	1	8	8	16
1	1	1	1	7	7	14	14	22
2	8	8	8	14	14	21	21	35
3	8	8	13	35	35	42	42	56
4	8	13	18	40	46	53	53	144
5	16	21	26	48	54	63	68	159
6	16	21	26	102	108	117	122	213
7	16	21	26	102	108	117	122	213

(c) Relative prefix sums R

	0	1	2	3	4	5	6	7
0	1	0	0	0	1	7	7	15
1	0	0	0	6	6	0	0	0
2	7	0	0	6	13	0	0	6
3	7	0	5	27	34	0	0	6
4	8	5	10	32	46	7	7	98
5	8	0	0	0	8	2	7	7
6	8	0	0	54	62	2	7	7
7	8	0	0	54	62	2	7	7

Fig. 2. Prefix sum techniques.

Corollary 1 (Range-sums). *Let $\mathcal{Q}$ be a d-dimensional range with lower corner $(s_1,\dots,s_d)$ and upper corner $(t_1,\dots,t_d)$ and let $[C]$ be equal to 1 iff C is true and 0 otherwise. The range-sum $|\mathcal{Q}|$ can be computed with the help of prefix sums as*

$$|\mathcal{Q}| = P_{t_1,\dots,t_d} + \sum_{\emptyset \neq D \subseteq \{1,2,\dots,d\}} (-1)^{|D|} \cdot P_D,$$

where $P_D = P_{(s_1-1)[1\in D]+(t_1)[1\notin D],\dots,(s_d-1)[d\in D]+(t_d)[d\notin D]}.$

For instance, for $d = 2$ dimensions, the non-empty subsets $D \subseteq \{1,2\}$ are $\{1\}$, $\{2\}$ and $\{1,2\}$, and for $D = \{1\}$ we have $P_D = P_{(s_1-1)\cdot 1 + t_1 \cdot 0, (s_2-1)\cdot 0 + t_2 \cdot 1} = P_{s_1-1,t_2}$. This gives $|\mathcal{Q}| = P_{t_1,t_2} - P_{s_1-1,t_2} - P_{t_1,s_2-1} + P_{s_1-1,s_2-1}$. Consider the query range in Fig. 2a with lower corner $(3,3)$ and upper corner $(7,4)$ (thick rectangle). The sum over this query range is $\mathcal{Q} = 16 + 6 + 77 = 99$. Using Corollary 1, this range sum can be computed from the four prefix sums in boldface in Fig. 2b as $\mathcal{Q} = P_{7,4} - P_{2,4} - P_{7,2} + P_{2,2} = 144 - 18 - 35 + 8 = 99$.

A problem with prefix sums is their dependency from many grid cells, which leads to large update costs and makes it difficult to exploit sparsity. Relative prefix sums [10] break these dependencies. The main idea is to partition a grid of size N into $\sqrt{N}$ blocks à $\sqrt{N}$ cells. Each cell in a block stores a local prefix sum, which depends only on the cells in that block. The *relative prefix sum* of a cell with coordinates $(t_1, \ldots, t_d)$ that is located in a block with lower corner $(a_1, \ldots, a_d)$ is then defined as follows:

$$R_{t_1,\ldots,t_d} = \begin{cases} P_{a_1,\ldots,a_d} & \text{if } t_1 = a_1 \wedge \ldots \wedge t_d = a_d, \\ P_{t_1,\ldots,t_d} - P_{a_1,\ldots,a_d} & \text{if } \exists_{i,j} : t_i = a_i \wedge t_j \neq a_j, \quad (1) \\ \sum_{c_1=a_1+1}^{t_1} \cdots \sum_{c_d=a_d+1}^{t_d} A_{c_1,\ldots,c_d} & \text{otherwise.} \end{cases}$$

Each block is identified by an *anchor cell*, i.e., the lower corner of the block with coordinates $(a_1, \ldots, a_d)$. The relative prefix sum at the anchor cell corresponds to the conventional prefix sum $P_{a_1,\ldots,a_d}$ (case 1). All cells $(c_1, \ldots, c_d)$ that share at least one coordinate with the anchor cell and differ in at least one coordinate are called *overlay cells*. At the overlay cells, the relative prefix sum corresponds to the prefix sum $P_{c_1,\ldots,c_d}$ minus the value of the anchor cell $P_{a_1,\ldots,a_d}$. The remaining cells are called *local cells* and store the prefix sum over the block cells excluding anchor and overlay cells. Figure 2c shows the relative prefix sum for our running example. For visibility we only use 4 blocks rather than $\sqrt{N} = 8$ blocks. Anchor cells are shown in boldface, whereas local cells are in gray. Relative prefix sums reduce the update costs from $\mathcal{O}(N)$ to $\mathcal{O}(\sqrt{N})$, while preserving constant query costs and construction costs of $\mathcal{O}(N)$.

The following corollary shows how to derive prefix sums from relative prefix sums so that Corollary 1 can be used to answer range sum queries.

Corollary 2 (Constructing prefix sums from relative prefix sums). *Let $R_{t_1,\ldots,t_d}$ denote the relative prefix sum of the cell with coordinates $(t_1, \ldots, t_d)$ and let $[C]$ be equal to 1 iff C is true and 0 otherwise. The prefix sum $P_{t_1,\ldots,t_d}$ can be calculated as follows:*

$$P_{t_1,\ldots,t_d} = \begin{cases} R_{a_1,\ldots,a_d} & \text{if } t_1 = a_1 \wedge \ldots \wedge t_d = a_d, \\ R_{t_1,\ldots,t_d} + R_{a_1,\ldots,a_d} & \text{if } \exists_{i,j} : t_i = a_i \wedge t_j \neq a_j, \\ R_{t_1,\ldots,t_d} + \sum_{\emptyset \neq E \subseteq \{1,2,\ldots,d\}} (-1)^{|E|-1} P_E & \text{otherwise,} \end{cases}$$

where $a_1, \ldots, a_d$ is the lower corner of the block containing $t_1, \ldots, t_d$ and $P_E = P_{(a_1)[1 \in E] + (t_1)[1 \notin E], \ldots, (a_d)[d \in E] + (t_d)[d \notin E]}$.

For the range-sum query in Fig. 2a we have $P_{7,4} - P_{2,4} - P_{7,2} + P_{2,2}$. To reconstruct prefix sums $P_{7,4}$ and $P_{2,4}$, the second case in Corollary 2 applies, i.e., $P_{7,4} = R_{7,4} + R_{5,5} = 98 + 46 = 144$ and $P_{2,4} = R_{2,4} + R_{1,3} = 10 + 8 = 18$ (cf. Figs. 2b and c). For prefix sums $P_{7,2}$ and $P_{2,2}$, the third case applies, i.e., $P_{7,2} = R_{7,2} + (R_{7,0} + R_{5,0}) + (R_{5,2} + R_{5,0}) - R_{5,0} = 6 + (15+1) + (34+1) - 1 = 56$ and $P_{2,2} = R_{2,2} + (R_{2,0} + R_{0,0}) + (R_{0,2} + R_{0,0}) - R_{0,0} = 0 + (0+1) + (7+1) - 1 = 8$. The final result is calculated as $P_{7,4} - P_{2,4} - P_{7,2} + P_{2,2} = 99$.

4   Sparse Prefix Sums

In this section, we first describe the construction of sparse prefix sums and the data structures used for the internal representation. Then, we describe an algorithm for answering range queries, followed by a complexity analysis.

4.1   Constructing Sparse Prefix Sums

Similar to relative prefix sums, the sparse prefix sum approach splits a grid A with N cells into $\sqrt{N}$ blocks. However, instead of storing dense arrays of cumulative sums, sparse prefix sums take advantage of sparsity in A and store only a subset of the sums. The construction of sparse prefix sums consists of two main steps: (1) computing local prefix sums and (2) computing anchor and overlay prefix sums. In the sequel, we describe and illustrate these two steps (cf. Fig. 4).

Step 1: Computing Local Prefix Sums. In the first step, we compute for each block the required local prefix sums, taking as input the data from a sparse array A. Different from relative prefix sums, we do not compute a dense array of local prefix sums. Instead, we take advantage of the sparsity in the data and identify those cells, for which the computation of local prefix sums can be avoided.

Definition 1 (Block slice). A block slice *is defined as the subset of all cells in a block B that share the same coordinate C in a dimension i, i.e., $\{(c_1, \ldots, c_2) \in B : c_i = C\}$.*

Block slices consist of all block cells that share the same coordinate in a specific dimension, e.g., rows and columns in two-dimensional blocks/arrays. The basic idea of the sparse representation of prefix sums is that we can omit the computation of local prefix sums for empty block slices and only store the remaining cells. Since only entire block slices are removed, we obtain a regular grid with a reduced coordinate space. Figures 3a and b illustrate this for the 3×3 local cells of the first block in our running example. The first column and the second row are completely empty, hence they are removed before the local prefix sums are computed.

After removing empty block slices, we compute the local prefix sums for all remaining cells in the reduced grid. This reduced block of prefix sums is

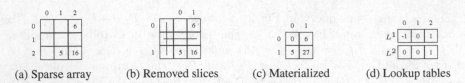

(a) Sparse array   (b) Removed slices   (c) Materialized   (d) Lookup tables

Fig. 3. Constructing a block's local prefix sums and lookup tables.

eventually materialized. For our example, the local prefix sums in the reduced space of 2×2 cells are shown in Fig. 3c. The value 27 in the cell $(1, 1)$ is computed as the sum $6 + 5 + 16 = 27$.

Definition 2 (Materialized local block). *A materialized local block consists of all local block cells that do not belong to an empty block slice. For these cells, local prefix sums are computed.*

To guarantee a constant look-up time for cells in the reduced coordinate space, given a query range in the original space, we introduce so-called block look-up tables.

Definition 3 (Block look-up table). *A block look-up table L^i for dimension i maps a coordinate c_i along dimension i to $L^i[c_i]$, such that $c_i - L^i[c_i]$ is the number of (omitted) empty block slices between 0 and c_i.*

Each block has one look-up table per dimension that maps block coordinates in the original grid to block coordinates in the materialized grid. Figure 3d shows the two look-up tables for the first block in our running example. They translate coordinates from the original 3×3 grid (Fig. 3b) into coordinates of the reduced 2×2 grid (Fig. 3c). L^1 maps the horizontal dimension and L^2 the vertical dimension. For instance, the first column is mapped to -1, the second to 0 and the third to 1. Intuitively, a value of -1 or a repetition of the same value in a look-up table indicates one omitted slice in that dimension. To access the prefix sum of the cell $(1, 2)$ in the original space, we have to access the cell $(L^1[1], L^2[2]) = (0, 1)$ in the reduced space. The situation of the running example after step 1 is shown in Fig. 4b.

Step 2: Computing Anchor and Overlay Prefix Sums. In the second step, we compute the values for the anchor cell and the overlay cells for each block (cf. case 1 and case 2 in Eq. 1). A naïve approach would be to compute the prefix sum for the entire grid and then discard the local cells. Since this would require excessive construction time and memory, we propose an algorithm that only materializes the sparse prefix sums plus a few additional values that are needed during construction process.

The key idea behind our algorithm for computing anchor and overlay cells is to exploit diagonals in the grid: prefix sums along a diagonal only depend on prefix sums along diagonals that are closer to the origin, while cells along

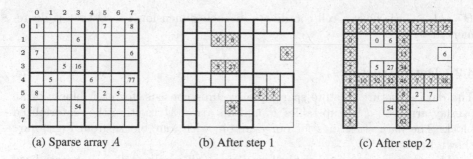

Fig. 4. Construction of sparse prefix sums from an array A (affected cells are hatched).

one diagonal are independent from each other. To enumerate diagonals, we sum the coordinates of the cells, i.e., the cell with coordinate $(s_1, \ldots, s_d)$ lays on the diagonal $s_1 + \ldots + s_d$. It is easy to see that all cells on a diagonal have the same sum of their coordinates. The blocks and the corresponding anchor and overlay cells are processed in increasing order of the sum $s_1 + \ldots + s_d$. This ensures that diagonals with a lower number are processed first. For instance, in Fig. 4c we first process the top-left block, followed by the top-right and lower-left block (same diagonal) in any order as they only depend on the top-left block. The last block is the lower-right that depends on both the top-right and lower-left block.

In our implementation, we additionally exploit the fact that neighboring overlay prefix sums share the same preceding prefix sums. These are either overlay prefix sums that have already been computed, or they lay on the upper border of a block and need to be computed. To avoid that the prefix sums of the upper border are computed multiple times, we compute them once when all overlay prefix sums in a block are available, and store them temporarily during the construction process (similar as the overlay prefix sums). After the construction, the upper block prefix sums can be discarded.

For the construction of the sparse prefix sum, we still need to compute the prefix sums for the overlay cells. Previous approaches compute prefix sums by a summation along any dimension individually. In contrast, we compute prefix sums (along the diagonals) incrementally from already existing prefix sums, as shown in the following lemma.

Lemma 1. *Each overlay prefix sum can be computed from preceding prefix sums using*

$$P_{t_1,\ldots,t_d} = A_{t_1,\ldots,t_d} + \sum_{\emptyset \neq D \subseteq \{1,2,\ldots,d\}} (-1)^{|D|-1} P_D,$$

where $P_D = P_{(t_1-1)[1\in D]+(t_1)[1\notin D],\ldots,(t_d-1)[d\in D]+(t_d)[d\notin D]}$.

Proof. It follows directly from solving the equation in Corollary 1 for $P_{t_1,\ldots,t_d}$. $\square$

Each overlay prefix sum is the sum of the value of the sparse array A at the corresponding cell plus the value of 2^{d-1} cells that, in a non-empty subset

$D \subseteq \{1, \ldots, d\}$, have a cell coordinate in each dimension $i \in D$ that is decremented by one.

4.2 Data Structures

The data structure to store sparse prefix sums consists, for each block, of an overlay array O, d lookup tables L, and an array M to store the materialized block. The data structures for our example with four blocks from Fig. 4c are shown in Fig. 5.

The overlay array O is of fixed size, since all overlay cells are materialized and we have the same number of overlay cells for each block. To store the overlay cells using a single one-dimensional array, we conceptually split the array into 2^d-1 groups. Each group intuitively represents a lower border of a block in a subset of all dimensions, i.e., group 0 is the anchor cell that is on the lower border in all dimensions, group 1 are the overlay border cells that are not on the lower border in the first dimension, group 2 are the overlay border cells that are not on the lower border in the second dimension, and so on. For instance, in Fig. 5 the first value in O is the anchor cell of a block, the next three values are the remaining overlay cells in the horizontal dimension (cf. Fig. 4c), and the last three in the vertical dimension. The offsets of the groups in array O are the same for all blocks, such that they can be simply precomputed once.

The sparse local cells of each block are stored as a one-dimensional array M in row-major order. The lookup tables L are used to translated coordinates from the block coordinate space into the coordinate space of M. This is the key idea behind sparse prefix sums to accomplish both constant query time and a sparse representation. For instance, in Fig. 5 the top-left block only requires materializing four of the nine inner block cells. The coordinate $(2, 1)$ is translated using $(L^1[2], L^2[1])$ into $(1, 0)$ and $M_{1,0} = 6$. The coordinate $(0, 2)$ is translated using $(L^1[0], L^2[2])$ into $(-1, 1)$ and any value for negative coordinates is equal to 0, i.e., the local prefix sum is 0.

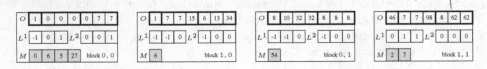

Fig. 5. Sparse prefix sum data structures.

4.3 Querying Sparse Prefix Sums

Sparse prefix sums are queried similarly to relative prefix sums, i.e., we reconstruct first relative prefix sums which in turn are transformed into prefix sums as shown below:

Range-sum $\xleftarrow{\text{Corollary 1}}$ | conventional prefix sum | $\xleftarrow{\text{Corollary 2}}$ | relative prefix sum | $\xleftarrow{\text{Corollary 3}}$ | sparse prefix sum |

Algorithm 1. Querying

Input: Query range $\mathcal{Q}$ from $(s_1, \ldots, s_d)$ to $(t_1, \ldots, t_d)$
Output: Range-sum over $|\mathcal{Q}|$

1 $q = 0$
2 **foreach** $\emptyset \neq D \subseteq \{1, \ldots, d\}$ **do**
3 $\quad (j_1, \ldots, j_d) \leftarrow ((s_1 - 1)[1 \in D] + (t_1)[1 \notin D], \ldots, (s_d - 1)[d \in D] + (t_d)[d \notin D])$
4 $\quad$ Determine block containing $(j_1, \ldots, j_d)$ with anchor $(a_1, \ldots, a_d)$, overlay cells O,
$\quad\quad$ materialized local block M, and look-up tables L
5 $\quad$ **if** $\exists_i : j_i = a_i$ **then**
6 $\quad\quad$ **if** $\exists_i : j_i \neq a_i$ **then**
7 $\quad\quad\quad | \quad P_{j_1, \ldots, j_d} \leftarrow O_{j_1 - a_1, \ldots, j_d - a_d} + O_{0, \ldots, 0}$
8 $\quad\quad$ **else**
9 $\quad\quad\quad \lfloor \quad P_{j_1, \ldots, j_d} \leftarrow O_{j_1 - a_1, \ldots, j_d - a_d}$
10 $\quad$ **else**
11 $\quad\quad (k_1, \ldots, k_d) \leftarrow (L^1[j_1 - a_1], \ldots, L^d[j_d - a_d])$
12 $\quad\quad$ **if** $\exists_i : k_i = -1$ **then**
13 $\quad\quad\quad | \quad p \leftarrow 0$
14 $\quad\quad$ **else**
15 $\quad\quad\quad \lfloor \quad p \leftarrow M_{k_1, \ldots, k_d}$
16 $\quad\quad$ **foreach** $\emptyset \neq E \subseteq \{1, \ldots, d\}$ **do**
17 $\quad\quad\quad (c_1, \ldots, c_d) \leftarrow (a_1[1 \in E] + j_1[1 \notin E], \ldots, a_d[d \in E] + j_d[d \notin E])$
18 $\quad\quad\quad$ **if** $\exists_i : j_i \neq a_i$ **then**
19 $\quad\quad\quad\quad | \quad P_{c_1, \ldots, c_d} \leftarrow O_{c_1 - a_1, \ldots, c_d - a_d} + O_{0, \ldots, 0}$
20 $\quad\quad\quad$ **else**
21 $\quad\quad\quad\quad \lfloor \quad P_{c_1, \ldots, c_d} \leftarrow O_{c_1 - a_1, \ldots, c_d - a_d}$
22 $\quad\quad\quad \lfloor \quad p \leftarrow p + (-1)^{|E| - 1} \cdot P_{c_1, \ldots, c_d}$
23 $\quad\quad \lfloor \quad P_{j_1, \ldots, j_d} \leftarrow p$
24 $\quad \lfloor \quad q \leftarrow q + (-1)^{|D|} P_{j_1, \ldots, j_d}$
25 **return** q

Corollary 3 (Constructing relative prefix sums from sparse prefix sums). *Let $(t_1, \ldots, t_d)$ be a cell in a block with anchor $(a_1, \ldots, a_d)$, overlay cells O, materialized local block M and block look-up tables L. The relative prefix sum $R_{t_1, \ldots, t_d}$ can be calculated as follows:*

$$R_{t_1, \ldots, t_d} = \begin{cases} O_{t_1 - (a_1 + 1), \ldots, t_d - (a_d + 1)} & \text{if } \exists_i : t_i = a_i, \\ 0 & \text{if } \exists_i : L^i[t_i - (a_i + 1)] = -1, \\ M_{L^1[t_1 - (a_1 + 1)], \ldots, L^d[t_d - (a_d + 1)]} & \text{otherwise.} \end{cases}$$

The algorithm for range-sum queries using sparse prefix sums is shown in Algorithm 1. It is based on the algorithm of relative prefix sums, which first reconstructs the prefix sums and then calculates range-sum queries using these prefix sums (cf. Corollary 1 and 2). In our sparse prefix sum algorithm the relative prefix sums are stored in a sparse representation and thus need to be reconstructed using Corollary 3. For each prefix sum to compute, we first determine the block that contains the cell (line 4). For better readability, we index O and M in the algorithm as grids rather than as linearized arrays. Lines 6–9 handle the cases where the cell is an overlay cell: we retrieve the cell value from O (cf. case 1 in Corollary 3) and proceed to produce the prefix sum. Lines 11–23 handle the cases where the coordinate is a local cell: we first translate, with the help of the look-up tables, the cell coordinates to the coordinates of the materialized local block, considering the case that the coordinate might be -1 (cf. case 2 and case 3 in Corollary 3). Then, from the reconstructed relative prefix sum the prefix sum is computed and the algorithm proceeds as for relative prefix sums.

4.4 Complexity Analysis

Theorem 1. *Sparse prefix sums have space complexity $\mathcal{O}(N^{1-\frac{1}{2d}} + SN^{\frac{1}{2}-\frac{1}{2d}})$, where N is the size of the grid, S is the number of non-empty cells and d is the dimensionality.*

Proof. The number of stored values is determined as the sum of the number of overlay cells, the number of values in the look-up tables and the number of values of the materialized blocks. Let $w^d = \sqrt{N}$ be the total number of cells in a block, i.e., $w = N^{\frac{1}{2d}}$.

Number of overlay cells: The number of local cells is $(w-1)^d$. Thus, the number of overlay cells in each block is equal to $w^d - (w-1)^d = (1+(w-1))^d - (w-1)^d$, which by the binomial theorem can be transformed to $\left(\sum_{k=0}^{d} \binom{d}{k}(w-1)^k \right) - (w-1)^d = \sum_{k=0}^{d-1} \binom{d}{k}(w-1)^k$. Clearly, $\sum_{k=0}^{d-1} \binom{d}{k}(w-1)^k < 2^d(w-1)^{d-1}$. Hence, the number of overlay cells cannot exceed $2^d(w-1)^{d-1}\sqrt{N} < 2^d w^{d-1}\sqrt{N} = 2^d N^{\frac{1}{2}-\frac{1}{2d}}\sqrt{N} = 2^d N^{1-\frac{1}{2d}}$, which is in $\mathcal{O}(N^{1-\frac{1}{2d}})$.

Size of look-up tables: The number of block look-up table values is equal to $d(w-1)\sqrt{N} < dN^{(\frac{1}{2}+\frac{1}{2d})}$, which is in $\mathcal{O}(N^{\frac{1}{2}+\frac{1}{2d}})$.

Number of non-empty local cells: The storage costs of a block's local cells with s sparse values cannot exceed $\min((w-1)^d, s^d)$. The worst case maximizes the storage costs per sparse value $\frac{\min((w-1)^d, s^d)}{s}$, which is maximized for $s = w - 1$. Thus, the storage costs for S sparse values cannot exceed $\frac{S}{(w-1)}(w-1)^d = S(w-1)^{d-1} = S(N^{\frac{1}{2d}}-1)^{d-1} < SN^{(\frac{1}{2}-\frac{1}{2d})}$ which is in $\mathcal{O}(SN^{(\frac{1}{2}-\frac{1}{2d})})$. Taken the three parts together, the overall storage costs are in $\mathcal{O}(N^{1-\frac{1}{2d}} + SN^{(\frac{1}{2}-\frac{1}{2d})})$. $\qquad\square$

The construction algorithm has time complexity $\mathcal{O}(N^{1-\frac{1}{2d}} + SN^{(\frac{1}{2}-\frac{1}{2d})})$. Step 1 has time complexity $\mathcal{O}(SN^{(\frac{1}{2}-\frac{1}{2d})})$, since it materializes at most that many cells. Step 2 has time complexity $\mathcal{O}(N^{1-\frac{1}{2d}})$, since it computes at most that many overlay prefix sums, and each overlay prefix sum is constructed from a constant number of preceding overlay prefixes.

The update complexity is $\mathcal{O}(\sqrt{N})$ as for relative prefix sums, as it is identical to relative prefix sums apart from the local block cells. All local block cells can be updated in $\mathcal{O}(\sqrt{N})$, since a block has at most $\mathcal{O}(\sqrt{N})$ materialized values.

An overview of the theoretical properties is given in Table 1.

Table 1. Complexity results for a grid of size N and d dimensions (treated as constant); S is the number of non-empty grid cells.

	Query costs	Update costs	Storage/construction costs
(Conventional) prefix sums (CPS)	$\mathcal{O}(1)$	$\mathcal{O}(N)$	$\mathcal{O}(N)$
Relative prefix sums (RPS)	$\mathcal{O}(1)$	$\mathcal{O}(\sqrt{N})$	$\mathcal{O}(N)$
Sparse prefix sums (SPS)	$\mathcal{O}(1)$	$\mathcal{O}(\sqrt{N})$	$\mathcal{O}(N^{(1-\frac{1}{2d})} + SN^{(\frac{1}{2}-\frac{1}{2d})})$

5 Experimental Evaluation

5.1 Setup and Datasets

In the experiments, we compare our sparse prefix sums technique (SPS) to conventional prefix sums (CPS) [11] and relative prefix sums (RPS) [10]. All algorithms are implemented in C++ by the same author and were compiled with GCC 4.9.2 using -O3 and run on a single core. The experiments run on a machine with an Intel(R) Xeon(R) CPU E5-2667 v3 @ 3.20 GHz and 100 GB of main memory.

We use the following datasets for the experiments: **OSM** – a spatial dataset from the OpenStreetMap project that records 2.9 billion two-dimensional GPS-coordinates (see Fig. 1). **ZIPF** – a synthetic dataset with 1000 cluster centers and a total of 10^7 points that are distributed according to a Zipf distribution between clusters in all dimensions, i.e., some clusters have much more points than others. The cluster centers are uniformly distributed, and the cluster members are normally distributed around the center. For this dataset we vary the dimensionality. **HRSL** [1] – a set of population datasets developed for eight countries for the year 2015 on grids with a resolution of roughly 30 meters.

5.2 Results

Impact of Grid Resolution. In the first experiment, we use the **OSM** dataset and create prefix sums at different resolutions to show the effect of the grid size N on the construction time, the memory usage, and the query time. The results for conventional prefix sums (CPS), relative prefix sums (RPS), and sparse prefix sums (SPS) are shown in Fig. 6 and confirm the theoretical results in Table 1. As expected, the storage costs of SPS are significantly smaller than for CPS and RPS, which have identical storage costs as both compute and store dense arrays. For instance, CPS and RPS require 64 GB for a grid with about 8 billion cells, whereas SPS only requires 5.1 GB. In terms of construction time, we observe that CPS is the fastest approach. SPS is only slightly slower than CPS, but clearly faster to construct than RPS (upon which SPS is based), since a large number of cells need not to be computed. The query time for all approaches is constant in the order of some microseconds. Compared to RPS, the query time of SPS

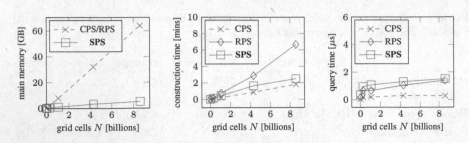

Fig. 6. Impact of grid resolution (**OSM** dataset).

is higher, because it has to store blocks in multiple pieces and the overhead of addressing these pieces causes an increase in query time.

Next, we repeat the same experiment, but for very high resolutions up to $300 \cdot 10^9$ cells. The results are shown in Fig. 7. Since for such high resolutions both CPS and RPS exceeded the memory capacity of our server, we were not able to run them; the memory usage in Fig. 7 is analytically computed, which is easy for dense arrays. In contrast, SPS can be constructed in the order of minutes and requires only 79 GB memory for the highest resolution, whereas the other two approaches would require more than 2TB. The query time of SPS remains roughly constant in the order of some microseconds.

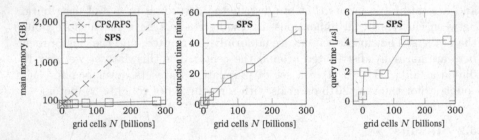

Fig. 7. Impact of a very high grid resolution (**OSM** dataset).

Impact of Dimensionality. Although we focus on low-dimensional data, for the sake of completeness we analyze the behavior of the three approaches for higher-dimensional data. We vary the number of dimensions in the **ZIPF** dataset, keeping the number of grid cells constant at $N = 268 \cdot 10^6$. The results are shown in Fig. 8. For higher-dimensional data, the memory usage of SPS quickly reaches the one of the competitors. This confirms our analytical results. The reason is that the number of border cells increases drastically with higher dimensions, thus a comparably much larger number of cells is materialized. The same observation holds for the construction time. Since there is not much gain in the number of non-materialized cells, the construction of the sparse representation is more expensive than the dense representation. We could achieve a similar construction

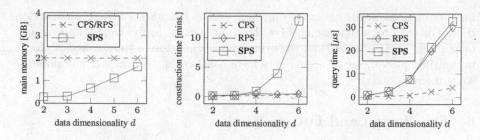

Fig. 8. Impact of dimensionality (**ZIPF** dataset).

Table 2. Storage costs, construction time and query time (**HRSL** datasets).

	Data properties		Storage		Construction			Query		
	Resolution	Sparsity	CPS/RPS	**SPS**	CPS	RPS	**SPS**	CPS	RPS	**SPS**
South Africa	66612 × 45748	99.55%	22.7 GB	1.7 GB	52 s	197 s	52 s	0.3 µs	0.5 µs	1.4 µs
Madagascar	28311 × 49159	99.78%	10 GB	0.7 GB	17 s	68 s	5 s	0.2 µs	0.6 µs	1.6 µs
Burkina Faso	28521 × 20442	99.75%	4.3 GB	0.5 GB	7 s	35 s	5 s	0.2 µs	0.6 µs	1.4 µs
Ivory Coast	23663 × 23147	99.77%	3.8 GB	0.3 GB	8 s	29 s	4 s	0.2 µs	0.5 µs	1.7 µs
Ghana	17639 × 23151	99.64%	2.9 GB	0.4 GB	5 s	25 s	10 s	0.2 µs	0.6 µs	1.4 µs
Malawi	11606 × 27931	99.44%	2.4 GB	0.4 GB	4 s	15 s	3 s	0.2 µs	0.7 µs	0.9 µs
Sri Lanka	8757 × 14103	96.89%	0.9 GB	0.3 GB	2 s	6 s	4 s	0.2 µs	0.5 µs	1.3 µs
Haiti	12473 × 7513	98.15%	0.6 GB	0.1 GB	1 s	3 s	1 s	0.2 µs	0.6 µs	1.0 µs

time by first computing CPS or RPS and then deriving SPS. However, such an
approach would have the same intermediate memory requirements as CPS/RPS.

Gridded Real-World Data. In the last experiment, we compare all three techniques on the eight real-world datasets in **HRSL**. The construction time, query time and storage costs for each dataset together with the resolution and the sparsity are summarized in Table 2. To facilitate the comparison, the performance measures are also plotted in Fig. 9. We can observe that the more sparse the data is the better is the memory improvement of SPS over the competitors. For instance, for the "South Africa" and "Madagascar" datasets, SPS not only yields

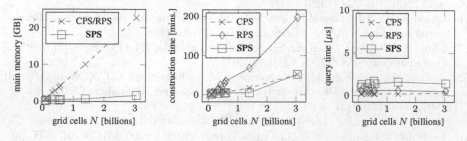

Fig. 9. Impact of grid resolution (**HRSL** datasets).

huge reductions in memory consumption, but it has also a faster construction time. For the query time we observe the same picture as in the previous experiments, i.e., SPS is slower than the other approaches due to the storage of blocks in multiple pieces, but all approaches have constant query times in the range of some microseconds.

6   Conclusion and Future Work

In this paper, we presented the sparse prefix sum technique. The basic idea is to start from relative prefix sums and then avoid materializing local rows/columns that are equal to zero in the data array. This leads to an order of magnitude reduction in storage costs for sparse low-dimensional arrays, while maintaining constant query time and a comparable construction time. As a result, the proposed approach makes it feasible to run microsecond-fast range sum queries on consumer hardware at vastly higher resolutions than before. We also show that, for high-precision grids based on satellite imagery, the approach reduces the storage costs by an order of magnitude.

In future work, we want to apply the proposed technique to dynamic data summaries and investigate how the approach can be improved to better deal with more data dimensions.

References

1. Facebook Connectivity Lab, Center for International Earth Science Information Network - CIESIN - Columbia University 2016. High Resolution Settlement Layer (HRSL). Source imagery for HRSL © 2016 DigitalGlobe. http://www.ciesin.columbia.edu/data/hrsl/. Accessed 01 Mar 2017
2. Agarwal, P.K., Erickson, J., et al.: Geometric range searching and its relatives. Contemp. Math. **223**, 1–56 (1999)
3. Bengtsson, F., Chen, J.: Space-efficient range-sum queries in OLAP. In: Kambayashi, Y., Mohania, M., Wöß, W. (eds.) DaWaK 2004. LNCS, vol. 3181, pp. 87–96. Springer, Heidelberg (2004). doi:10.1007/978-3-540-30076-2_9
4. Chan, C.Y., Ioannidis, Y.E.: Hierarchical prefix cubes for range-sum queries. In: VLDB, pp. 675–686 (1999)
5. Chazelle, B.: Filtering search: a new approach to query-answering. SIAM J. Comput. **15**(3), 703–724 (1986)
6. Chazelle, B.: A functional approach to data structures and its use in multidimensional searching. SIAM J. Comput. **17**(3), 427–462 (1988)
7. Chun, S., Chung, C., Lee, S.: Space-efficient cubes for OLAP range-sum queries. Decis. Support Syst. **37**(1), 83–102 (2004)
8. de Berg, M., van Kreveld, M., Overmars, M., Schwarzkopf, O.C.: Computational geometry. Computational Geometry, pp. 1–17. Springer, Heidelberg (2000). doi:10.1007/978-3-662-04245-8_1
9. Geffner, S., Agrawal, D., Abbadi, A.: The dynamic data cube. In: Zaniolo, C., Lockemann, P.C., Scholl, M.H., Grust, T. (eds.) EDBT 2000. LNCS, vol. 1777, pp. 237–253. Springer, Heidelberg (2000). doi:10.1007/3-540-46439-5_17

10. Geffner, S., Agrawal, D., El Abbadi, A., Smith, T.R.: Relative prefix sums: an efficient approach for querying dynamic OLAP data cubes. In: ICDE, pp. 328–335 (1999)
11. Ho, C., Agrawal, R., Megiddo, N., Srikant, R.: Range queries in OLAP data cubes. In: SIGMOD Conference, pp. 73–88 (1997)
12. Kang, H., Min, J., Chun, S., Chung, C.: A compression method for prefix-sum cubes. Inf. Process. Lett. **92**(2), 99–105 (2004)
13. Liang, W., Wang, H., Orlowska, M.E.: Range queries in dynamic OLAP data cubes. Data Knowl. Eng. **34**(1), 21–38 (2000)
14. Riedewald, M., Agrawal, D., El Abbadi, A.: pCUBE: update-efficient online aggregation with progressive feedback and error bounds. In: SSDBM, pp. 95–108 (2000)
15. Riedewald, M., Agrawal, D., El Abbadi, A.: Dynamic multidimensional data cubes. In: Multidimensional Databases, pp. 200–221 (2003)
16. Riedewald, M., Agrawal, D., Abbadi, A.E., Pajarola, R.: Space-efficient data cubes for dynamic environments. In: Kambayashi, Y., Mohania, M., Tjoa, A.M. (eds.) DaWaK 2000. LNCS, vol. 1874, pp. 24–33. Springer, Heidelberg (2000). doi:10.1007/3-540-44466-1_3
17. Takaoka, T.: Efficient algorithms for the maximum subarray problem by distance matrix multiplication. Electr. Notes Theor. Comput. Sci. **61**, 191–200 (2002)
18. Viola, P.A., Jones, M.J.: Rapid object detection using a boosted cascade of simple features. In: CVPR (1), pp. 511–518 (2001)

Targeted Feedback Collection Applied to Multi-Criteria Source Selection

Julio César Cortés Ríos[✉], Norman W. Paton, Alvaro A.A. Fernandes,
Edward Abel, and John A. Keane

School of Computer Science, University of Manchester,
Oxford Road, Manchester M13 9PL, UK
juliocesar.cortesrios@manchester.ac.uk

Abstract. A multi-criteria source selection (MCSS) scenario identifies, from a set of candidate data sources, the subset that best meets a user's needs. These needs are expressed using several criteria, which are used to evaluate the candidate data sources. A MCSS problem can be solved using multi-dimensional optimisation techniques that trade-off the different objectives. Sometimes we may have uncertain knowledge regarding how well the candidate data sources meet the criteria. In order to overcome this uncertainty, we may rely on end users or crowds to annotate the data items produced by the sources in relation to the selection criteria. In this paper, we introduce an approach called Targeted Feedback Collection (TFC), which aims to identify those data items on which feedback should be collected, thereby providing evidence on how the sources satisfy the required criteria. TFC targets feedback by considering the confidence intervals around the estimated criteria values. The TFC strategy has been evaluated, with promising results, against other approaches to feedback collection, including active learning, using real-world data sets.

Keywords: Data integration · Source selection · Feedback collection · Pay-as-you-go · Multi-objective optimisation

1 Introduction

The number of available data sources is increasing at an unprecedented rate [8]. Open data initiatives and other technological advances, like publishing to the web of data or automatically extracting data from tables and web forms, are making the source selection problem a critical topic. In this context, it is crucial to select those data sources that satisfy user requirements on the basis of well-founded decisions.

Regarding the properties that the data sources must exhibit, there have been studies of the source selection problem considering specific criteria, such as accuracy, cost and freshness [5,18]. In this paper, we deploy a multi-criteria approach that can be applied to diverse criteria in order to accommodate a wider variety of user requirements and preferences, while considering the trade-off between

© Springer International Publishing AG 2017
M. Kirikova et al. (Eds.): ADBIS 2017, LNCS 10509, pp. 136–150, 2017.
DOI: 10.1007/978-3-319-66917-5_10

the required criteria. In this approach, from a collection of sources, S, the problem is to identify a subset of the sources S' from which R data items can be obtained that reflect the user's preferences. These preferences are represented by a collection of weighted criteria; for example, the criteria could be of the form *accuracy:0.4, freshness:0.3, relevance:0.3*, indicating that *freshness* and *relevance* are of equal importance to the user, and that *accuracy* is more important still.

To solve the multi-dimensional source selection problem, where each dimension represents a different criterion relating to the data sources, a multi-dimensional optimisation technique is used to provide a solution that takes into account the user's preferences (represented as weights) in relation to the criteria. This Multi-Criteria Source Selection problem (MCSS) has been addressed before (e.g. [16,17]). This paper addresses the MCSS problem using an approach where the objective is to retrieve an optimal number of items from each supplier given the weighted criteria [21] that model a user's requirements.

To inform the source selection process, where criteria estimates are likely to be unreliable, we need to annotate the candidate data sources to obtain their criteria values; this is an essential step, as we need to know how each source scores for each criterion. Given that there may be many sources and criteria, the source annotation process can become expensive. In this paper, we focus on pay-as-you-go approaches, and collect feedback in the form of true and false positive annotations on data items, that indicate whether or not these data items satisfy a specific criterion. Such feedback could come from end users or crowd workers, and has been obtained in previous works [1,2,7,19].

Having an efficient way to collect the feedback required to improve our knowledge of the data sources is important, as there are costs involved. Hence, we need to carefully identify the data items on which to ask for feedback in order to maximise the effect of the new evidence collected, and to minimise the amount of feedback we need to collect. Some recent work has focused on targeting feedback, especially in *crowdsourcing* (e.g. [4,9,14]); here we complement such work by providing an approach to feedback collection for multi-criteria source selection.

In this paper, we build upon the same statistical foundation as [19], which also targets feedback in a way that takes into account the margins of error in criteria estimates. However, in [19] the approach targeted feedback in a setting where there was a trade-off between two fixed criteria (precision and recall), which was explored using single-dimensional, constrained optimisation. In contrast, here there may be arbitrary numbers of criteria, these criteria are weighted, and the search for solutions involves multi-dimensional, constrained optimisation. So, in comparison with [19], the problem solved in this paper is harder in the sense that a solution must satisfy a variable number of weighted criteria of different types (and not only in terms of data quality).

The following contributions are reported in this paper: (i) a strategy for targeted feedback collection for use with MCSS in which there are arbitrary numbers of weighted criteria; (ii) an algorithm that implements the strategy, using the confidence intervals around the criteria estimates to identify those sources that require more feedback to improve the results of MCSS; and (iii) an experimental assessment of our approach using real-world data to show that TFC can consistently and considerably reduce the amount of feedback required to achieve high-quality solutions.

2 Problem Description

MCSS is a complex problem when the number of criteria is large and the user can declare preferences over these criteria. Concretely, the MCSS problem can be defined as: given a set of candidate sources $S = \{s_1, \ldots, s_m\}$, a set of user criteria $C = \{c_1, \ldots, c_n\}$ with weights $W = \{w_1, \ldots, w_n\}$, and a target number of data items R, identify a subset S' of S that satisfies the requirements expressed by the user while maximising the combined criteria. The solution is presented as a collection $\mathbf{X}$ with m elements, indicating how many data items each source in S contributes to the solution. Sources in $S \setminus S'$ do not contribute.

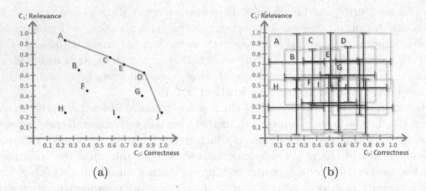

(a)   (b)

Fig. 1. Multi-criteria data source selection without (a) and with (b) uncertainty.

Consider the example presented in Fig. 1(a), in which there are 10 data sources $S = \{A, B, C, D, E, F, G, H, I, J\}$, and 2 data property criteria to balance: relevance (c_1) and correctness (c_2) with the following weights $W = \{w_1 = 0.5, w_2 = 0.5\}$. The user requires a particular number of data items (R) from a subset of sources in S that maximise both criteria and reflect the weights W (in this case, the user considers the two criteria to be of identical importance).

This problem can be solved by using linear programming or other multi-dimensional optimisation techniques. In our case, we are considering an additional factor, the presence of uncertainty in the source criteria estimates. This uncertainty is caused by incomplete evidence, and can be reduced by annotating data items produced by each source to determine if they satisfy the conditions for each criterion. We developed the TFC strategy to identify the data items that reduce this uncertainty to support better solutions for the MCSS problem.

To solve the MCSS problem we can apply an optimisation technique, such as that presented in Subsect. 3.3, and obtain a solution $\mathbf{X}$ which is a vector containing the values of the decision variables x after the optimisation for all the sources. This solution indicates how many data items each data source from S contributes. In Fig. 1(a) we draw a line to highlight the sources in $S' = \{A, C, D, E, J\}$, that in our example contribute data items to the solution.

We now consider the case where we have uncertain knowledge about the criteria values. This brings an even more complex problem. In Fig. 1(b), instead of dots representing the data sources in the multi-criteria space, we have bi-dimensional intervals representing the uncertainty around each source's criteria values. The real criterion value may be expected to lie within this area, but, in the absence of more evidence, we do not know exactly where. Therefore, now the question is: how can we cost-effectively select data items to collect feedback on, in order to reduce the uncertainty in a way that benefits the optimisation technique in solving the MCSS problem? We propose TFC, an approach that minimises the number of data items on which we need to collect feedback and determines the point beyond which the collection of more feedback can be expected to have no effect on which sources were selected.

3 Technical Background

3.1 Data Criteria

A data criterion is a metric that can be applied to a data set to evaluate how the data set fares on that criterion. There are many different criteria that can be applied to the data source selection problem. For instance, in [15,17] accuracy (degree of correctness) and freshness (how recent is the information) were used.

In this paper, we evaluate the estimated value of a data criterion $\hat{c}$ as the ratio between the elements satisfying the notion for a given metric (and for which feedback has been collected) or true positives, tp, and all the elements that have been annotated (which is the sum of the true and false positives, fp). For example, to evaluate the relevance of a source we divide the number of relevant data items over the total number of data items labelled for that source. We use the following formula to compute the ratio of expected data items over all the data items on which feedback has been collected for a source s:

$$\hat{c}_s = \frac{|tp_s|}{|tp_s| + |fp_s|} \tag{1}$$

3.2 Confidence Interval, Overlap and Sample Size

Our strategy is based on classifying the candidate sources into those that are suitable (given a collection of data criteria) to include in a solution, and those sources that are not. This classification is done by analysing the overlapping of the confidence intervals around the criteria value estimates for each source. A *confidence interval* is the range formed by flanking the estimated value (based on the available evidence or feedback) with a margin of error for a required confidence level and represents the space in which the true value is expected to be contained. This confidence interval is associated with a given *confidence level*, which is the percentage of all possible samples that can be expected to include the real value. Following this definition, the larger the number of data items we have labelled for a source, the greater the confidence in the estimated values, and hence, the smaller the confidence intervals around these estimates. We use the following formulae to compute the confidence intervals for a source s [3] (assuming the data is normally distributed as the real distribution is unknown):

$$se_s = \sqrt{\frac{\hat{c}_s \cdot (1 - \hat{c}_s)}{L_s}} \tag{2}$$

$$fpc_s = \sqrt{\frac{T_s - L_s}{T_s - 1}} \tag{3}$$

$$e_s = z_{cL} \cdot se_s \cdot fpc_s \tag{4}$$

To compute the upper and lower bounds of the confidence interval we have:

$$upCI = \min(\hat{c}_s + e_s, 1.0) \tag{5}$$

$$lowCI = \max(\hat{c}_s - e_s, 0.0) \tag{6}$$

where s is a source in the set of candidate data sources S, se_s is the *standard error*, fpc_s is the *finite population correction factor* used to accommodate data sources of any size assuming that they have a finite number of data items, L_s is the number of feedback instances collected for s, T_s is the total number of data items produced by s, $\hat{c}_s$ is the estimated data criterion, and $lowCI$ and $upCI$ are the lower and upper bounds of the confidence interval, respectively. The result is the *margin of error e_s* around our estimate, e.g. $\hat{c}_s \pm e_s$, for a given confidence level cL and its corresponding z–score z_{cL}.

An important part of our strategy relies on the confidence intervals surrounding the criteria estimates for each source, and how these confidence intervals overlap. The approach [10] is to determine not only if two confidence intervals overlap, but also the amount of overlapping. Analysis of this overlapping helps in determining whether two intervals are significantly different or else can be considered as equivalent. In our approach we only consider intervals which are *significantly overlapping*, i.e. if the values are significantly different (thereby providing strong evidence). The estimated values are significantly different if:

$$\hat{c}_{s_1} - \hat{c}_{s_2} > z_{cL} \cdot \sqrt{se_{s_1}^2 + se_{s_2}^2}, \tag{7}$$

and there is no overlap between confidence intervals if:

$$\hat{c}_{s_1} - \hat{c}_{s_2} > z_{cL} \cdot (se_{s_1} + se_{s_2}). \tag{8}$$

The TFC strategy uses the notion of *sampling size* to compute the number of data items required to obtain a representative sample of the entire population given a confidence level and a margin of error. Feedback on the sample elements is then obtained to establish an initial understanding of the underlying data quality. The sample size sS for a finite population T [6,12] is also used to estimate the number of elements required during each feedback collection episode:

$$sS_T = \frac{ssInf}{1 + \frac{ssInf-1}{|T|}} \qquad (9)$$

which is based on the formula for the sample size of a very large (infinite) population for a given margin of error e and a desired confidence level cL:

$$ssInf = \frac{z_{cL}^2 \cdot (\hat{c}_s \cdot (1 - \hat{c}_s))}{e^2} \qquad (10)$$

3.3 Multi-criteria Optimisation

Regarding the multi-dimensional or multi-criteria optimisation techniques, we consider that the data criteria are evaluated using linear functions and therefore linear-programming techniques can be used to find the optimal solution that balances all the criteria and user preferences (represented as weights). To solve the optimisation problem, we have selected *Min-sum*, a Multi-objective linear programming (MOLP) algorithm, whose objective is to maximise the overall weighted utility of the solution considering the minimum deviation λ from the objectives, and the trade-off between all the objectives. This can be represented as a linear programming problem with a collection of objective functions **Z** and their associated constraints, and the general goal of maximising the overall weighted utility of the solution. The solution is represented as the vector **X** containing the values of all the decision variables x after the optimisation.

First we need to obtain the ideal Z_k^* and negative ideal Z_k^{**} solutions (best and worst possible solutions respectively) for each criterion k by using single objective optimisation. These solutions are found by optimising each criterion with respect to the following single objective function Z:

$$Z_k = \frac{\sum_{i=1}^{m} x_i \cdot \hat{c}_{k_{s_i}}}{R} \quad k = 1, 2, \ldots, n, \qquad (11)$$

where m is the number of candidate sources available, n is the number of user criteria, x_i is the number of data items used from source s_i, $\hat{c}_{k_{s_i}}$ is the value of the criterion k for source s_i, and R is the number of data items requested.

The objective function in Eq. 11 is solved as both maximisation and minimisation objective functions for each criterion k. These functions are constrained as follows. The number of data items chosen from each source in S, x_i, cannot exceed the maximum number of data items $|s|$ produced by that source:

$$x_i \leq |s_i| \quad i = 1, 2, \ldots, m \qquad (12)$$

The total number of data items chosen must equal the amount requested:

$$\sum_{i=1}^{m} x_i = R \tag{13}$$

And the minimum value for the decision variables x is 0 (non-negativity):

$$x_i \geq 0 \quad i = 1, 2, \ldots, m \tag{14}$$

The ideal and negative ideal solutions for each criterion k are then computed to obtain the range of possible values. These solutions, along with the constraints from Eqs. 12–14, and the user preference weights w are used to find a solution that minimises the sum of the criteria deviations. Each per-criterion deviation measures the compromise in a solution with respect to the corresponding ideal value given the user weights.

The weighted deviation for each criterion D_k is computed by comparing how far the current solution is from the ideal solution, as follows:

$$D_k = \frac{w_k \cdot (Z_k^* - Z_k)}{Z_k^* - Z_k^{**}} \quad k = 1, 2, \ldots, n, \tag{15}$$

And finally, the optimisation model consists in minimising the sum of criteria deviations λ (measure of the overall deviation from the objective) by considering the constraints in Eqs. 12–15 as follows:

$$\min \quad \lambda, \lambda = D_1 + D_2 + \ldots + D_n, \tag{16}$$

4 Targeting Feedback Using Multi-dimensional Confidence Intervals

We now define the TFC strategy. Consider the MCSS example problem described in Sect. 2, where the goal is to select, from the available candidates, the data sources that provide the maximum combined relevance and correctness. For this goal, some budget was allocated to fund feedback on b data items. We assume no up-front knowledge of the relevance or correctness.

To solve this problem, we need some initial estimates about the values of the criteria for the candidate data sources as shown in Fig. 1(b). We obtain these by collecting feedback on a representative random sample of data items (Eq. 9). To compute the criteria estimates, we use Eq. 1, and calculate the associated margin of error with Eqs. 4, 5 and 6, to obtain the confidence intervals for each criterion for each data source as shown in Fig. 2(a).

Given these initial estimates, we address the goal of finding the combination of sources that maximises the desired criteria (relevance and correctness in the example) while considering the trade-off between them. We can formulate this objective as a Min-sum model using Eq. 16 with the criteria estimates as our coefficients and the number of data items from each source as the decision variables. Min-sum then finds the combination of data items returned by each source that yields the maximum overall weighted utility oU for the optimisation goal (maximum combined relevance and correctness).

By applying Min-sum over the candidate sources, we find the subset of sources forming a non-dominated solution $Ss = \{A, C, D, G, J\}$. This preliminary solution is illustrated in Fig. 2(a). In the same figure, a different subset of sources $So = \{B, E, I\}$ is identified that are not part of Ss but that have confidence intervals for the optimisation criteria that overlap with sources in Ss (graphically, the areas defined by the confidence intervals of B, E and I overlap those defined by the the confidence intervals of the sources in the non-dominated solution, whereas F and H do not). This overlap is computed using Eqs. 7 and 8. It suggests that, in addition to the sources in the non-dominated solution, we need to collect more feedback on B, E, and I in order to decide whether they belong to the solution or not. Our strategy then collects more feedback on a new set, $S' = Ss \cup So = \{A, B, C, D, E, G, I, J\}$. The sources in S' benefit from additional feedback either by reducing the uncertainty as to whether or not they should contribute to the solution, or by refining their criteria estimates.

Having decided on the data sources we need to collect more feedback on, we determine how much feedback should be obtained. This is obtained with Eq. 9, which computes the sample size over a population which, in this case, are the unlabelled data items produced by all the sources in S'.

Having decided on the number of data items that we need to collect feedback on, we collect the feedback and use it to refine our criteria estimates; this is done by recalculating the estimates and margin of error for each criterion for each source. We follow this approach for the sources which are either part of a preliminary solution or are candidates to become part of the solution.

This refinement continues while we have enough budget b for additional feedback collection, there is still some overlap between the confidence intervals of sources contributing to the solution and those from non-contributing sources, and there are still unlabelled data items.

It is important to notice that in Fig. 2(a), some sources are not considered for further feedback collection (viz., F and H), since their confidence intervals do not overlap with those of any of the sources contributing to the solution, and therefore, they have no statistical possibility of being part of it, unless there is a need for more data items than those produced by sources in S'. By filtering out these outliers, TFC focuses on those sources that can be part of the solution.

The strategy leads to a state where the confidence intervals for sources in the solution do not overlap with the intervals of other sources, as shown in Fig. 2(b). The result is a solution with low error estimates and a set of data sources that were excluded from additional feedback collection as they have a low likelihood of being part of an improved solution.

An important feature of our strategy is that it can be applied to problems with multiple criteria, varied user preferences (weights) for each criterion, and over a large number of data sources of variable size and quality.

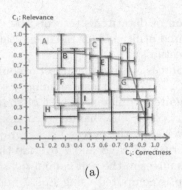
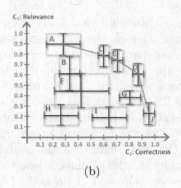

(a)   (b)

Fig. 2. Confidence intervals with overlapping (a) and without overlapping (b)

5   Algorithm

In this section, we describe our algorithm for the TFC strategy applied to the MCSS problem. The pseudo-code for the algorithm is given in Fig. 3.

The inputs for the algorithm are: S: the collection of sources from which we need to select a subset that together satisfy the user requirements considering the criteria and specific preferences; C: the collection of criteria modelled as described in Subsect. 3.1; U: the set of unlabelled data items produced by sources in S; W: the collection of criteria weights representing the user's preferences; b: the allocated budget for the total number of items on which feedback can be obtained; R: the total number of user requested data items; and for statistical estimations cL: the confidence level; and e: the initial margin of error. The output is a vector $\mathbf{X}$ with the number of data items each source contributes.

To solve the MCSS problem, based on the criteria estimates refined by our TFC approach, we first need to obtain the sample size for the number of data items that need to be annotated to achieve a statistically representative view of all candidate data items (line 3). We compute this sample size with Eq. 9 using the number of unlabelled data items produced by sources in S' to represent the sample population. The confidence level and margin of error determine the sample size of the data items on every source.

The function *collectFeedback* (line 4) represents the feedback collection process, which takes as arguments the set of sources considered for feedback S', the set of unlabelled data items U, and the number of additional data items on which feedback is required sS. This function randomly selects from U at most sS data items on which feedback needs to be collected. The remaining budget is updated accordingly depending on the number of data items identified (line 5).

The criteria values can be estimated for each candidate data source in S once the feedback is collected and assimilated. The function *estCriteriaValues* (line 6) uses the candidate data sources S, the sets of labelled and unlabelled data items L and U, the collection of criteria C, and a given confidence level and margin of error cL and e, to compute the collection of estimated criteria values $\hat{C}$ for each data source. These estimates rely on the data items already labelled and are computed with Eq. 1, as described in Subsect. 3.1. The estimates obtained are used to build the confidence intervals (Eqs. 5 and 6) around each criterion estimate (and for each source), by computing the margin of error with Eq. 4. The confidence intervals are then analysed for overlapping one dimension at a time. We follow this approach to handle multiple dimensions without considering simultaneously for the statistical computations. An example of how the confidence intervals may look at this early stage of the process is shown in Fig. 1 from Sect. 2, where there is high overlapping between the confidence intervals and no clear candidate sources.

At this point, with initial estimated criteria values for all the candidate sources, the MCSS problem can be solved by applying an optimisation model as described in Subsect. 3.3 (Min-sum) to obtain a solution that maximises the overall weighted utility oU. The *solveMCSS* function (line 7) represents this step and requires the collection of candidate data sources S, the set of estimated criteria $\hat{C}$, the set of weights representing the user preferences W, and the total number of user requested data items R. The output from this optimisation is a vector $\mathbf{X}$ with the number of data items each candidate source contributes. The set S' is initialised before processing the candidate sources (line 8).

```
Input: set of sources S
Input: set of data property criteria C
Input: set of unlabelled data items U
Input: set of weights for the data property criteria W
Input: a budget size b
Input: a number of user requested data items R
Input: a confidence level cL
Input: a margin of error e
Output: vector X with no. of items contributed per source
 1: L ← {}, X ← {}, S' ← S, sS ← 1
 2: while b > 0 and sS > 0 and S' <> {} do
 3:     sS ← computeSampleSize(S', U, b, cL, e)
 4:     L ← collectFeedback(S', U, sS)
 5:     b ← b − sS
 6:     Ĉ ← estCriteriaValues(S, L, U, C, cL, e)
 7:     X ← solveMCSS(S, Ĉ, W, R)
 8:     S' ← {}
 9:     for s ∈ S do
10:         oL ← false
11:         for c ∈ C do
12:             oL ← isSignificantlyOverlapping(s_c, X) or oL
13:         end for
14:         if X[s] > 0 or oL then
15:             S' ← S' ∪ s
16:         end if
17:     end for
18: end while
19: return X
```

Fig. 3. TFC algorithm

Having the confidence intervals for each criterion and data source, and the sources that contribute to a preliminary solution $\mathbf{X}$, we can analyse the overlap between these intervals. This analysis is performed in the *isSignificantlyOverlapping* function (line 12), which is called with the estimate for each criterion c in C applied to each data source s in S. The function also requires the solution for the MCSS problem $\mathbf{X}$ to determine which intervals from sources contributing to the solution significantly overlap with intervals from non-contributing sources. The overlapping analysis uses the concepts defined in Subsect. 3.2, in particular Eqs. 7 and 8 to determine if two intervals are significantly overlapping or not. As we evaluate this overlapping at source level (not criterion level), when at least one criterion is evaluated with significant overlap the source s is therefore considered for feedback collection (condition: *or oL* in line 12).

The next step is for each source contributing to the solution or for each non-contributing source that has some significant overlap with sources contributing to the solution (line 14), to be added to the set S' which holds the sources on which feedback needs to be collected (line 15). S' is used in the next cycle to compute a new sample size sS over the remaining unlabelled data items. After a few rounds of feedback collection the scenario can be as in Fig. 2(a), where there is still some overlapping but the sources contributing to the solution are now mostly identified.

The iteration continues while any of the following conditions hold (line 2): (i) There is overlapping between confidence intervals of sources that contribute to the solution and sources that do not contribute. (ii) The number of data items on which to collect additional feedback obtained by using Eq. 9 is greater than zero. In other words, we have still some data items left for feedback collection. (iii) The remaining budget b is greater than zero.

When the loop exits, the solution $\mathbf{X}$ (line 19) is in the form of a collection of counts of the number of data items to be used from each candidate source in S. Figure 2(b) presents a potential image at this stage, where no overlapping exists between confidence intervals of sources contributing and not contributing to the solution. Note that when the loop exists because there is no longer any overlap between the confidence intervals of sources that contribute to the solution and sources that do not contribute, this indicates that the selected sources should not change if additional feedback is collected. This is, thus, a well-founded approach to deciding when additional feedback is unlikely to be fruitful.

6   Evaluation: TFC vs. Random and Uncertainty Sampling

In this section we present the experimental results for evaluating the TFC strategy against two competitors: random and uncertainty sampling. Random acts as a baseline. Uncertainty sampling is a general technique that is applicable to the setting we are exploring. To the best of our knowledge there are no specific contributed solutions to this problem in the research literature.

The *random* sampling does not target specific data items for feedback. This baseline competitor considers, as candidates for feedback, all unlabelled items produced by the sources, providing an even distribution of the feedback collected.

Uncertainty sampling is a technique that follows the active learning paradigm, which is based on the hypothesis that if a learning algorithm is allowed to choose the information from which it is learning then it will perform better and with less training [20]. In the uncertainty sampling technique, an active learner poses questions to an oracle over the instances for which it is less certain of the correct label [11]. Often the uncertainty is represented by a probabilistic model that represents the degree of uncertainty we have in the instances. In this paper, the uncertainty is represented by a heuristic that considers the weights of the criteria and the margins of error for the estimated criterion of a source. Feedback is collected first on those data items whose originating source has the largest margin of weighted error, thus taking into account the importance the user places on the criterion. The uncertainty is computed using this formula

$$u_t = \max(w(\hat{c}_{k_s}) \cdot e(s_t, \hat{c}_{k_s})); k = 1, 2, \ldots, n \qquad (17)$$

where t is a data item produced by the source s, u is the uncertainty value on which the items are ranked, w is the data criterion weight, e is the margin of error (Eq. (4)), $\hat{c}_{k_s}$ is the data criterion, and n is the number of criteria. For feedback collection we target first those items with the highest uncertainty, considering all criteria and candidate sources.

6.1   Experimental Setup

The evaluation uses a data set about food (world.openfoodfacts.org/data). This data set contains nutritional information about world food products in an open database format. The information has been gradually collected from vendor's websites and added to the database by unpaid contributors. An additional data set (about UK real estate data) was used to evaluate our approach but since the results were very similar to those presented here, we have not included them as well.

For these experiments, we consider 86,864 different data items produced by 117 virtual sources (where each virtual source represents a contributor to the database). Each data item has the following attributes: code, url, creator, product_name, quantity, origins, countries, serving_size, additives, nutrition_score; all of which were stored as text strings. The targeting approaches were tested with 2, 4 and 6 data criteria, and varied weights among them (user's preferences). The data criteria considered were (in order): correctness, relevance, usefulness, consistency, conciseness and interpretability. The weights corresponding to each tested scenario and for each data criterion are presented in Table 1.

Table 1. Criteria weights for experimental scenarios

	2 criteria ($\mathbf{w_1}$)	4 criteria ($\mathbf{w_2}$)	6 criteria ($\mathbf{w_3}$)
Accuracy (c_1)	0.5	0.4	0.3
Relevance (c_2)	0.0	0.2	0.1
Usefulness (c_3)	0.0	0.1	0.2
Consistency (c_4)	0.5	0.3	0.1
Conciseness (c_5)	0.0	0.0	0.2
Interpretability (c_6)	0.0	0.0	0.1

The experiments were repeated 20 times to reduce the fluctuations due to the random sampling, and the average values were reported. All the statistical computations assume that the data for every source is normally distributed, and are based on a 95% confidence level ($z - score = 1.96$) with an error of 0.05.

For these experiments, the feedback collected is simulated by sampling over the ground truth, in order to evaluate the performance of the approaches without considering additional factors like worker reliability in crowdsourcing [13].

The ground truth was obtained by modelling a typical user's intention over the food data and evaluating this intention across the 6 data criteria in Table 1.

In these experiments, we evaluate the maximum overall weighted utility oU by applying the Min-sum model from Eq. 16 to solve the MCSS problem. oU is a measure of the utility of a solution considering the user's preferences.

6.2   Results

The plots presented in Fig. 4 show the oU for the 3 targeting strategies compared on 3 different scenarios for which the weights are given in Table 1.

In Fig. 4(a) we compare the averaged oU for the 3 targeting strategies with incremental levels of feedback for 2 criteria. The dotted line represents a reference solution achieved *without uncertainty* (100% of the data items labelled). In these results, TFC found a solution above 0.8 oU with only 2.5% of the data items labelled, in comparison with 0.32 and 0.43 oU achieved by the random and uncertainty sampling approaches, respectively, for the percentage of feedback collected. As this scenario considers only 2 criteria ($\mathbf{W}_1$) the solution is hard to find, because the number of potential solutions is larger than when we have more criteria to balance, in other words, if the number of constraints increases (by having more criteria to select the data items), the number of potential solutions decreases which, in turn, reduces the complexity of the optimisation problem.

In Fig. 4(b) TFC still clearly outperforms its competitors, in a scenario with 4 data criteria. The averaged overall weighted utility oU for the reference solution is not as high as in the previous scenario due to the reduction in the number of potential solutions, which is caused by imposing more restrictions (more criteria) in the optimisation problem. This reduces the difference between the 3 strategies but, even so, TFC reaches the reference solution with 6% of the data items labelled while the other approaches have reached barely above the half of the reference oU at the same point. In terms of the improvement, by using TFC the solution, with 2.5% of labelled data items, has 0.7 oU, while random and uncertainty sampling achieve 0.33 and 0.39 respectively.

Figure 4(c) shows the averaged oU for the scenario with 6 criteria. In this case, as we increase even more the number of constraints, the optimisation algorithm finds solutions with lower combined oU hence the smaller difference between the 3 strategies. The advantage of the TFC approach is smaller but it still reaches the reference solution with less feedback than the competitors. For instance, with 2% of labelled data items, TFC allows a solution with oU of 0.6, compared with 0.31 and 0.39 for random and uncertainty sampling respectively.

In the three figures described, the return on investment is clearly favourable for the TFC approach as the overall weighted utility oU of the solution achieved by solving the MCSS problem is always larger with TFC, particularly for small amounts of feedback, which is aligned to the objective of reducing the feedback required to obtain effective when following a pay-as-you-go approach.

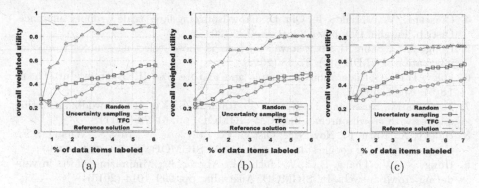

Fig. 4. Results summary for MCSS experiments for (a) 2, (b) 4, and (c) 6 criteria.

7   Conclusions

This paper presented TFC, a strategy for targeting data items for feedback, to enable cost-effective MCSS. TFC was developed to address the problem of incomplete evidence about the criteria that inform source selection. Key features of the approach are that: (i) Feedback is collected in support of multi-criteria optimisation, in a way that takes into account the impact of the uncertainty on the result of the optimisation. (ii) Feedback is collected not for individual sources in isolation, but rather taking into account the fact that the result is a set of sources. (iii) Feedback is collected on diverse types of criteria, of which there may be an arbitrary number, and user preferences in the form of weights are taken into account during the targeting. (iv) Feedback collection stops when the collection of further feedback is expected not to change which sources contribute to a solution (i.e. there is no significant overlap between the criteria estimates for the selected and rejected sources). (v) Experimental results, with real world data, show substantial improvements in the cost-effectiveness of feedback, compared with a baseline (random) solution and an active learning technique. Future work will evaluate the TFC approach using feedback collected using crowdsourcing, and under a complex and well-known problem domain, i.e. supplier selection.

Acknowledgement. Julio César Cortés Ríos is supported by the Mexican National Council for Science and Technology (CONACyT). Data integration research at Manchester is supported by the UK EPSRC, through the VADA Programme Grant.

References

1. Belhajjame, K., Paton, N.W., Embury, S.M., Fernandes, A.A.A., Hedeler, C.: Incrementally improving dataspaces based on user feedback. Inf. Syst. **38**(5), 656–687 (2013)
2. Bozzon, A., Brambilla, M., Ceri, S.: Answering search queries with crowdsearcher. In: WWW 2012, Lyon, France, pp. 1009–1018, 16–20 April 2012
3. Bulmer, M.G.: Principles of Statistics. Dover Publications, New York (1979)

4. Crescenzi, V., Merialdo, P., Qiu, D.: Crowdsourcing large scale wrapper inference. Distrib. Parallel Databases **33**(1), 95–122 (2015)
5. Dong, X.L., Saha, B., Srivastava, D.: Less is more: selecting sources wisely for integration. PVLDB **6**(2), 37–48 (2012)
6. Foley, D.H.: Considerations of sample and feature size. IEEE Trans. Inf. Theor. **18**(5), 618–626 (1972)
7. Franklin, M., Kossmann, D., Kraska, T., Ramesh, S., Xin, R.: Crowddb: answering queries with crowdsourcing. In: ACM SIGMOD, pp. 61–72 (2011)
8. Halevy, A., Korn, F., Noy, N.F., Olston, C., Polyzotis, N., Roy, S., Whang, S.E.: Goods: organizing google's datasets. In: ACM SIGMOD, pp. 795–806 (2016)
9. Hung, N.Q.V., Thang, D.C., Weidlich, M., Aberer, K.: Minimizing efforts in validating crowd answers. In: SIGMOD, Australia, pp. 999–1014 (2015)
10. Knezevic, A.: Overlapping confidence intervals and statistical significance. Stat-News, Cornell University, Cornell Statistical Consulting Unit 73 (2008)
11. Lewis, D.D., Gale, W.A.: A sequential algorithm for training text classifiers. In: ACM-SIGIR, pp. 3–12 (1994)
12. Lewis, J.R., Sauro, J.: When 100% really isn't 100%: improving the accuracy of small-sample estimates of completion rates. JUS **3**(1), 136–150 (2006)
13. Liu, X., Lu, M., Ooi, B.C., Shen, Y., Wu, S., Zhang, M.: CDAS: a crowdsourcing data analytics system. PVLDB **5**(10), 1040–1051 (2012)
14. Mozafari, B., Sarkar, P., Franklin, M.J., Jordan, M.I., Madden, S.: Scaling up crowd-sourcing to very large datasets: a case for active learning. PVLDB **8**(2), 125–136 (2014)
15. Pipino, L.L., Lee, Y.W., Wang, R.Y.: Data quality assessment. Commun. ACM **45**(4), 211–218 (2002). Supporting community and building social capital, USA
16. Rekatsinas, T., Deshpande, A., Dong, X.L., Getoor, L., Srivastava, D.: Sourcesight: enabling effective source selection. In: SIGMOD Conference, San Francisco, CA, USA, pp. 2157–2160 (2016). http://doi.acm.org/10.1145/2882903.2899403
17. Rekatsinas, T., Dong, X.L., Getoor, L., Srivastava, D.: Finding quality in quantity: the challenge of discovering valuable sources for integration. In: CIDR (2015)
18. Rekatsinas, T., Dong, X.L., Srivastava, D.: Characterizing and selecting fresh data sources. In: SIGMOD, pp. 919–930 (2014)
19. Ríos, J.C.C., Paton, N.W., Fernandes, A.A.A., Belhajjame, K.: Efficient feedback collection for pay-as-you-go source selection. In: SSDBM, pp. 1:1–1:12 (2016)
20. Settles, B.: Active learning. Synth. Lect. Artif. Intell. Mach. Learn. **6**(1), 1–114 (2012)
21. Ting, S.C., Cho, D.I.: An integrated approach for supplier selection and purchasing decisions. Supply Chain Manag. Int. J. **13**(2), 116–127 (2008)

Graph Databases

Cost Model for Pregel on GraphX

Rohit Kumar[1,2]([✉]), Alberto Abelló[2], and Toon Calders[1,3]

[1] Department of Computer and Decision Engineering,
Université Libre de Bruxelles, Brussels, Belgium
`rohit.kumar@ulb.ac.be`
[2] Department of Service and Information System Engineering,
Universitat Politécnica de Catalunya (BarcelonaTech), Barcelona, Spain
[3] Department of Mathematics and Computer Science, Universiteit Antwerpen,
Antwerp, Belgium

Abstract. The graph partitioning strategy plays a vital role in the overall execution of an algorithm in a distributed graph processing system. Choosing the best strategy is very challenging, as no one strategy is always the best fit for all kinds of graphs or algorithms. In this paper, we help users choosing a suitable partitioning strategy for algorithms based on the Pregel model by providing a cost model for the Pregel implementation in Spark-GraphX. The cost model shows the relationship between four major parameters: (1) input graph (2) cluster configuration (3) algorithm properties and (4) partitioning strategy. We validate the accuracy of the cost model on 17 different combinations of input graph, algorithm, and partition strategy. As such, the cost model can serve as a basis for yet to be developed optimizers for Pregel.

1 Introduction

Large graphs with millions of nodes and billions of edges are becoming quite common now. Social media graphs, road network graphs, and relationship graphs between buyers and products are some of the examples of large graphs generated and processed regularly [3]. With the increase in size of these graphs, the classical approach of graph processing is becoming insufficient [7,8]. Hence, to address these shortcomings, *vertex-centric programming models* [10] have been proposed to transform the way graph problems are managed. Pregel [11] is one such programming models which supports distributed (parallel) graph computations. Many distributed graph computing (DGC) systems like PowerGraph [4] and Spark-GraphX [15] provide implementations of the Pregel model for graph computations. DGC systems distribute the graph computation by partitioning the graph over different nodes of a cluster.

There are many partitioning strategies proposed in literature [4,12,14] for performing efficient graph computations on DGC systems. Most of the DGC systems provide the same programing model and offer similar features and strategies to use. Depending on the internal implementation of these strategies and algorithms, the systems can give different performance. Even once a user has

© Springer International Publishing AG 2017
M. Kirikova et al. (Eds.): ADBIS 2017, LNCS 10509, pp. 153–166, 2017.
DOI: 10.1007/978-3-319-66917-5_11

decided a system to use, there are not enough guidelines on which partitioning strategy to use for which application or graph. Verma et al. in [13] attempts to address this question with an experimental comparison of different partitioning strategies on three different DGC systems resulting in a set of rules. However, there is no clear theoretical justification of why one partitioning strategy performs better than another depending on a particular combination of graph and algorithm. Moreover, the paper does not consider the cluster properties which according to our cost model, is one of the parameters in deciding the best partitioning strategy. In this paper, we address this question by providing a cost model for the Pregel implementation in GraphX. Cost models are used in the database community for query plan evaluation. We contend that DGC systems should be able to choose the best partitioning strategy for a given graph and algorithm using our cost model in iterative graph computations.

Concretely, in this paper, we make the following contributions: *(i)* we formulate a cost model to capture the different dominating factors involved in the Pregel model (Sect. 3); *(ii)* we validate our cost model on GraphX by estimating the computation time and comparing it with real execution time (Sect. 4). To the best of our knowledge this is the first work in which a cost model based approach has been proposed for Pregel to help users to choose the best partitioning strategy. Similar cost models could be obtained for Pregel on other DGC systems.

2   Background

In this section, we present background information on (1) the Pregel model, and (2) the different partitioning strategies we used in the experiments.

2.1   Pregel Model

In order to render graph computations more efficient, new graph programming models such as Pregel have been introduced [11]. In Pregel, graph algorithms are expressed as iterative vertex-centric computations which can be easily and transparently distributed automatically. We illustrate this principle with the following graph algorithm CC for computing connected components in a graph: we start with assigning to each vertex a unique identifier. In the first step each vertex sends a message with its unique identifier to all its neighbors. Subsequently, for each vertex the minimum is computed of all incoming identifiers. If this minimum is lower than its own identifier, the vertex updates its internal state with this new minimum and sends a message to its neighbors to notify them of its new minimum. This process continues until no more messages are sent. It is easy to see that this iteration will terminate and that the result will be that each vertex holds the minimal identifier over all vertices in its connected component, which can then serve as an identifier of that connected component.

As we can see in this example, a user of Pregel only has to provide the following components:

- Initialization: one initial message per vertex. In the case of CC, this initial message contains the unique identifier of that vertex;
- Function to combine all incoming messages for a vertex. In our example, the combine function takes the minimum over all incoming identifiers.
- A function called the vertex program to update the internal state of the vertex if the minimum identifier received is less than the current identifier of the vertex.
- A function to send the vertex current identifier to its neighbors. In CC, the internal state of a vertex is updated only if the vertex receives a identifier smaller than it is already storing. Only in that case messages are sent to its neighbors with this updated minimum.

Figure 1 illustrate this programming model; every iteration of running the vertex program and combining the messages that will be input for the next iteration is called a super-step. In the first super-step every vertex is activated and executes its vertex program. In Fig. 1, the vertex programs are called "tasks" and the blue lines represent messages sent between vertices. In the second super-step in this figure, vertex 1 does not receive any message and hence will not be active in super-step 2. Vertex 2 receives two messages which are combined and the vertex program is executed. Similarly, vertex 3 receives one message and executes its vertex program. The time it takes for each task could be different and hence there is a synchronization barrier after every super-step. Finally, in super-step 4 no messages are generated and the computation stops.

The main benefit of the Pregel programming model is that it provides a powerful language in which many graph algorithms can be expressed in a natural way. At the same time, however, the programs are flexible enough to allow for automatically and transparently distributing their execution as we will see in next section.

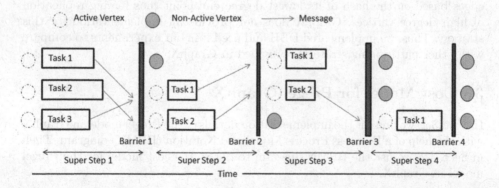

Fig. 1. An example of Pregel model consisting of three vertices.

2.2 Partitioning

There are two kinds of partitioning strategies for distributed graph processing:
(1) vertex-cut [4] and (2) edge-cut [1,6]. In vertex-cut partitioning the edges are
assigned to partitions and thus the vertices can span partitions i.e. vertices are
replicated or mirrored across partitions. In edge-cut, the vertices are partitioned
and the edge can span across partitions i.e. edge is replicated or mirrored across
partitions. GraphX utilizes the vertex-cut partitioning strategy. In vertex-cut
partitioning, the goal of a partitioning strategy is to partition the edges such
that the load (number of edges) in every partition is balanced and vertex *replication* (number of mirrors of vertex) is minimum. Average *replication factor* is
a common metric to measure the effectiveness of vertex-cut partitioning.

The simplest vertex-cut partitioning strategy is to partition edges using a
hash function. GraphX [15] has two different variants for this: *Random Vertex
Cut* (RVC) and *Canonical Random Vertex Cut* (CRVC). Given a hash function
h, RVC assigns an edge (u, v) based on the hash of the source and destination
vertex (i.e. $A(u, v) = h(u, v) \bmod k$). CRVC partitions the edge regardless of
the direction and hence an edge (u, v) and (v, u) will be assigned to the same
partition. CRVC or RVC provides a good load balance due to the randomness in
assigning the edges but do not grantee any upper bound on the replication factor.
There is another strategy which uses two-dimensional sparse matrix and is similar to grid partitioning [5], EdgePartition2D [2]. In EdgePartition2D partitions
are arranged as a square matrix, and for an edge it picks a partition by choosing
column on the basis of the hash of the source vertex and row on the basis of
the hash of the destination vertex. It ensures a replication factor of $(2\sqrt{N} - 1)$
where N is the number of partitions. In practice, these approaches result in large
number of vertex replications and do not perform well for a power-law graphs.

Recently, a *Degree-Based Hashing* (DBH) algorithm [14] was introduced with
improved grantees on replication factor for power-law graphs. DBH partitions
edges based on the hash of its lowest degree end point thus forcing replication
of high degree vertices. GraphX does not provide an implementation for this
strategy. Thus, we implemented DBH and used it in our experiments to compare
with other partitioning strategies provided in GraphX.

3 Cost Model for Pregel GraphX

In Sect. 3.1, we present the implementation details of the Pregel model in GraphX
with the help of a Business Process Model and Notation (BPMN) diagram. Then
in Sect. 3.2, we use the BPMN diagram to derive the cost model for the Pregel
model in GraphX.

3.1 Pregel Model in GraphX

GraphX is built on top of Apache Spark which uses a distributed data structure
called Resilient Distributed Datasets (RDD) [16]. A graph in GraphX is represented as a pair of vertex and edge property collections namely *VertexRDD* and

EdgeRDD. The *VertexRDD* contains all the vertices of the graph and acts as the master copy, which runs the UPDATEVERTEX program. The *EdgeRDD* contains all the edge attributes and the vertex ids of the source and destination vertices. During Pregel execution, a materialized view (*EdgeTripletRDD*) is created by joining *VertexRDD* and *EdgeRDD* for the set of active vertices. The RDDs are partitioned across the cluster nodes and the computation happens in a shared-nothing architecture. The *VertexRDD* is partitioned randomly based on the hash of the vertex id and the *EdgeRDD* is partitioned using the graph partitioning strategy provided (vertex-cut strategies discussed in Sect. 2.2). *EdgeTripletRDD* is partitioned using the same partitioner used by *EdgeRDD*.

The Pregel computation in GraphX consists of four phases: Initialization, Apply, Gather and Reduce. The Initialization happens only once and the other three repeat in a loop until the program stops or a given maximal number of super-steps is exceeded. The Initialization phase, is executed by the driver/master as a single instance. The other three phases run in multiple instances. Each instance is processing of one partition of either the *VertexRDD* or *EdgeRDD*. After the Initialization phase the Apply phase runs one instance per partition of the *VertexRDD* and updates the vertices state. Then the Gather phase runs one instance per partition of the *EdgeRDD* to fetch the latest copy of the vertex state from *VertexRDD* and generate messages for next super-step. The Gather phase does a local reduce of the messages as well by combining all the messages generated for the same vertex on each instance. Finally, the reduce phase does a global reduce by combining of all the messages generated for the same vertex at vertex partitions. The reduce phase runs one instance per partition of the *VertexRDD*. Figure 2 shows all the phases and precedences. Please note, unlike the ideal Pregel model where every vertex could execute the vertex program in parallel and send and receive messages in parallel, in GraphX the parallelization is at the level of an instance or partition. For example, the vertex program of CC algorithm in GraphX will run during the Apply phase in parallel for every partition of the *VertexRDD*. Inside one partition of a *VertexRDD*, the vertex program will run in sequence for all the vertices.

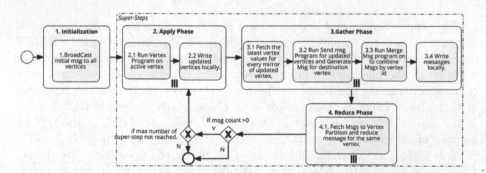

Fig. 2. BPMN diagram representing the Pregel computation model.

3.2   The Cost Model Formulation

For the sake of simplicity of the cost model we make following assumptions:

1. All the nodes in the cluster have the same characteristics, i.e. they have same processing speed, IO and network bandwidth. This assumption does not reduce the applicability of the model, since extending it to heterogeneous nodes is straight forward.
2. Resource scheduling is not considered and hence, we assume all the instances run in parallel. This assumption is a natural choice to maximize performance as it offers maximum parallelization. To ensure this we just need to make sure that we keep the number of partitions to be equal to the number of available workers in the cluster.

From the BPMN diagram in Fig. 2, it is clear that the cost of the Pregel job is the sum of the costs of four phases. We represent the cost of the Initialization phase as a function $cInit$ which depends on: the vertices (V), the algorithm (A) which determines the cost of creating the initial message and its size, and finally, the number of vertex partitions to which the initial message will be sent. We combine the remaining three: Apply, Gather and Reduce phases, in function $cSuperStep$, representing the cost of the subsequent super-steps. Let s be the number of super-steps. Hence, we can represent the cost of the Pregel model ($cPregel$) as shown in Eq. (1). For a super-step i the cost $cSuperStep$ depends on: currently active vertices (V_i), currently active edges (E_i) and the messages (M_{i-1}) generated in previous super-step. How a vertex or an edge becomes active depends on the algorithm (A). We define A_v, A_s, and A_m as three functions for UPDATEVERTEX, SENDMSG, and MERGEMSG programs respectively. Additionally, $cSuperStep$ also depends on how V_i and E_i is partitioned (i.e., vertex partitioning strategy (P_v) and edge partitioning strategy (P_e)).

$$cPregel(V, E, s, A, P_e, P_v) := cInit(V, A, |P_v|)$$
$$+ \sum_{i=1}^{s} cSuperStep(V_i, E_i, A, M_{i-1}, P_e, P_v) \tag{1}$$

The Apply, Gather and Reduce phases run in sequence and hence the cost of one super-step is the sum of the cost of each phase. But, as shown in the BPMN diagram there are multiple instances of each phase. As per our assumption, we have all the instances running in parallel in the cluster. Hence, we denote the cost of running one phase as the maximum cost among all the instances of that phase. There are tasks inside each phase which run sequentially except in the case of Reduce phase where there is only one task. Let $|P_v|$ and $|P_e|$ be number of vertex and edge partitions respectively, and q ($0 \leq q \leq |P_v|$) and k ($0 \leq k \leq |P_e|$) as corresponding index of vertex or edge partition. We define, $E_i^k \subset E_i$ as set of active edges on a partition k; V_i^k as set of vertices at super-step i in edge partition k which is either a source or destination vertex of an active edge E_i^k; $V_i^q \subset V_i$ as set of active vertices in vertex partition q; M_i^k as set of messages generated in super-step i in edge partition k; $M_i^q \subset M_i$ as set of messages received in super-step i in vertex partition q. We represent the cost of each super-step as shown in Eq. (2).

$$cSuperStep(V_i, E_i, A, M_{i-1}, P_e, P_v) := \max_{0 \le q \le |P_v|} \{cApply(V_i^q, M_{i-1}^q, A_v, P_e, P_v)\}$$

$$+ \max_{0 \le k \le |P_e|} \{cGather(E_i^k, M_i^k, V_i^k, A_s, A_m, P_e)\}$$

$$+ \max_{0 \le q \le |P_v|} \{cReduce(M_i^q, V_i^q, A_m, P_e, P_v)\}$$

$$(2)$$

As shown in Fig. 2, the Apply phase has two tasks:

- The first task is to run the UPDATEVERTEX program on the active vertices. It runs sequentially for every vertex in the local partition. Hence, the total cost of the first task is defined as the sum of the cost of running the UPDATEVERTEX program for every active vertex in the partition, which depends on the vertex state, the input message and the algorithmic characteristics. We capture all this as a function $cVertexProg$ and assume its cost is known to the user defining the algorithm.
- The second task is to write the updated vertex attributes to file so that it can be sent to required edge partitions. It consists of creating $|P_e|$ different file segments, one for each edge partition. The writing is buffered, so each write task writes in an internal memory buffer of size B_s, and when the buffer is full, the content is flushed to the file segment. For example, in Fig. 3a the mapper node having the vertex partition 1 with vertices a, b, c, d will create two files. As one vertex can have its replication in more than one edge partition, it needs to be written in more than one file segment. Let $V_i^{*q} \subseteq V_i^q$ be the set of vertices which updated their state after the first task. We define $replication(v)$ as the number of replication of vertex v in edge partitions and $sizeOf(v)$ as the size of vertex object v in bytes. Hence, the total blocks written would be equal to the size of every vertex object times its replication. Let B_w be the cost of writing one block and B_s be the size of one block, hence the total cost for this task would be $B_w \times \frac{Total\ bytes\ written}{B_s}$.

Apart from the cost of the above mentioned task we define α_1 as a constant to capture some housekeeping tasks done by Spark (like task scheduling) for this phase. We use α_2 and α_3 as separate constant costs for the other two phases. The cost of Apply phase is given as the sum of the cost of the two task and the constant α_1 in Eq. (3).

$$cApply(V_i^q, M_{i-1}^q, A_v, P_e, P_v) := \sum_{v \in V_i^q} cVertexProg(v, M_{i-1}^q(v), A_v)$$

$$+ \beta_w \times \left\lceil \frac{\sum_{v \in V_i^{*q}} sizeOf(v) \times replication(v)}{B_s} \right\rceil + \alpha_1$$

$$(3)$$

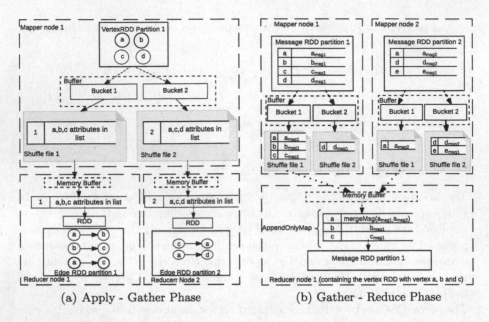

(a) Apply - Gather Phase   (b) Gather - Reduce Phase

Fig. 3. Data shuffle between the phases. Dashed arrows represent in-memory data transfer, Solid arrows represent memory to local disk write and dotted arrows represent remote disk to memory read.

The Gather phase consists of four tasks :

- The first task consists of reading the file segments created in the previous phase. For simplicity, we focus only on the remote reads as local reads are quite fast and do not affect the overall cost significantly. Each file will be read and deserialized to create or update an AppendOnlyMap (an internal data structure used by Spark to create an RDD). In this case there is only one key in the map (the partition id) and the value is a list with vertex attributes. For example, as shown in Fig. 3a there is only one record in the map with key "1" and value a list of vertex attributes of a, b and c. The AppendOnlyMap is then converted into an RDD and combined with *EdgeRDD* to generate *EdgeTripletRDD*. As the number of records in the map is just one, the cost of this task is due to the size of the list. Let V_i^* be the set of all vertices which got updated in previous phase, then the list of vertices read in this task is given as $V_i^k \cap V_i^*$. We represent the total cost of this task as total bytes read multiplied by the cost of reading and deserializing one byte (β_r).
- The second task consists of running the SENDMSG program on every active edge. It depends on the attributes of the source and destination vertices and the algorithm definition A_s. We capture this cost as a function *cSendProg*. Hence, the total cost for this task is given as the sum of running the *cSendProg* for every active edge.
- The third task consist of running the MERGEMSG program to combine all the messages generated for a vertex $v \in V_i^k$. We define the cost of

running MERGEMSG program which combines two messages as $cMergeProg$. It depends on the algorithm definition A_m. We define $M_i^k(v)$ as the set of messages generated for a vertex v. MERGEMSG will run $|M_i^k(v)| - 1$ times.

- The final task is the shuffle write task, which consists of writing to disk the final list of reduced messages $\widehat{M_i^k}$ as shown in Fig. 3b. The writing will be buffered as in the Apply phase, but the number of records written will be equal to the number of final messages ($|\widehat{M_i^k}|$). One message can belong only to one shuffle file, hence the total blocks written would be size of all messages divided by the block size.

The cost of the Gather phase is defined as the sum of the cost of the four tasks and the constant α_2 given in Eq. (4).

$$
\begin{aligned}
cGather(E_i^k, M_i^k, V_i^k, A_s, A_m, P_e) := & \; \beta_r \times \sum_{v \in V_i^k \cap V_i^*} sizeOf(v) \\
& + \sum_{(u,v) \in E_i^k} cSendProg(u, v, A_s) \\
& + cProcess(M_i^k, V_i^k, A_m) \\
& + \beta_w \times \left\lceil \frac{\sum_{m \in \widehat{M_i^k}} sizeOf(m)}{B_s} \right\rceil + \alpha_2
\end{aligned}
\tag{4}
$$

where,

$$
cProcess(M_i^k, V_i^k, A_m) := \sum_{v \in V_i^k} \left(|M_i^k(v)| - 1 \right) \times cMergeProg(A_m)
\tag{5}
$$

The Reduce phase consists of only one task which is to fetch the messages generated in the previous phase and reduce the messages for the same vertex into one message. For example, as shown in Fig. 3b a_{msg1} and a_{msg2} are fetched from two mappers and reduced into one message for vertex a. Unlike the read in the Gather phase, in this phase the number of records in the AppendOnlyMap will be equal to the numbers of messages. For example, as shown in Fig. 3 there is one record in the shuffle file for the Gather phase where as upto 3 records in the shuffle file for the Reduce phase. The size of each message record is constant, hence the cost of the read is dominated by the number of records and not the size of the record. We define γ as the constant cost of reading and updating the AppendOnlyMap per record. Thus, we can define cost for the read task as γ times number of records fetched. The reducing of the messages can start as soon as there are two messages for the same vertex. As Spark uses parallel threads to read data and process data, there will be an overlap in the execution of these tasks. Hence, in a multi-core system, as soon as first block of messages is read, it can start processing the messages while in parallel keep fetching remaining blocks. Let C be the number of cores in a cluster node; hence C threads can fetch data in parallel. Let b be the number of blocks of messages received in this phase and M^b represent the set of messages in the bth block. Then, the overall cost of this phase is given as the sum of the cost of fetching the first block plus the cost of processing all messages (if processing is slower than fetching) or the cost of fetching remaining blocks plus processing the last block (if fetching is slower than processing) as expressed in Eq. (6).

$$cReduce(M_i^q, V_i^q, A_m, P_e, P_v) := \gamma \times |M^1|$$
$$+ max \{cProcess(M_i^q, V_i^q, A_m),$$
$$\frac{\gamma}{C} \times \sum_{2 \leq j \leq b} |M^j| + C \times cProcess(M^b, V_i^q, A_m)\}$$
$$+ \alpha_3$$

$$(6)$$

For a single core node, the fetching of data and processing can not run in parallel, hence Eq. (6) simplifies to the sum of the cost of fetching all messages and processing them as given in Eq. (7).

$$cReduce(M_i^q, V_i^q, A_m, P_e, P_v) := \gamma \times |M_i^q|$$
$$+ cProcess(M_i^q, V_i^q, A_m) + \alpha_3$$

$$(7)$$

4   Experimental Validation of the Cost Model

In this section, we describe the experimental setup to obtain the cluster specific variables (α_1, α_2, α_3, β_r, β_w and γ) in the cost model and then share the results of the validation of the cost model on different configurations.

4.1   Experiment Configuration and Setup

There are four main parameters which affect the execution of a GraphX Pregel job: (1) Cluster setup, (2) Input Graph, (3) Partitioning Strategy, and (4) Graph Algorithm to be executed. In our experiments, we always keep the cluster setup constant and vary the other three. All experiments are done on a cluster with a master node and 5 worker nodes. All nodes are Linux systems with Intel Xeon E5-2630L v2 a 2.40 GHz processor, 1 TB SATA-3 Hard disk, 128 GB RAM, and 4 GB Ethernet. We deployed Spark 2.0.2 in cluster mode with each worker node having 1 executor with 1 thread and 45 GB RAM assigned to it.

Input Graph: We used three real world datasets: the CollegeMsg network is a directed graph of messages sent between users on a Facebook-like platform at UC-Irvine; Higgs activity time (Higgs) is a dataset which provides information about activity on Twitter during the discovery of the Higgs boson particles (both datasets were taken from the SNAP repository [9]); Apart from this, we also use a re-tweet network collected from information about activity on Twitter during the Punjab Election 2017 (twitter) in India collected by ourselves for 3 days.

Partitioning Strategy: We use three partitioning strategies in the experiments: EdgePartition2D; Canonical Random Vertex Partitioning(CRVC) (both strategies provided by the default GraphX API) and our own implementation of Degree Based hashing (DBH). As explained earlier, these partitioning strategies only partition the *EdgeRDD*. For *VertexRDD* we used the default random Hash Based partitioner provided by Spark. The number of partitions was equal to 5 in all experiments.

Graph Algorithm: We used the classical PageRank and Connected Component algorithms in our experiments.

4.2   Estimating Cluster Specific Variables

Monitoring the factors in the cost model is not straightforward. Hence, we applied following simplifications to approximate the value of the constant parameters:

1. We used the same code provided in GraphX for the Page Rank and Connected component algorithms but just added additional counters on each of the three GraphX functions to keep a count of how many times the UPDATEVERTEX, SENDMSG and MERGEMSG programs were executed in each task of a super-step.
2. The execution time of the three functions is very small and difficult to monitor precisely. A more accurate measurement of these functions allows for a more accurate estimation of the cluster constants in the formula, hence we introduced a constant time delay of 1 millisecond in all three functions. This constant time delay is only for accurate estimation of the cluster parameters and does not affect the cost model accuracy. Let $count(f)$ be the number of times a program f is executed in an instance. This enables us to approximate:
   - $\sum cVertexProg(v, M_{i-1}^q(v), A_v) = count(\text{UPDATEVERTEX}) \times 1\,\text{ms}$
   - $\sum cSendProg(u, v, A_s) = count(\text{SENDMSG}) \times 1\,\text{ms}$
   - $\sum \left(|M_i^k(v)| - 1\right) \times cMergeProg(A_m) = count(\text{MERGEMSG}) \times 1\,\text{ms}$
3. We kept the number of edge partitions, vertex partitions and number of nodes in the cluster equal, so that every node in the cluster is processing only one partition of the *VertexRDD* and *EdgeRDD* (i.e. $|P_e| = |P_v| = N$).
4. Every node has only one core assigned to it (i.e. $C = 1$), hence we can use Eq. 7 for the reduce phase.

We used `twitter` graph data with the CRVC partitioning strategy and the Page Rank algorithm to estimate the constants $\alpha_1, \alpha_2, \alpha_3, \beta_r, \beta_w$ and γ of the cost model. We used the SPARK UI API (a monitoring service provided by Spark) to get the run time of each phase separately and other factors of the cost model. Since we used a shared cluster while running the experiments, we repeated the experiments 10 times and took the minimum execution time of a super-step as the baseline cost of that super-step, assuming that higher time to execute the same super-step is due to the interferences with parallel executions of other processes on the cluster. *cInit* is a constant one time cost for a graph and algorithm and do not change based on the partitioning strategy hence we do not estimate this cost for every partitioning strategy.

We estimated the value of α_1 and β_w from Eq. 3 by substituting the values of all other factors. For every super-step, we replaced *cApply* by the execution time of the phase, $\sum cVertexProg(v, M_{i-1}^q(v), A_v)$ by *count(*UPDATEVERTEX*)* and the number of blocks written by total bytes written divided by 32 MB (the default value of B_s in Spark), for the task which took the maximum time for this phase. Substituting these values, results in a linear equation of the form $Y = \beta_w \times X + \alpha_1$ where $Y = cApply - count($UPDATEVERTEX$)$ and X is the number of blocks written. We got the value of X and Y for all the super-steps and obtained α_1 and β_w by ordinary least square (OLS) method. The result of the linear curve fitting is show in Fig. 3. We get $\alpha_1 = 1.366$ ms and $\beta_w = 100.77$ ms/block with a R-squared value of 0.9815. We believe the deviation(outliers) from the line is due to discretization of the write bytes into number of buckets as for some cases the last bucket would be almost full and for some it will be almost empty resulting in different write time. Similarly, we estimated α_2 and β_r from Eq. 4 by replacing β_w with 100.77; *cGather* by the stage execution time. For the right hand side parameters of the equation we substituted values for the longest running task. Hence, we replaced $\sum_{v \in V_i^k \cap V_i^*} sizeOf(v)$ by the volume of remote bytes read by the task, $cProcess(M_i^k, V_i^k, A_m)$ by *count(*MERGEMSG*)*, $\sum_{(u,v) \in E_i^k} cSendProg(u, v, A_s)$ by *count(*SENDMSG*)* and the number of blocks written by the volume of total bytes written by the task divided by 32 MB. Substituting these values, results in a linear equation of the form $Y = \beta_r \times X + \alpha_2$, where $Y = cGather - count($MERGEMSG$) - count($SENDMSG$) - \beta_w \times$ #blocks and X is remote bytes read. After applying OLS we get $\alpha_2 = 43.214$ ms

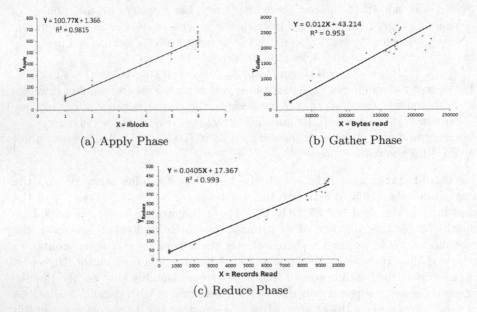

(a) Apply Phase   (b) Gather Phase

(c) Reduce Phase

Fig. 4. Using Linear curve fitting to estimate the variables in the cost model

and $\beta_r = 0.012$ ms/byte with a R-squared value of 0.953 as shown in Fig. 4a. Similarly, from Eq. 7 we get a linear equation of the form $Y = \gamma \times X + \alpha_3$ where, $Y = cReduce - count(\text{MERGEMSG})$ and X is the number of message records. We get $\alpha_3 = 17.367$ ms and $\gamma = 0.0405$ ms/record with R-squared value of 0.993 as shown in Fig. 4c.

4.3   Cost Model Validation

We used 3 different graph data, 3 different edge partitioning strategy and 2 different graph algorithms in our experiments resulting in 18 different combinations of graph, partitioning strategy and algorithm. In order to validate the cost model, we estimated the cluster constants $\alpha_1, \alpha_2, \alpha_3, \beta_r, \beta_w$ and γ in the cost model for graph $=$ `twitter`, partitioning strategy $=$ CRVC and algorithm $=$ Page Rank (Sect. 4.2), then we used other 17 combinations of graph, partitioning strategy and algorithm to estimate the execution cost. We replace the values of $\alpha_1, \alpha_2, \alpha_3, \beta_r, \beta_w$ and γ in the cost model and predict the job execution time by measuring other attributes required by the cost model. Then we estimate the accuracy of the cost model by comparing with the actual execution time of all the super-steps. We report the prediction accuracy in Table 1. We get 96.9% average accuracy in predicting the job execution time in 17 different combination with minimum accuracy of 94.6% and maximum accuracy of 99.8%.

Table 1. Prediction accuracy(%) of the cost model for different combinations of dataset, partitioning strategy and graph algorithm.

Dataset	Algorithm	Partition strategy		
		EdgePartition2D	CRVC	DBH
CollegeMsg	PageRank	96.4	97.9	97.7
	CC	97.6	96.1	96.7
twitter	PageRank	97.7	-	99.3
	CC	98.9	98.7	97.1
Higgs	PageRank	94.6	97.2	99.8
	CC	97.9	95.9	94.9

5   Concluding Remarks

We presented a cost model to estimate the execution cost of Pregel-based algorithms on Spark GraphX and evaluated on different combinations of input graph, algorithm and partitioning strategy. We see from the cost model that the overall execution time depends on different factors such as: the execution time of each function (i.e., UPDATEVERTEX, SENDMSG and MERGEMSG); the cluster configuration (such as data transfer between different nodes). The cost model depends on many variables which are not known before hand and hence, for an optimizer, they will need to be estimated. In future work, we will experiment by varying the different dominating factors in the cost model, to see how they determine the best partitioning strategy.

Acknowledgement. This work was supported by the Fonds de la Recherche Scientifique-FNRS under Grant(s) no. T.0183.14 PDR. The student is also part of IT4BI DC program.

References

1. Barnard, S.T.: Parallel multilevel recursive spectral bisection. In: Proceedings of the 1995 ACM/IEEE Conference on Supercomputing, p. 27. ACM (1995)
2. Çatalyürek, Ü.I.T.V., Aykanat, C., Uçar, B.: On two-dimensional sparse matrix partitioning: models, methods, and a recipe. SIAM J. Sci. Comput. (2010)
3. Ching, A., Edunov, S., Kabiljo, M., Logothetis, D., Muthukrishnan, S.: One trillion edges: graph processing at facebook-scale. VLDB (2015)
4. Gonzalez, J.E., Low, Y., Gu, H., Bickson, D., Guestrin, C.: Powergraph: distributed graph-parallel computation on natural graphs. In: OSDI (2012)
5. Jain, N., Liao, G., Willke, T.L.: Graphbuilder: scalable graph ETL framework. In: GRADES (2013)
6. Karypis, G., Kumar, V.: Multilevel graph partitioning schemes. In: ICPP, vol. 3 (1995)
7. Kumar, R., Calders, T.: Information propagation in interaction networks. In: Proceedings of the 20th International Conference on Extending Database Technology, EDBT 2017 (2017)
8. Kumar, R., Calders, T., Gionis, A., Tatti, N.: Maintaining sliding-window neighborhood profiles in interaction networks. In: Appice, A., Rodrigues, P.P., Santos Costa, V., Gama, J., Jorge, A., Soares, C. (eds.) ECML PKDD 2015. LNCS, vol. 9285, pp. 719–735. Springer, Cham (2015). doi:10.1007/978-3-319-23525-7_44
9. Leskovec, J., Krevl, A.: SNAP Datasets: stanford large network dataset collection, http://snap.stanford.edu/data
10. Lumsdaine, A., Gregor, D., Hendrickson, B., Berry, J.: Challenges in parallel graph processing. Parallel Process. Lett. (2007)
11. Malewicz, G., Austern, M.H., Bik, A.J., Dehnert, J.C., Horn, I., Leiser, N., Czajkowski, G.: Pregel: a system for large-scale graph processing. In: SIGMOD (2010)
12. Petroni, F., Querzoni, L., Daudjee, K., Kamali, S., Iacoboni, G.: HDRF: stream-based partitioning for power-law graphs. In: CIKM. ACM (2015)
13. Verma, S., Leslie, L.M., Shin, Y., Gupta, I.: An experimental comparison of partitioning strategies in distributed graph processing. Proc. VLDB Endow. (2017)
14. Xie, C., Yan, L., Li, W.J., Zhang, Z.: Distributed power-law graph computing: theoretical and empirical analysis. In: Advances in Neural Information Processing Systems (2014)
15. Xin, R.S., Gonzalez, J.E., Franklin, M.J., Stoica, I.: Graphx: a resilient distributed graph system on spark. In: GRADES. ACM (2013)
16. Zaharia, M., Chowdhury, M., Das, T., Dave, A., Ma, J., McCauley, M., Franklin, M.J., Shenker, S., Stoica, I.: Resilient Distributed Datasets: A Fault-tolerant Abstraction for In-memory Cluster Computing. USENIX Association (2012)

Historical Traversals in Native Graph Databases

Konstantinos Semertzidis$^{(\boxtimes)}$ and Evaggelia Pitoura

Department of Computer Science and Engineering,
University of Ioannina, Ioannina, Greece
{ksemer,pitoura}@cs.uoi.gr

Abstract. Since most graph data, such as data from social, citation and computer networks evolve over time, it is useful to be able to query their history. In this paper, we focus on supporting traversals of such graphs using a native graph database. We assume that we are given the history of an evolving graph as a sequence of graph snapshots representing the state of the graph at different time instances. We introduce models for storing such snapshots in the graph database and we propose algorithms for supporting various types of historical reachability and shortest path queries. Finally, we experimentally evaluate and compare the various models and algorithms using both real and synthetic datasets.

Keywords: Graph database · Historical traversals · Reachability · Path computation

1 Introduction

Recently, increasing amounts of graph structured data are made available from a variety of sources, such as social, citation, computer, hyperlink and biological networks. Almost all such real-world networks evolve over time. Querying the evolution of such graphs is an important and challenging problem.

In this paper, we assume that we are given the history of an evolving graph in the form of a sequence of graph snapshots representing the state of the graph at different time instances. Our focus is on efficiently storing and querying these snapshots using a native graph database. Native graph databases offer an attractive means for storing and processing big graph datasets.

To store the sequence of graph snapshots in a graph database, we propose models based on associating with each node and edge, its lifespan, i.e., the time intervals, during which the node and edge is valid. The multi-edge approach (ME) uses a different edge type for each of the time instances during which the edge was valid. The single-edge approaches use a single edge annotated with a complex type for representing the lifespan of the edge. We consider two single-edge approaches, one that models the lifespan as an ordered list of time instances (SETP), and one that uses an interval representation (SETI).

We also introduce historical graph traversals that consider paths that existed in a sufficient number of graph snapshots. We exploit variants of two types of historical traversals, reachability and shortest paths. Historical reachability

© Springer International Publishing AG 2017
M. Kirikova et al. (Eds.): ADBIS 2017, LNCS 10509, pp. 167–181, 2017.
DOI: 10.1007/978-3-319-66917-5_12

queries ask whether two nodes are connected in some time instance, in all time instances, or in a sufficient number of time instances. Historical shortest path queries ask for the shortest path between two nodes posing requirements on the lifespan of such paths. We present algorithms for processing historical queries for both the multi-edge and the single-edge approaches.

We have implemented our approach in the Neo4j graph database and present experimental results using both real and synthetic datasets. For very short-lived edges, using multiple edges to represent lifespans, seems to work well by taking advantage of the built-in traversal methods of the native graph database. However, for all other cases, using the interval-based approach to represent lifespans (SETI) proves more efficient both in terms of processing time and storage. We also present a case study regarding connectivity among authors of different conferences through time.

Related Work: There has been recent interest on analytical processing and mining of evolving graphs, including among others developing models [15], discovering communities [2], and computing measures such as PageRank [3]. There has been also research on building graph engines tailored to supporting analytical processing in dynamic graphs, such as Kineograph [5] and Chronos [7]. However, our focus here is on query processing.

There has been some work on historical query processing. The common assumption is that the graph is either kept in main memory or is stored in disk, but not in a native graph database. Most research assumes as a first step the reconstruction of the relevant snapshots. Then, queries are processed through an online traversal on each of the snapshots. Various optimizations for reducing the storage and snapshot reconstruction overheads have been proposed. Optimizations include the reduction of the number of snapshots that need to be reconstructed by minimizing the number of deltas applied [12], using a hierarchical index of deltas and a memory pool [11], avoiding the reconstruction of all snapshots [17], and improving performance by parallel query execution and proper snapshot placement and distribution [14]. Other research considers in-memory processing of specific types of historical queries [1,4,9,13,18,19].

Very few works [4,6,8,20] are built on top of a native graph database. In particular [4] proposes an approach for storing time-varying networks in the Neo4j graph database using a hierarchical time index to support snapshots with different granularity (e.g., months and days). They do not discuss historical traversal queries, but, instead consider retrieving specific snapshots. In [6] the authors focus on graph data with structural changes, and present time logs that capture when an event has occurred (i.e. add/remove of edge/node) in the history of the graph. Although, their indexes are used to retrieve fast a state of the graph in a given period they are not designed for supporting historical traversal queries. A short discussion of the storage models ME and SETP is made in position paper [20]. Finally, the work in [8] targets specific types of graphs with static structure but frequent changes in node and edges properties. Our focus here is on structural updates and reachability and path queries.

Paper outline: The rest of this paper is structured as follows. In Sect. 2, we formally define historical traversal queries. We introduce three approaches for storing the graph snapshots in graph databases in Sect. 3 and algorithms for processing historical traversal queries in Sect. 4. In Sect. 5, we experimentally evaluate the different approaches. Section 6 concludes the paper.

2 Traversals in Historical Graphs

In this section, we first define historical graphs and then introduce traversal queries on them.

2.1 Historical Graphs

Graphs are used to represent relationships between entities where entities are modeled as nodes and relationships as edges between them. Labels may be assigned to both edges to capture different types of relationships and to nodes to capture node attributes. Formally, a node and edge labeled graph G is a tuple $(V, E, \lambda^V, \lambda^E)$ where V is a set of nodes, $E \subseteq V \times V$ is a set of edges, $\lambda^V : V \rightarrow L^V$ and $\lambda^E : E \rightarrow L^E$ are labeling functions that map a node and an edge to a label from a set L^V of node labels and a set L^E of edge labels respectively. Some graph databases also support an extension of labeled graphs, termed *property graphs*, where instead of labels, a set of property-value pairs is associated with nodes and edges. Such pairs are also sometimes called *attributes*.

Most real world graphs evolve over time. New nodes and edges are added, and existing nodes or edges are deleted. We assume that time is discrete and use successive integers to denote successive points in time. We use $G_t = (V_t, E_t, \lambda_t^V, \lambda_t^E)$ to denote the *graph snapshot* at time instance t, that is the sets of node, edge and labeling functions that exist at time instance t.

Definition 1 (HISTORICAL GRAPH).
A historical graph $\mathcal{G}_{[t_i, t_j]}$ in time interval $[t_i, t_j]$ is a sequence $\{G_{t_i}, G_{t_i+1}, \ldots, G_{t_j}\}$ of graph snapshots.

An example is shown in Fig. 1 which depicts a historical graph $\mathcal{G}_{[1,5]}$ consisting of five graph snapshots $\{G_1, G_2, G_3, G_4, G_5\}$. Nodes and edge labels are omitted for simplicity. Note that the granularity of creating time instances and graph snapshots may vary. For example, we may create a new graph snapshot at every second, hour or day.

We use the term *lifespan* (*ls*) to refer to the period of time that a graph element, that is, a node, edge, or labeling function, existed. Lifespans are sets of time intervals. Set of time intervals are also known as *temporal elements* [10]. For example, the lifespan $ls((u_1, u_3))$ of edge (u_1, u_3) in Fig. 1 is equal to $\{[1, 1], [3, 4]\}$ meaning that edge (u_1, u_3) existed in graph snapshots G_1, G_3 and G_4. We also define a useful operation on interval sets, called *time join* [18]. Given two sets of intervals $\mathcal{I}$ and $\mathcal{I}'$, their time join $\mathcal{I} \otimes \mathcal{I}'$ is the set of time intervals that includes the time instances that exist in both $\mathcal{I}$ and $\mathcal{I}'$. For example, $\{[1, 3], [5, 10], [12, 13]\} \otimes \{[2, 7], [11, 15]\}$ is equal to $\{[2, 3], [5, 7], [12, 13]\}$.

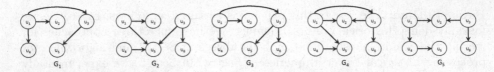

Fig. 1. Example of a historical graph

2.2 Historical Traversal Queries

A graph traversal allows the navigation of the structure of the graph and is a fundamental graph query. In an abstract form, a traversal query Q can be expressed as a path query $Q = u \xrightarrow{\alpha} v$, where α specifies conditions on the paths that we wish to traverse and u, v denote the starting and ending points of these paths. The starting and ending points can be specific nodes or properties of the nodes, or a mix of both. The expression α involves constraints on the properties (or, labels) of the nodes and edges in the path. For example, we may look for paths connecting two people in a social network with edges labeled as "friends".

Traversals retain the paths from u to v that satisfy α. In general, there are may be many such paths, even an infinite number, if there are cycles in the dataset. Thus, besides maintaining all possible paths, various other semantics may be associated with the evaluation of traversals. Common ones are retaining only the shortest paths, or only paths that consist of no-repeated nodes or edges.

We now define traversal queries on historical graphs. First, let us define the lifespan of a path. Let $\mathcal{G}_{[t_i,t_j]}$ be a historical graph and $p = u_1 u_2 \ldots u_m$ be a path of m nodes where $u_k \in \cup_{t_l=t_i}^{t_j} V_{t_l}$, $1 \leq k \leq m$. We define the lifespan, $ls(p)$, of path p as follows: $ls(p) = ls((u_1, u_2)) \otimes ls((u_2, u_3)) \ldots \otimes ls((u_{m-1}, u_m))$. For example, the lifespan of path $u_1 u_3 u_6$ of $\mathcal{G}_{[1,5]}$ in Fig. 1 is $\{[3, 4]\}$.

Definition 2 (HISTORICAL TRAVERSAL QUERY).
A traversal query Q_H on a historical graph, $\mathcal{G}_{[t_i,t_j]}$, called a historical traversal query, is a tuple $(Q, \mathcal{I}, L)$ where Q is a traversal query $Q = u \xrightarrow{\alpha} v$, $\mathcal{I}$ is a set of time intervals and L is a positive integer. For a path p, let $D(p) = ls(p) \otimes \mathcal{I} \otimes [t_i, t_j]$. Q_H retains the paths p from u to v in $\mathcal{G}_{[t_i,t_j]}$ that satisfy α and for which in addition $D(p)$ contains at least L time instances.

Intuitively, we ask that the paths retained by a historical query exist in at least L of the graph snapshots. At one extreme a path must appear at least once in the graph history, in which case $L = 1$. At the other extreme, a path must appear in all time instances, in which case L must be equal with the number of time instances in $\mathcal{I} \otimes [t_i, t_j]$.

A traversal query may produce different outputs. For example, the output may be the set of the retained paths, or, the set of nodes, or edges on the retained paths. Furthermore, in the special case of reachability queries, the output of the traversal is boolean, i.e., true if there exists a path, and false otherwise. Without loss of generality, in this paper we focus on reachability shortest path queries. Additional semantics may be associated with the output of historical traversals.

For example, for reachability queries, we may ask that two nodes are reachable in at least one time instance (*disjunctive queries*), in all time instances (*conjunctive queries*), or in at least-k time instances. For shortest path queries, we may ask, for example, for the earliest shortest path (ESP), for the shortest among the paths that existed in all time instances (stable shortest path (SSP)), or, for the shortest among the paths that existed in at least-k snapshots (KSP).

3   Storing Historical Graphs

In this section, we present different approaches for representing a historical graph in a native graph database. The basic idea is to augment each graph element with its lifespan. For edges and nodes, lifespans are stored as labels (i.e., property, attribute) of the corresponding edge and node. Based on the type of labels used, we have two different approaches.

Multi-edge Representation. The *multi-edge approach* (ME) utilizes a different edge type between two nodes u and v for each time instance of the lifespan of the edge (u, v). The multi-edge representation of the historical graph $\mathcal{G}_{[1,5]}$ of Fig. 1 is depicted in Fig. 2. For instance, to represent a relationship between nodes u_1, u_3 with lifespan $\{[1, 1], [3, 4]\}$, we use three edges with different labels to connect u_1 and u_3. Since all native graph databases provide efficient traversal of edges having a specific label, the ME approach provides an efficient way of retrieving the graph snapshot G_t corresponding to time instance t. Similarly, multiple labels are associated with each node.

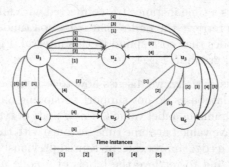

Fig. 2. ME representation of the historical graph of Fig. 1 (nodes labels are not shown for clarity)

Single-edge Representation. The *single-edge approach* uses a single edge between any two nodes appropriately labeled with the lifespan of the edge. To represent the lifespan of an edge or node, we consider two different approaches. In the *single-edge with time points approach* (SETP), the lifespan of a node or edge is modeled using a label that is a sorted list of the time instances in their lifespan. The SETP representation of the historical graph $\mathcal{G}_{[1,5]}$ of Fig. 1 is shown in Fig. 3(a).

For example, the lifespan of edge (u_1, u_3) is now represented by a single edge having as label $[1, 3, 4]$. In the *single-edge with time intervals approach* (SETI), we use Ls and Le, each one an ordered list of m elements, where m is the number of time intervals in the lifespan of the edge or node. In particular, $Ls[i]$, $1 \le i \le m$, denotes the start of the i-th interval in the lifespan, while $Ls[i]$ the end of the interval. An example is shown in Fig. 3(b). With the single-edge approaches, retrieving the graph snapshot G_t at time instance t requires further processing of the related labels.

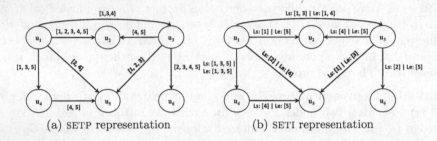

(a) SETP representation   (b) SETI representation

Fig. 3. Single-edge representations of the historical graph of Fig. 1 (nodes labels are not shown for clarity)

Indexing. For faster retrieval of specific graph snapshots, we build an index within the graph database by creating a new node type T where each node of the given type has a unique value that corresponds to a specific time instance. A T node that denotes a time instance t is connected with all nodes that existed at time instance t. To retrieve the nodes that exist in a time interval, we get the neighbors of the T nodes that correspond to this interval. Figure 4(a) shows the index of the historical graph in Fig. 1.

Time-varying labels. Finally, we discuss how to store labels that change over time. Current graph databases do not support versioning on labels and thus we need to create for each unique label value l, a new node of type l. We connect all nodes or edges that have value l at some time instance with the node representing l using one of the three edge approaches presented previously. Doing so, we only store each label once and to retrieve the labels of a node u in a time interval, we retrieve all the nodes type of l that are connected to u by edges that refer to the time instances in the interval. In Fig. 4(b), we depict an example of storing the time-varying labels of two nodes u_1, u_2 using SETP.

4   Processing Historical Traversal Queries

In this section, we focus on processing historical traversal queries in native graph databases. For simplicity, we consider a single interval I, but the algorithms easily extend to sets of time intervals.

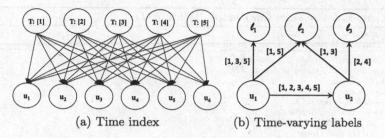

(a) Time index   (b) Time-varying labels

Fig. 4. (a) Time index of the historical graph of Fig. 1 and (b) an example of time-varying labels

Multi-edge Representation. A basic functionality provided by all native graph databases is a TRAVERSALBFS method that implements a BFS traversal of all edges of a specific type (i.e., with a specific label) starting from a source node. At each step, TRAVERSALBFS returns either the current traversed node or all the previously traversed nodes in a form of a path. One approach for retrieving the paths that exist between two nodes u and v during a time interval I is to invoke TRAVERSALBFS starting from u once for each time instance t in I and then combine these results. Another approach is to process paths an edge-at-a-time. Starting from u for each time instance t in I we traverse only the edges of type t until we reach v. Which of the two approaches is more efficient depends on the type of the traversal query under consideration.

For reachability queries where we ask that two nodes are reachable, without posing any requirement on the lifespan of the paths that connect them, the approach that uses the built-in TRAVERSALBFS is more efficient. We invoke TRAVERSALBFS for each time instance t in I until: (a) for disjunctive queries: the first time instance that we find v, (b) for conjunctive queries: the first time instance that we do not find v and (c) for the least-k queries: when we find v in at least k time instances, or the remaining time instances are not enough to get reachability at k time instances.

For queries that require that the paths exist in at least $L > 1$ time instances, using the TRAVERSALBFS method is in general expensive, since we retrieve all paths at each time instance, even those paths that appear only in a single time instance. Thus, TRAVERSALBFS is used only for the earliest shortest path (ESP) queries, where it returns the shortest path that connects u to v in the first time instance. For stable (SSP) and at least-k (KSP) shortest path queries, we use the second approach. We traverse the edge type that refers to the first time instance in I and we continue the traversal only if for each edge (w, x) there are all (SSP) or at least k (KSP) type of edges (w, x) that refer to other time instances in I.

Single-edge Representation. For the *single-edge approaches*, we cannot use the TRAVERSALBFS, since we need to post-process the lifespan label of each edge to determine the time instances where the edges were active. Thus, we implemented our own TRAVERSALBFS algorithm which traverses edges that are

Algorithm 1. (SETP-SETI) Conjunctive-BFS(u, v, I)

Require: nodes u, v, interval I
Ensure: True if v is reachable from u in all time instances in I and false otherwise

---
1: create a queue N, create a queue INT
2:   enqueue u onto N, enqueue I onto INT
3:   **while** $N \neq \emptyset$ **do**
4:       $n \leftarrow N.dequeue()$
5:       $i \leftarrow INT.dequeue()$
6:       **for each** $e \in n.getEdges()$ **do**
7:           $I_e \leftarrow$ TIME_JOIN(e, i)
8:           **if** $I_e = \emptyset$ **then**
9:               continue
10:           **end if**
11:           $w \leftarrow r.getOtherNode(n)$
12:           **if** $w = v$ **then**
13:               $R \leftarrow R \cup I_e$
14:               **if** $R \sqsupseteq I$ **then**
15:                   **return** true
16:               **end if**
17:               continue
18:           **end if**
19:           **if** $\mathcal{IN}(w) \not\sqsupseteq I_e$ **then**
20:               $\mathcal{IN}(w) \leftarrow \mathcal{IN}(w) \cup I_e$
21:               enqueue w onto N
22:               enqueue I_e onto INT
23:           **end if**
24:       **end for**
25:   **end while**
26:   **return** false

---

alive in the given interval. We present in Algorithm 1, the algorithm for processing conjunctive reachability queries. Algorithm 1 can be used for processing all other types of historical queries with only small modifications.

Since a node v may be reachable from u through different paths at different graph snapshots, we maintain an interval set R with the part of $ls(u \to v) \cup I_e$ covered so far (line 13), where I_e is the intersection of the lifespan of an edge with a given interval. The traversal ends when R covers the whole query time interval I (lines 14–16).

To retrieve I_e, we use method TIME_JOIN (line 7) and getOtherNode(n) which given a node n that is attached to an edge, returns the other node (line 11). In SETP, TIME_JOIN retrieves the lifespan label from the edge and using an intersection algorithm for sorted lists it returns the intersection of edge lifespan and I. In SETI, TIME_JOIN retrieves the edge lifespan labels L_s and L_e and for each $[s', e'] \in I$ s.t. $\exists\ i$ s.t $\max(L_s[i], s') \geq \min(L_e[i], e')$ it returns the overlapping time instances $\{[s', e'] \cap [L_s[i], L_e[i]]\}$.

To speed-up traversal, we perform a number of pruning tests. The traversal stops when we traverse an edge that is not alive in the query interval (lines 7–10). Still an edge may be traversed multiple times, if it participates in multiple paths from source to target. To reduce the number of such traversals, we provide additional pruning by recording for each node w, an interval set $\mathcal{IN}(w)$ with the parts of the query interval for which it has already been traversed. If the query reaches w again looking for interval $I_e \subseteq I$ and $\mathcal{IN}(w) \sqsupseteq I$, the traversal is pruned (lines 19–23).

Indexing. The time index can be used similarly in all approaches to prune some computations. For example, for the least-k reachability query that asks whether nodes u and v are reachable in at least k time instances, we can first check using the index whether both nodes were active in at least k common time instances. If they were not active, we do not need to traverse the graph. Otherwise, we traverse the graph using a subinterval of I that contains only the instances when both nodes were active.

5   Experimental Evaluation

In this section, we present an experimental comparison of the different approaches for supporting historical traversal queries in a native graph database. We used the Neo4j[1] graph database that supports fast processing of graph data and implemented all algorithms using the Neo4j Java API.

We use two real and one synthetic dataset. In particular, we use DBLP[2] in time interval [1959, 2016] where each graph snapshot corresponds to one year. At each graph snapshot, a node represents an author and an edge a co-authorship relation between two authors in the corresponding year. We also use a FB [21] dataset which consists of 871 daily snapshots where at each snapshot a node represents a user and an edge represents a relation between two users. The synthetic dataset was generated using a preferential attachment graph generator [16], where a new snapshot is created after 10,000 nodes. The dataset characteristics are summarized in Table 1(a). The FB dataset and the default synthetic dataset are insert-only, i.e., contain no node/edge deletions.

We ran our experiments on a system with a quad-core Intel Core i7-3820 3.6 GHz processor, with 64GB memory. We only used one core in all experiments.

Table 1. Dataset and Graph database characteristics

Dataset	# Nodes	# Edges	# Snapshots
DBLP	1,167,854	5,364,298	58
FB	61,967	905,565	871
Synthetic	1,000,000	1,999,325	100

(a) Dataset characteristics

Dataset		GDB Size (MB)	Index Size (MB)	Time (sec)
DBLP	ME	353		39
	SETP	528.84	131.37	22
	SETI	546.55		23
FB	ME	6,000		631
	SETP	400	830	65
	SETI	31.98		33
Synthetic	ME	4,500		1,620
	SETP	513	1,700	145
	SETI	253		86

(b) Graph database size and creation time

5.1   Size and Load Time

We stored all datasets in three different database instances (GDBs) using the three different representations, namely, ME, SETP, and SETI introduced in Sect. 3.

[1] https://neo4j.com/.
[2] http://dblp.uni-trier.de.

Also, in each GDB we stored a time index on the lifespan of the nodes. Table 1(b) shows the size and construction time of each graph database instance. Multi-edge approaches use a different edge type for each time instance, which leads to larger sizes. This difference in size is more evident in the FB dataset, since most edges in the DBLP dataset have short lifespans, because many co-authorships appear only once or span very few years. To load the datasets into the graph databases we used the CSV importing system of Neo4j. Again, ME requires more time to be loaded since it has to create more edges than the other models.

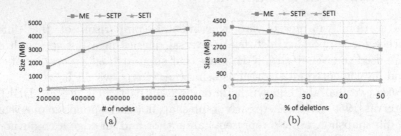

Fig. 5. Size (a) for varying number of nodes and (b) percentage of deletions

In Fig. 5(a), we report graph database sizes for varying number of nodes (and thus snapshot) using the synthetic dataset. As shown, the single-edge approaches are much smaller than the multi-edge in all cases, as expected. We also vary the percentage of edge deletes. For each edge, we randomly remove 10% to 50% of the time instances in its lifespan. Figure 5(b) presents the results. We observe that the size of ME decreases; since removing a time instance leads to less edges types. The number of removals in the lifespan (stored as lists) in SETP leads to slower size reduction. SETI size is increasing since removing time instances leads to more subintervals and thus to larger L_s and L_e lifespan structures. Overall, that single-edges are the best choice in terms of size efficiency for storing large graphs. Among them, SETI is more space-efficient, especially, when there are few subintervals in the lifespan.

5.2 Query Processing

We now focus on query processing. We report the average execution time of 200 historical traversal queries where the source and target nodes are chosen uniformly at random with the restriction that both nodes are present in the graph at the beginning and the end of the query interval. For the FB and the synthetic dataset, the query interval is chosen randomly. However, in DBLP dataset which is more active in the last two decades, we use $I = [2011, 2016]$ as default query interval. For larger intervals we increase it using earlier years for starting time instances. For the *at least-k* queries we set k to be equal to $|I|/2$.

Reachability Queries. In Figs. 6 and 7, we depict the average query times for DBLP and FB. A general remark that holds independently of the graph representation model and the dataset is that disjunctive queries are faster than conjunctive queries, since they stop once an instance where the nodes are reachable is found. Conjunctive queries are in turn faster than at least-k queries, since they stop once an instance where the nodes are not reachable is found.

The main difference between the two datasets is that in DBLP edges represent co-authorships, consequently, in general, their lifespans include very few years, in most cases, just 1 or 2. In FB, lifespans are larger, and since we have no deletions, include just one interval. The ME approach is very fast for short-lived edges and is a clear winner for reachability queries in DBLP. For FB which contains a large number of multiple edge types, the response time of ME increases linearly with the size of the query interval. An exception is disjunctive reachability queries, where traversal stops once an instance where a path exists is found and ME remains competitive.

Among the single edge approaches, SETP outperforms SETI only when the lifespan includes very few time instances (as in DBLP). In this case, the time join between the lifespan and any interval is fast. Furthermore, in this case, SETI includes many small intervals. When lifespans become larger and more continuous (as in FB), SETI outperforms SETP.

To study further the effect of lifespans on query performance, we experimented using the synthetic dataset with different percentage of deletions and with a query interval of length 10 in Fig. 8. We observe that ME and SETI are competitive in conjunctive and disjunctive queries whereas in at least-k queries SETI is the winner. ME takes advantage of the use of the native TRAVERSALBFS method.

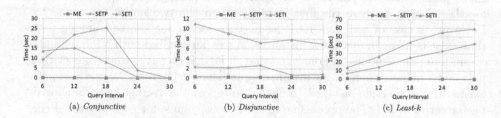

| (a) Conjunctive | (b) Disjunctive | (c) Least-k |

Fig. 6. Query time for historical reachability queries in DBLP

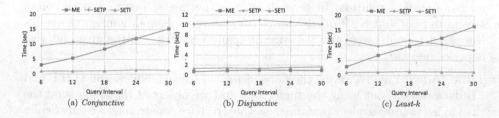

| (a) Conjunctive | (b) Disjunctive | (c) Least-k |

Fig. 7. Query time for historical reachability queries in FB

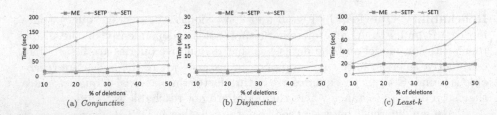

(a) *Conjunctive*   (b) *Disjunctive*   (c) *Least-k*

Fig. 8. Query time for historical reachability queries in the synthetic dataset

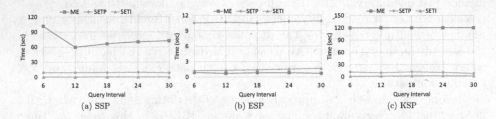

(a) SSP   (b) ESP   (c) KSP

Fig. 9. Query time for historical shortest path queries in FB

SETI performs well in all type of queries and it is starting to slow down when the percentage of deletions is getting higher and the number of intervals in the lifespan gets large.

Path Queries. We also evaluated the performance of historical path queries. ESP queries perform similar to disjunctive reachability queries, since we seek for the shortest path in the first instance when the two nodes are connected. However, in case of SSP and KSP we need to locate the shortest among paths that exist in all or in at least-k instances. We experimented with a large number of random pair of nodes and observed that in DBLP no paths that connect these pairs exist in more than 6 time instances. Furthermore, in most cases, these paths existed in just a single instance. In Fig. 9, we report the average time for shortest path queries in FB. The processing in ME is costly since for each traversed edge that connects u to v the traversal algorithm has to check if there are also other type of edges that refer to all (or k) time instances that u to v. Thus, we set a limit of 120 s for each path query type. KSP queries in ME exceed the time limit for computing a solution. In general, SETI is the fastest one and SETP comes second in SSP and KSP queries, since they traverse a small number of queries compared to multi-edge and the edge lifespan verification in the given interval is performed fast.

Time Index. Finally, we ran the same historical traversal queries in DBLP and FB datasets without using the time index and we observed that in general the time index improves query performance. Due to space constraints, we only depict the change in performance for conjunctive queries in Fig. 10(a)(b). In particular, in DBLP dataset we observe high performance as long as the query interval is

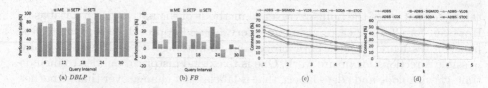

Fig. 10. (a)(b) Time index performance boost for conjunctive queries and (c)(d) percentage of connected pair of nodes in various conferences

increasing since there are not many connected pairs in all time instances and thus indexing returns the negative answers very fast. However, in FB dataset where there are nodes that are connected in whole interval even for larger ones, we notice that indexing is more helpful in ME and SETP since we do not pay the cost for traversing the graph for pairs that are not connected. SETI performance in FB does not increase very much since traversal algorithms run very fast by pruning edges that are not active in the interval. The same trend is observed in historical path queries and thus results are omitted.

5.3 Case Study

In this study, we use historical queries to study connectivity between authors at difference conferences in DBLP. We selected 4 database (ADBIS, SIGMOD, VLDB, ICDE) and 2 theory (SODA, STOC) conferences. For each conference, we randomly selected 500 pair of nodes representing authors that have at least one publication in the conference and examined whether they are reachable in at least k years in the interval [1959, 2016]. We depict the results in Fig. 10(c) where we observe that theory conferences have the most reachable pairs of nodes which indicates that they consist of more well-connected communities compared to database conferences. As expected, the percentage of nodes that are reachable decreases as k increases. We also conducted a second study to show connectivity between ADBIS authors and authors in the other 5 conferences. As show in Fig. 10(d), somehow surprisingly ADBIS authors are more connected with authors in the theory conferences than with authors in the database conferences. Not surprisingly, connectivity between authors of the same conference is larger than connectivity among ADBIS and other conferences.

6   Conclusions

In this paper, we study the problem of storing and querying the history of an evolving graph in a native graph database. We have proposed different approaches for storing such graphs based on associating with each node and edge a lifespan, i.e., a set of time intervals indicating when they were valid. We have also proposed algorithms for processing various types of traversal queries using the proposed storage models. For very short-lived edges, using multiple edges to represent lifespans, one for each time instance, seems to work well by taking advantage of the built-in traversal methods of the native graph databases. However, for all other cases, using an interval-based approach to represent lifespans proves more efficient both processing and storage wise. There are many possible directions for future work. One is to extend historical queries to include time-varying node and edges labels, that is labels, that change over time. Another direction is to provide support for historical graph queries inside the native graph database.

References

1. Akiba, T., Iwata, Y., Yoshida, Y.: Dynamic and historical shortest-path distance queries on large evolving networks by pruned landmark labeling. In: WWW, pp. 237–248 (2014)
2. Backstrom, L., Huttenlocher, D.P., Kleinberg, J.M., Lan, X.: Group formation in large social networks: membership, growth, and evolution. In: KDD, pp. 44–54 (2006)
3. Bahmani, B., Chowdhury, A., Goel, A.: Fast incremental and personalized pagerank. PVLDB 4(3), 173–184 (2010)
4. Cattuto, C., Quaggiotto, M., Panisson, A., Averbuch, A.: Time-varying social networks in a graph database: a Neo4j use case. In: GRADES, p. 11 (2013)
5. Cheng, R., Hong, J., Kyrola, A., Miao, Y., Weng, X., Wu, M., Yang, F., Zhou, L., Zhao, F., Chen, E.: Kineograph: taking the pulse of a fast-changing and connected world. In: EuroSys, pp. 85–98 (2012)
6. Durand, G.C., Pinnecke, M., Broneske, D., Saake, G.: Backlogs and interval timestamps: building blocks for supporting temporal queries in graph databases. In: EDBT/ICDT Workshops (2017)
7. Han, W., Miao, Y., Li, K., Wu, M., Yang, F., Zhou, L., Prabhakaran, V., Chen, W., Chen, E.: Chronos: a graph engine for temporal graph analysis. In: EuroSys, p. 1 (2014)
8. Huang, H., Song, J., Lin, X., Ma, S., Huai, J.: Tgraph: a temporal graph data management system. In: CIKM, pp. 2469–2472 (2016)
9. Huo, W., Tsotras, V.J.: Efficient temporal shortest path queries on evolving social graphs. In: SSDBM, p. 38 (2014)
10. Jensen, C.S., Snodgrass, R.T.: Temporal element. In: Liu, L., Tamer Özsu, M. (eds.) Encyclopedia of Database Systems, p. 2966. Springer, Heidelberg (2009). doi:10.1007/978-0-387-39940-9_1419
11. Khurana, U., Deshpande, A.: Efficient snapshot retrieval over historical graph data. In: ICDE, pp. 997–1008 (2013)
12. Koloniari, G., Souravlias, D., Pitoura, E.: On graph deltas for historical queries. In: WOSS (2012)
13. Labouseur, A.G., Birnbaum, J., Olsen Jr., P.W., Spillane, S.R., Vijayan, J., Hwang, J.H., Han, W.S.: The G* graph database: efficiently managing large distributed dynamic graphs. Distrib. Parallel Databases 33, 479–514 (2014)
14. Labouseur, A.G., Olsen, P.W., Hwang, J.H.: Scalable and robust management of dynamic graph data. In: VLDB, pp. 43–48 (2013)
15. Leskovec, J., Kleinberg, J.M., Faloutsos, C.: Graphs over time: densification laws, shrinking diameters and possible explanations. In: KDD, pp. 177–187 (2005)
16. Newman, M.E.J.: The structure and function of complex networks. SIAM Rev. 45(2), 167–256 (2003)

17. Ren, C., Lo, E., Kao, B., Zhu, X., Cheng, R.: On querying historical evolving graph sequences. PVLDB **4**(11), 726–737 (2011)
18. Semertzidis, K., Lillis, K., Pitoura, E.: Timereach: historical reachability queries on evolving graphs. In: EDBT, pp. 121–132 (2015)
19. Semertzidis, K., Pitoura, E.: Durable graph pattern queries on historical graphs. In: ICDE, pp. 541–552 (2016)
20. Semertzidis, K., Pitoura, E.: Time traveling in graphs using a graph database. In: EDBT/ICDT Workshops (2016)
21. Viswanath, B., Mislove, A., Cha, M., Gummadi, P.K.: On the evolution of user interaction in Facebook. In: WOSN, pp. 37–42 (2009)

Formalising openCypher Graph Queries
in Relational Algebra

József Marton[1(✉)], Gábor Szárnyas[2,3], and Dániel Varró[2,3]

[1] Database Laboratory, Budapest University of Technology and Economics,
Budapest, Hungary
marton@db.bme.hu
[2] Fault Tolerant Systems Research Group,
MTA-BME Lendület Research Group on Cyber-Physical Systems,
Budapest University of Technology and Economics, Budapest, Hungary
{szarnyas,varro}@mit.bme.hu
[3] Department of Electrical and Computer Engineering,
McGill University, Montreal, Canada

Abstract. Graph database systems are increasingly adapted for storing and processing heterogeneous network-like datasets. However, due to the novelty of such systems, no standard data model or query language has yet emerged. Consequently, migrating datasets or applications even between related technologies often requires a large amount of manual work or ad-hoc solutions, thus subjecting the users to the possibility of vendor lock-in. To avoid this threat, vendors are working on supporting existing standard languages (e.g. SQL) or standardising languages.

In this paper, we present a formal specification for openCypher, a high-level declarative graph query language with an ongoing standardisation effort. We introduce relational graph algebra, which extends relational operators by adapting graph-specific operators and define a mapping from core openCypher constructs to this algebra. We propose an algorithm that allows systematic compilation of openCypher queries.

1 Introduction

Context. Graphs are a well-known formalism, widely used for describing and analysing systems. Graphs provide an intuitive formalism for modelling many real-world scenarios, as the human mind tends to interpret the world in terms of objects (*vertices*) and their respective relationships to one another (*edges*) [15].

The *property graph* data model [17] extends graphs by adding labels/types and properties for vertices and edges. This gives a rich set of features for users to model their specific domain in a natural way. Graph databases are able to store property graphs and query their contents with complex graph patterns, which otherwise would be are cumbersome to define and/or inefficient to evaluate on traditional relational databases [21].

Neo4j[1], a popular NoSQL property graph database, offers the Cypher query language to specify graph patterns. Cypher is a high-level declarative query

---

[1] https://neo4j.com/.

© Springer International Publishing AG 2017
M. Kirikova et al. (Eds.): ADBIS 2017, LNCS 10509, pp. 182–196, 2017.
DOI: 10.1007/978-3-319-66917-5_13

language which allows the query engine to use sophisticated optimisation techniques. Neo Technology, the company behind Neo4j initiated the openCypher project [13], which aims to deliver an open specification of Cypher.

Problem and Objectives. The openCypher project provides a formal specification of the *grammar* of the query language and a set of acceptance tests that define the semantics of various language constructs. This allows other parties to develop their own openCypher-compatible query engine. However, there is no mathematical formalisation for the language. In ambiguous cases, developers are advised to consult Neo4j's Cypher documentation or to experiment with Neo4j's Cypher query engine and follow its behaviour. Our goal is to provide a formal specification for the core features of openCypher.

Contributions. In this paper, we use a formal definition of the property graph data model [7] and relational graph algebra, which operates on multisets (bags) [6] and is extended with additional graph-specific operators. Using these foundations, we construct a concise formal specification for the core features in the openCypher grammar. This specification is detailed enough to serve as a basis for an openCypher compiler [23].

2 Data Model and Running Example

Data model. A *property graph* is defined as $G = (V, E, st, L, T, \mathcal{L}, \mathcal{T}, P_v, P_e)$, where V is a set of vertices, E is a set of edges and $st : E \rightarrow V \times V$ assigns the source and target vertices to edges. Vertices are labelled and edges are typed:

- L is a set of vertex labels, $\mathcal{L} : V \rightarrow 2^L$ assigns a *set of labels* to each vertex.
- T is a set of edge types, $\mathcal{T} : E \rightarrow T$ assigns a *single type* to each edge.

To define properties, let $D = \cup_i D_i$ be the union of atomic domains D_i and let ε represent the NULL value.

- P_v is a set of vertex properties. A vertex property $p_i \in P_v$ is a partial function $p_i : V \rightarrow D_i \cup \{\varepsilon\}$, which assigns a property value from a domain $D_i \in D$ to a vertex $v \in V$, if v has property p_i, otherwise $p_i(v)$ returns ε.
- P_e is a set of edge properties. An edge property $p_j \in P_e$ is a partial function $p_j : E \rightarrow D_j \cup \{\varepsilon\}$, which assigns a property value from a domain $D_j \in D$ to an edge $e \in E$, if e has property p_j, otherwise $p_j(e)$ returns ε.

In the context of this paper, we define a *relation* as a bag (*multiset*) of tuples: a tuple can occur more than once in a relation [6]. Given a property graph G, relation r is a *graph relation* if the following holds:

$$\forall A \in \text{sch}(r) : \text{dom}(A) \subseteq V \cup E \cup D,$$

where the schema of r, $\text{sch}(r)$, is a list containing the attribute names, $\text{dom}(A)$ is the domain of attribute A, V is the vertices of G, and E is the edges of G.

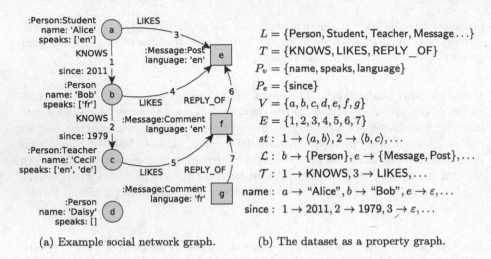

(a) Example social network graph.   (b) The dataset as a property graph.

Fig. 1. Social network example represented graphically and formally. To improve readability, we use letters for vertex identifiers and numbers for edge identifiers.

Property access. When defining relational algebra expression on graph relations, it is often required (e.g. in projection and selection operators) to access a certain property of a vertex/edge. Following the notation of [7], if x is an attribute of a graph relation, we use $x.p$ to access the corresponding value of property p. Also, $\mathcal{L}(v)$ returns the labels of vertex v and $\mathcal{T}(e)$ returns the type of edge e.

Running example. An example graph inspired by the LDBC Social Network Benchmark [5] is shown on Fig. 1(a), while Fig. 1(b) presents the formalised graph. The graph vertices model four **Persons** and three **Messages**, with edges representing **LIKES**, **REPLY_OF** and **KNOWS** relations. In social networks, the **KNOWS** relation is symmetric, however, the property graph data model does not allow undirected edges. Hence, we use directed edges with an arbitrary direction and model the symmetric semantics of the relation in the queries.

3   The openCypher Query Language

Cypher is the a high-level declarative graph query language of the Neo4j graph database. It allows users to specify graph patterns with a syntax resembling an actual graph, which makes the queries easy to comprehend. The goal of the openCypher project [13] is to provide a specification of the Cypher language. In the following, we introduce features of the language using examples.

3.1   Language Constructs

Inputs and Outputs. openCypher queries take a *property graph* as their input, however the result of a query is not a graph, but a *graph relation*.

Vertex and Path Patterns. The basic building blocks of queries are patterns of vertices and edges. List. 3.1 shows a query that returns all vertices that model a Person. The query in List. 3.2 matches Person and Message pairs connected by a LIKES edge and returns the person's name and the message language. List. 3.3 describes person pairs that know each other directly or have a friend in common, i.e. from person p1, the other person p2 can be reached using one or two hops.

```
MATCH (p:Person)
RETURN p
```

List. 3.1. Getting vertices

```
MATCH (p:Person)-[:LIKES]->(m:Message)
RETURN p.name, m.language
```

List. 3.2. Pattern matching

```
MATCH
  (p1:Person)-[ks:KNOWS*1..2]-
  (p2:Person)
RETURN p1, p2
```

List. 3.3. Variable length path

```
MATCH (p:Person) WITH p
UNWIND p.speaks AS language
RETURN language,
  count(DISTINCT p.name) as cnt
```

List. 3.4. Grouping

Filtering. Pattern matches can be filtered in two ways as illustrated in List. 3.5 and List. 3.6. (1) Vertex and edge patterns in the MATCH clause might have vertex label/edge type constraints written in the pattern after a colon, and (2) the optional WHERE subclause of MATCH might hold predicates.

```
MATCH (p1:Person)-[k:KNOWS]-(p2:Person)
WHERE k.since < 2000
RETURN p1.name, p2.name
```

List. 3.5. Filtering for edge property

```
MATCH (p:Person)
WHERE p.name ='Bob'
RETURN p.speaks
```

List. 3.6. Filtering

Unique and Non-unique Edges. A MATCH clause defines a graph pattern. A query can be composed of multiple patterns spanning multiple MATCH clauses. For matches of a pattern within a single MATCH clause, edges are required to be unique. However, matches for multiple MATCH clauses can share edges. This means that in matches returned by List. 3.7, k1 and k2 are required to be different, while in matches returned by List. 3.8, k1 and k2 are allowed to be equal. For vertices, this restriction does not apply.[2] This is illustrated in List. 3.9, which returns adjacent persons who like the same message.

---

[2] Requiring uniqueness of edges is called *edge isomorphic matching*. Other query languages and execution engines might use *vertex isomorphic matching* (requiring uniqueness of vertices), *isomorphic matching* (requiring uniqueness of both vertices and edges) or *homomorphic matching* (not requiring uniqueness of either) [8].

```
MATCH (p1)-[k1:KNOWS]-(p2),
      (p2)-[k2:KNOWS]-(p3)
RETURN p1, k1, p2, k2, p3
```

List. 3.7. Different edges

```
MATCH (p1)-[k1:KNOWS]-(p2)
MATCH (p2)-[k2:KNOWS]-(p3)
RETURN p1, k1, p2, k2, p3
```

List. 3.8. Non-unique edges

```
MATCH (m:Message)<-[:LIKES]-(p1:Person)--(p2:Person)-[:LIKES]->(m)
RETURN p1, p2, m
```

List. 3.9. Triangle

Creating the Result Set. The result set[3] of a query is basically given in the RETURN clause, which can be de-duplicated using the DISTINCT modifier, sorted using the ORDER BY subclause. Skipping rows after sorting and limiting the result set to a certain number of records can be achieved using SKIP and LIMIT modifiers.

List. 3.10 illustrates these concepts by returning the name of the persons. The result set is restricted to the second and third names in alphabetical order.

```
MATCH (p:Person)
RETURN DISTINCT p.name
  ORDER BY p.name
    SKIP 1 LIMIT 2
```

```
1  MATCH
2    ()-[:LIKES]->(m:Message)<-[:LIKES]-(),
3    (m)<-[:REPLY_OF]-(r)
4  RETURN r
```

List. 3.10. Deduplicate and sort

List. 3.11. Multiple patterns

Combining Patterns. Multiple patterns (in the same or in different) MATCH clauses are combined together based on their common variables. List. 3.11 illustrates this by showing two patterns on lines 2 and 3. The first pattern describes a message m that has at least two likes. The second pattern finds replies to m.

Aggregation. openCypher specifies aggregation operators for performing calculations on multiple tuples.[4] Unlike in SQL queries, the *grouping criteria* is determined implicitly in the RETURN as well as in and WITH clauses. Each expression of the expression list in WITH and RETURN are forced to contain either (1) no aggregate functions or (2) a single aggregate function at the outermost level. The grouping key is the tuple built from expressions of type (1).[5] The query of List. 3.4 counts the number of persons commanding each language.

---

[3] The term *result set* refers to the *result collection*, which can be a set, a bag or a list.
[4] The avg, count, max, min, percentileCont, percentileDisc, stdDev, stdDevP, sum functions return a single scalar value, while collect returns a list.
[5] Decision on grouping semantics is due after the camera ready submission deadline. The semantics presented in this paper is one of the possible approaches.

```
MATCH (p:Person) WITH p
UNWIND p.speaks AS lang
RETURN p.name, lang
```

p.name	lang
Alice	en
Bob	fr
Cecil	en
Cecil	de

List. 3.12. Unwind   **Fig. 2.** Output of the unwind query.

Unwinding a List. The UNWIND construct takes an attribute and multiplies each tuple by appending the list elements one by one to the tuple, thus modifying the schema of the query part. By applying UNWIND to the speaks attribute List. 3.12 lists persons along the languages they speak. Figure 2 shows the output of this query. As Cecil speaks two languages, he appears twice in the output. Note that "Daisy" speaks no languages, thus no tuples belong to her in the output.

3.2   Query Structure

In openCypher a query is composed as the UNION of one or more single queries. Each single query must have the same resulting schema, i.e. the resulting tuples must have the same arity and the same name at each position.

Single Queries. A single query is composed of one or more query parts written subsequently. Query parts that form a prefix of a single query have one result set with the schema of the last query part's schema in that prefix.

Query Parts. Clause sequence of a query part matches the regular expression as follows: MATCH*((WITH UNWIND?)|UNWIND|RETURN). They begin with an arbitrary number of MATCH clauses, followed by either (1) WITH and an optional UNWIND, (2) a single UNWIND, or (3) a RETURN in case of the last query part.[6]

The RETURN and WITH clauses use similar syntax and have the same semantics, the only difference being that RETURN should be used in the last query part while WITH should only appear in the preceding ones. These clauses list expressions whose value form the tuples, thus they determine the schema of the query parts.

Example. An openCypher single query composed of two query parts is shown on List. 3.13 along with its result on Fig. 3. It retrieves the language of messages that were written in a language no other message uses. If that message was a reply, the language of the original message is also retrieved.

The first query part spans lines 1–3 and the second spans lines 4–6. The result of the first query part is a single tuple ⟨"fr", 1⟩ with the schema ⟨singleLang, cnt⟩.

---

[6] In openCypher, the filtering WHERE operation is a subclause of MATCH and WITH. When used in WITH as illustrated on line 3 of List. 3.13, WHERE is similar to the HAVING construct of SQL with the major difference that, in openCypher it is also allowed when no aggregation was specified in the query.

```
1 MATCH (m1:Message)
2 WITH m1.language AS singleLang, count(*) AS cnt
3 WHERE cnt = 1
4 MATCH (m2:Message) WHERE m2.language = singleLang
5 OPTIONAL MATCH (m2)-[:REPLY_OF]->(m3:Message)
6 RETURN m2.language as reply, m3.language as orig
```

List. 3.13. Single query with multiple query parts

reply	orig
fr	en

Fig. 3. Result.

The second query part takes this result as an input to retrieve messages of the given languages and in case of a reply the original message in m3. The result of these two query parts together produces the final result whose schema is determined by the RETURN of the last query part (line 6).

4   Mapping openCypher to Relational Graph Algebra

In this section, we present relational graph algebra using the examples of Sect. 3.1 and provide a mapping that allows compilation from openCypher to this algebra.

Table 1. Number of operands, properties and result schemas of relational graph algebra operators. A unary operator α is idempotent (i), iff $\alpha(x) = \alpha(\alpha(x))$ for all inputs. A binary operator β is commutative (c), iff $x \ \beta \ y = y \ \beta \ x$ and associative (a), iff $(x \ \beta \ y) \ \beta \ z = x \ \beta \ (y \ \beta \ z)$. For schema transformations, append is denoted by $\|$, while removal is marked by $\setminus$. L represents a (possibly empty) set of vertex labels and T represents a (possibly empty) set of edge types.

#ops.	notation	name	props.	schema
0	$\bigcirc_{(v:L)}$	get-vertices	—	$\langle v \rangle$
1	$\updownarrow \ {}^{(w:L)}_{(v)} [e:T] (r)$	expand-both	—	$\mathrm{sch}(r) \| \langle e, w \rangle$
	$\neq_{\mathrm{variables}} (r)$	all-different	i	$\mathrm{sch}(r)$
	$\omega_{xs \to x}(r)$	unwind	—	$\mathrm{sch}(r) \setminus \langle xs \rangle \| \langle x \rangle$
	$\sigma_{\mathrm{condition}}(r)$	selection	i	$\mathrm{sch}(r)$
	$\pi_{x_1, x_2, \ldots}(r)$	projection	i	$\langle x_1, x_2, \ldots \rangle$
	$\gamma^{c_1, c_2, \ldots}_{x_1, x_2, \ldots}(r)$	grouping	i	$\langle x_1, x_2, \ldots \rangle$
	$\delta(r)$	duplicate-elimination	i	$\mathrm{sch}(r)$
	$\tau_{\downarrow x_1, \uparrow x_2, \ldots}(r)$	sorting	i	$\mathrm{sch}(r)$
	$\lambda^{\mathrm{limit}}_{\mathrm{skip}}(r)$	top	—	$\mathrm{sch}(r)$
2	$r \cup s$	union	—	$\mathrm{sch}(r)$
	$r \uplus s$	bag union	c, a	$\mathrm{sch}(r)$
	$r \bowtie s$	natural join	c, a	$\mathrm{sch}(r) \| (\mathrm{sch}(s) \setminus \mathrm{sch}(r))$
	$r ⟕ s$	left outer join	—	$\mathrm{sch}(r) \| (\mathrm{sch}(s) \setminus \mathrm{sch}(r))$

4.1   An Algebra for Formalising Graph Queries

We present both standard operators of relational algebra [4] and operators for graph relations. Table 1 provides an overview of the operators of relational graph algebra. We follow the openCypher query language and present a mapping from the language constructs to their algebraic equivalents[7], summarized in Table 2. The corresponding rows of the table (e.g. ①) are referred to in the text.

Basic Operators. The *projection* operator π keeps a specific set of attributes in the relation: $t = \pi_{x_1,\ldots,x_n}(r)$. Note that the tuples are not deduplicated (i.e. filtered to sets), thus t will have the same number of tuples as r. The projection operator can also rename the attributes, e.g. $\pi_{x1 \to y1}(r)$ renames x1 to y1.

The *selection* operator σ filters the incoming relation according to some criteria. Formally, $t = \sigma_\theta(r)$, where predicate θ is a propositional formula. Relation t contains all tuples from r for which θ holds.

Vertices and Patterns. ①–② The *get-vertices* [7] nullary operator $\bigcirc_{(v:l_1 \wedge \ldots \wedge l_n)}$ returns a graph relation of a single attribute v that contains vertices that have *all* of labels $l_1, \ldots, l_n$. Using this operator, the query in List. 3.1 is compiled to

$$\bigcirc_{(p:Person)}$$

③–⑥ The *expand-out* unary operator $\uparrow_{(v)}^{(w:l_1 \wedge \ldots \wedge l_n)} [e : t_1 \vee \ldots \vee t_k](r)$ adds new attributes e and w to each tuple iff there is an edge e from v to w, where e has *any* of types $t_1, \ldots, t_k$, while w has *all* labels $l_1, \ldots, l_n$.[8] More formally, this operator appends the $\langle e, w \rangle$ to a tuple iff $st(e) = \langle v, w \rangle$, $l_1, \ldots, l_n \in \mathcal{L}(w)$ and $\mathcal{T}(e) \in \{t_1, \ldots, t_k\}$. Using this operator, the query in List. 3.2 can be formalised as

$$\pi_{p.name,m.language} \uparrow_{(p)}^{(m:Message)} [\_e1 : \mathsf{LIKES}] \bigcirc_{(p:Person)}$$

Similarly to the expand-out operator, the *expand-in* operator $\downarrow$ appends $\langle e, w \rangle$ iff $st(e) = \langle w, v \rangle$, while the *expand-both* operator $\updownarrow$ uses edge e iff either $st(e) = \langle v, w \rangle$ or $st(e) = \langle w, v \rangle$. We also propose an extended version of this operator, $\uparrow_{(v)}^{(w)} [e*_{min}^{max}]$, which may use between min and max hops. Using this extension, List. 3.3 is compiled to

---

[7] Patterns in the openCypher query might contain anonymous vertices and edges. In the algebraic form, we denote this with names starting with an underscore, such as _v1 and _e2.

[8] Label and type constraints can be omitted for the get-vertices operator and the expand operators. For example, $\bigcirc_{(v)}$ returns all vertices, while $\uparrow_{(v)}^{(w)} [e](r)$ traverses all outgoing edges e from vertices v to w, regardless of their labels/types.

$$\pi_{\text{p1,p2}} \not\equiv_{\text{ks}} \updownarrow \, {}^{\text{(p2:Person)}}_{\text{(p1)}} \, [\text{ks} : \text{KNOWS}*^2_1] \, \bigcirc_{\text{(p1:Person)}}$$

Combining and Filtering Pattern Matches. ⑦–⑪ In order to express the uniqueness criterion for edges (illustrated in Sect. 3.1) in a compact way, we propose the *all-different* operator. The all-different operator $\not\equiv_{E_1,\dots,E_n} (r)$ filters r to keep tuples where variables in $\cup_i E_i$ are pairwise different. Note that the operator is actually a shorthand for the following selection:

$$\not\equiv_{E_1,\dots,E_n} (r) = \sigma_{\underset{e_1 \neq e_2}{\underset{e_1,e_2 \in \cup_i E_i}{\wedge \, r.e_1 \neq r.e_2}}} (r)$$

Using the all-different operator, query in List. 3.7 is compiled to

$$\pi_{\text{p1,k1,p2,k2,p3}} \not\equiv_{\text{k1,k2}} \uparrow \, {}^{\text{(p3)}}_{\text{(p2)}} \, [\text{k2} : \text{KNOWS}] \uparrow \, {}^{\text{(p2)}}_{\text{(p1)}} \, [\text{k1} : \text{KNOWS}] \, \bigcirc_{\text{(p1)}}$$

⑦–⑧ The result of the *natural join* operator $\bowtie$ is determined by creating the Cartesian product of the relations, then filtering those tuples which are equal on the attributes that share a common name. The combined tuples are projected: from the attributes present in both of the two input relations, we only keep the ones in r and drop the ones in s. Thus, the join operator is defined as

$$r \bowtie s = \pi_{R \cup S} \left(\sigma_{r.A_1 = s.A_1 \wedge \dots \wedge r.A_n = s.A_n} (r \times s) \right),$$

where $\{A_1, \dots, A_n\} = R \cap S$ is the set of attributes that occur both in R and S. In order to allow pattern matches to share the same edge, they must be included in different MATCH clauses as shown on List. 3.8 which is compiled to

$$\pi_{\text{p1,p2,p3}} \left(\left(\updownarrow \, {}^{\text{(p2)}}_{\text{(p1)}} \, [\_\text{e1} : \text{KNOWS}] \, \bigcirc_{\text{(p1)}} \right) \bowtie \left(\updownarrow \, {}^{\text{(p3)}}_{\text{(p2)}} \, [\_\text{e2} : \text{KNOWS}] \, \bigcirc_{\text{(p2)}} \right) \right)$$

The query in List. 3.11 with two patterns in one MATCH clause is compiled to:

$$\pi_r \not\equiv_{\_\text{e1},\_\text{e2},\_\text{e3}} \left(\left(\downarrow \, {}^{(\_\text{v2})}_{(\text{m})} \, [\_\text{e2} : \text{LIKES}] \uparrow \, {}^{(\text{m:Message})}_{(\_\text{v1})} \, [\_\text{e1} : \text{LIKES}] \, \bigcirc_{(\_\text{v1})} \right) \right.$$
$$\left. \bowtie \left(\downarrow \, {}^{(\text{r})}_{(\text{m})} \, [\_\text{e3} : \text{REPLY\_OF}] \, \bigcirc_{(\text{m:Message})} \right) \right)$$

⑨–⑪ The *left outer join* operator produces $t = r \bowtie\!\!\!\!\!\!\times s$ combining matching tuples of r and s according to a given matching semantics.[9] In case there is no matching tuple in s for a particular tuple $e \in r$, e is still included in the result, with tuple attributes that exclusively belong to relation s having a value of ε.

---

[9] Matching semantics might use value equality of attributes that share a common name (similarly to natural join) to use an arbitrary condition (similarly to θ-join).

Result and Subresult Operations. (16) The *duplicate-elimination* operator δ eliminates duplicate tuples in a bag.

(17) The *grouping* operator γ groups tuples according to their value in one or more attributes and aggregates the remaining attributes.

We generalize the grouping operator to explicitly state the *grouping criteria* and allow for complex aggregate expressions. This is similar to the SQL query language where the grouping criteria is explicitly given in GROUP BY.

We use the notation $\gamma_{e_1,e_2,\ldots}^{c_1,c_2,\ldots}$, where $c_1, c_2, \ldots$ in the superscript form the *grouping criteria*, i.e. the list of expressions whose values partition the incoming tuples into groups. For each and every group this aggregation operator emits a single tuple of expressions listed in the subscript, i.e. $\langle e_1, e_2, \ldots \rangle$. Given attributes $\{a_1, \ldots, a_n\}$ of the input relation, c_i is an arithmetic expression built from a_j attributes using common arithmetic operators, while e_i is an expression built from a_j using common arithmetic operators and grouping functions.

We have discussed the aggregation semantics of openCypher in Sect. 3.1. The formal algorithm for determining the grouping criteria is given in Algorithm 1. Building on this algorithm and the grouping operator, List. 3.4 is compiled to

$$\gamma_{\text{language},\text{count\_distinct}(\text{p.name})\rightarrow\text{cnt}}^{\text{language}} \omega_{\text{p.speaks}\rightarrow\text{language}} \bigcirc (\text{p:Person})$$

Data: E is the list of expressions in the RETURN or WITH clause
1 **Function** DetermineGroupingCriteria(E)
2      $G \leftarrow \{\}$ // initial set of grouping criteria
3   **foreach** $e \in E$ **do**
4   **if** *e has an aggregate function call at its outermost level* **then**
5   // do nothing as this is an aggregation
6   **else if** *e contains aggregate function call* **then**
7   // aggregation allowed only at the outermost level
8   **raise** SemanticError(Illegal use of aggregation function)
9   **else**
10             $G \leftarrow G \cup \{e\}$ // append to the grouping key
11   **end**
12   **end**
13   **return** G

Algorithm 1: Determine grouping criteria from return item list.

Unwinding and List Operations. (19) The *unwind* [3] operator $\omega_{\text{xs}\rightarrow\text{x}}$ takes the list in attribute xs and multiplies each tuple adding the list elements one by one to an attribute x, as demonstrated in Fig. 2. Using this operator, the query in List. 3.12 can be formalised as

$$\pi_{\text{p.name},\text{lang}} \omega_{\text{p.speaks}\rightarrow\text{lang}} \pi_{\text{p}} \bigcirc (\text{p:Person})$$

(20) The *sorting* operator τ transforms a bag relation of tuples to a list of tuples by ordering them. The ordering is defined for selected attributes and with a certain direction for each of them (ascending $\uparrow$/descending $\downarrow$), e.g. $\tau_{\uparrow x1,\downarrow x2}(r)$.

(21) The *top* operator λ_l^s (adapted from [11]) takes a list as its input, skips the first s tuples and returns the next l tuples.[10]

Using the sorting and top operators, the query of List. 3.10 is compiled to:

$$\lambda_2^1 \tau_{\uparrow \text{p.name}} \delta \, \pi_{\text{p.name}} \bigcirc (\text{p:Person})$$

Combining Results. The $\cup$ operator produces the *set union* of two relations, while the $\uplus$ operator produces the *bag union* of two operators, e.g. $\{\langle 1,2\rangle, \langle 3,4\rangle\} \uplus \{\langle 1,2\rangle\} = \{\langle 1,2\rangle, \langle 1,2\rangle, \langle 3,4\rangle\}$. For both the union and bag union operators, the schema of the operands must have the same attributes.

4.2 Mapping openCypher Queries to Relational Graph Algebra

In this section, we give the mapping algorithm of openCypher queries to relational graph algebra and also give a more detailed listing of the compilation rules for the query language constructs in Table 2. We follow a bottom-up approach to build the relational graph algebra expression.

1. Process each single query as follows and combine their result using the union operation. As the union operator is technically a binary operator, the union of more than two single queries are represented as a left-deep tree of UNION operators.
3. For each query part of a single query, denoted by t, the relational graph algebra tree built from the prefix of query parts up to—but not including—the current query part, process the current query part as follows.
   1. A single pattern is turned left-to-right to a get-vertices for the first vertex and a chain of expand-in, expand-out or expand-both operators for inbound, outbound or undirected relationships, respectively.
   2. Comma-separated patterns in a single MATCH are connected by natural join.
   3. Append an all-different operator for all edge variables that appear in the MATCH clause because of the non-repeating edges language rule.
   4. Process the WHERE subclause of a single MATCH clause.
   5. Several MATCH clauses are connected to a left-deep tree of natural join. For OPTIONAL MATCH, left outer join is used instead of natural join. In case there is a WHERE subclause, its condition becomes part of the join condition, i.e. it will never filter on the input from previous MATCH clauses.
   6. If there is a positive or negative pattern deferred from WHERE processing, append it as a natural join or a combination of left outer join and selection operator filtering on no matches were found, respectively.

---

[10] SQL implementations offer the OFFSET and the LIMIT/TOP keywords.

7. If this is not the first query part, combine the current query part with the relational graph algebra tree of the preceding query parts by appending a natural join here. Its left operand will be t and its right operand will be the relational graph algebra tree built so far from the current subquery.
8. Append grouping, if RETURN or WITH clause has grouping functions inside.
9. Append a projection operator based on the RETURN or WITH clause. This operator will also handle the renaming (i.e. AS).
10. Append a duplicate-elimination operator, if the RETURN or WITH clause has the DISTINCT modifier.
11. Append a selection operator if WITH had the optional WHERE subclause.

4.3 Summary and Limitations

In this section, we presented a mapping that allows us to express the example queries of Sect. 3.1 in graph relational algebra. We extended relational algebra by adapting operators ($\bigcirc$, $\uparrow$, τ, λ), precisely specifying grouping semantics (γ) and defining the all-different operator ($\neq$). Finally, we proposed an algorithm for compiling openCypher graph queries to graph relational algebra.

Our mapping does not completely cover the openCypher language. As discussed in Sect. 3, some constructs are defined as legacy and thus were omitted. The current formalisation does not include expressions (e.g. conditions in selections) and maps. Compiling data manipulation operations (such as CREATE, DELETE, SET, and MERGE) to relational algebra is also subject of future work.

5 Related Work

Property Graph Data Models. The TinkerPop framework aims to provide a standard data model for property graphs, along with Gremlin, a high-level imperative graph traversal language [16] and the Gremlin Structure API, a low-level programming interface.

EMF. The Eclipse Modeling Framework is an object-oriented modelling framework widely used in model-driven engineering. Henshin [1] provides a visual language for defining patterns, while Epsilon [9] and VIATRA Query [2] provide high-level declarative (textual) query languages, the Epsilon Pattern Language and the VIATRA Query Language.

SPARQL. Widely used in semantic technologies, SPARQL is a standardised declarative graph pattern language for querying RDF [24] graphs. SPARQL bears close similarity to Cypher queries, but targets a different data model and requires users to specify the query as triples instead of graph vertices/edges [14]. G-SPARQL [19] extended the SPARQL language for attributed graphs, resulting in a language with an expressive power similar to openCypher.

Table 2. Mapping from openCypher constructs to relational algebra. Variables, labels, types and literals are typeset as «v» . The notation ⟨p⟩ represents patterns resulting in a relation p, while ⟦r⟧ denotes previous query fragment resulting in a relation r. To avoid confusion with the "`..`" language construct (used for ranges), we use ⋯ to denote omitted query fragments.

Language construct	Relational algebra expression	
Vertices and patterns. ⟨p⟩ denotes a pattern that contains a vertex «v».		
`(«v»)`	$\bigcirc_{(v)}$	①
`(«v»:«l1»:⋯:«ln»)`	$\bigcirc_{(v:l1\wedge\cdots\wedge ln)}$	②
`⟨p⟩-[«e»:«t1»|⋯|«tk»]->(«w»)`	$\uparrow\,^{(w)}_{(v)}[e:t1\vee\cdots\vee tk]\,(p)$, where e is an edge	③
`⟨p⟩<-[«e»:«t1»|⋯|«tk»]-(«w»)`	$\downarrow\,^{(w)}_{(v)}[e:t1\vee\cdots\vee tk]\,(p)$, where e is an edge	④
`⟨p⟩<-[«e»:«t1»|⋯|«tk»]->(«w»)`	$\updownarrow\,^{(w)}_{(v)}[e:t1\vee\cdots\vee tk]\,(p)$, where e is an edge	⑤
`⟨p⟩-[«e»*«min»..«max»]->(«w»)`	$\uparrow\,^{(w)}_{(v)}[e*^{max}_{min}]\,(p)$, where e is a list of edges	⑥
Combining and filtering pattern matches		
`MATCH ⟨p1⟩, ⟨p2⟩, ⋯`	$\neq_{\text{edges of p1, p2, }\cdots}(p1\bowtie p2\bowtie\cdots)$	⑦
`MATCH ⟨p1⟩` `MATCH ⟨p2⟩`	$\neq_{\text{edges of p1}}(p1)\bowtie\neq_{\text{edges of p2}}(p2)$	⑧
`OPTIONAL MATCH ⟨p⟩`	$\{\langle\rangle\}\bowtie\neq_{\text{edges of p}}(p)$	⑨
`OPTIONAL MATCH ⟨p⟩ WHERE ⟨condition⟩`	$\{\langle\rangle\}\bowtie_{\text{condition}}\neq_{\text{edges of p}}(p)$	⑩
`⟦r⟧ OPTIONAL MATCH ⟨p⟩`	$\neq_{\text{edges of r}}(r)\bowtie\neq_{\text{edges of p}}(p)$	⑪
`⟦r⟧ WHERE «condition»`	$\sigma_{\text{condition}}(r)$	⑫
`⟦r⟧ WHERE («v»:«l1»:⋯:«ln»)`	$\sigma_{\mathcal{L}(v)=l1\wedge\cdots\wedge\mathcal{L}(v)=ln}(r)$	⑬
`⟦r⟧ WHERE ⟨p⟩`	$r\bowtie p$	⑭
Result and subresult operations. Rules for `RETURN` also apply to `WITH`.		
`⟦r⟧ RETURN «x1» AS «y1», ⋯`	$\pi_{x1\rightarrow y1,\cdots}(r)$	⑮
`⟦r⟧ RETURN DISTINCT «x1» AS «y1», ⋯`	$\delta\left(\pi_{x1\rightarrow y1,\cdots}(r)\right)$	⑯
`⟦r⟧ RETURN «x1», «aggr»(«x2»)`	$\gamma^{x1}_{x1,\text{aggr}(x2)}(r)$ (see Sec. 3.1)	⑰
`⟦r⟧ WITH «x1»` `⟦s⟧ RETURN «x2»`	$\pi_{x2}\left(\left(\pi_{x1}(r)\right)\bowtie s\right)$	⑱
Unwinding and list operations		
`⟦r⟧ UNWIND «xs» AS «x»`	$\omega_{xs\rightarrow x}(r)$	⑲
`⟦r⟧ ORDER BY «x1» ASC, «x2» DESC, ⋯`	$\tau_{\uparrow x1,\downarrow x2,\cdots}(r)$	⑳
`⟦r⟧ SKIP «s» LIMIT «l»`	$\lambda^{s}_{l}(r)$	㉑
Combining results		
`⟦r⟧ UNION ⟦s⟧`	$r\cup s$	㉒
`⟦r⟧ UNION ALL ⟦s⟧`	$r\uplus s$	㉓

SQL. In general, relational databases offer limited support for graph queries: recursive queries are supported by PostgreSQL using the `WITH RECURSIVE` keyword and by the Oracle Database using the `CONNECT BY` keyword. Graph queries are supported in the SAP HANA prototype [18], through a SQL-based language [10].

Cypher. Due to its novelty, there are only a few research works on the formalisation of (open)Cypher. The authors of [7] defined *graph relations* and introduced the GETNODES, EXPANDIN and EXPANDOUT operators. While their work focused on optimisation transformations, this paper aims to provides a more complete and systematic mapping from openCypher to relational algebra.

In [8], graph queries were defined in a Cypher-like language and evaluated on Apache Flink. However, formalisation of the queries was not discussed in detail.

Comparison of Graph Query Frameworks. Previously, we published the Train Benchmark, a framework for comparing graph query frameworks across different technological spaces, such as property graphs, EMF, RDF and SQL [21].

6   Conclusion and Future Work

In this paper, we presented a formal specification for a subset of the openCypher query language. This provides the theoretical foundations to use openCypher as a language for graph query engines.

As future work, we plan to provide a formalisation based on graph-specific theoretical query frameworks, such as [12]. We will also give the formal specification of the operators for incremental query evaluation, which requires the definition of *maintenance operations* that keep the result in sync with the latest set of changes [22]. Our long-term research objective is to design an openCypher-compatible *distributed, incremental graph query engine* [20].[11]

Acknowledgements. Gábor Szárnyas and Dániel Varró were supported by the MTA-BME Lendület Research Group on Cyber-Physical Systems and the NSERC RGPIN-04573-16 project. The authors would like to thank Gábor Bergmann and János Maginecz for their comments on the draft of this paper.

References

1. Arendt, T., Biermann, E., Jurack, S., Krause, C., Taentzer, G.: Henshin: advanced concepts and tools for in-place EMF model transformations. In: Petriu, D.C., Rouquette, N., Haugen, Ø. (eds.) MODELS 2010. LNCS, vol. 6394, pp. 121–135. Springer, Heidelberg (2010). doi:10.1007/978-3-642-16145-2_9
2. Bergmann, G., Horváth, Á., Ráth, I., Varró, D., Balogh, A., Balogh, Z., Ökrös, A.: Incremental evaluation of model queries over EMF Models. In: Petriu, D.C., Rouquette, N., Haugen, Ø. (eds.) MODELS 2010. LNCS, vol. 6394, pp. 76–90. Springer, Heidelberg (2010). doi:10.1007/978-3-642-16145-2_6
3. Botoeva, E., et al.: OBDA beyond relational DBs: a study for MongoDB. In: Proceedings of the 29th International Workshop on Description Logics (2016)
4. Elmasri, R., Navathe, S.B.: Fundamentals of Database Systems, 3rd edn. Addison-Wesley-Longman, Boston (2000)

---

[11] Our prototype, *ingraph*, is available at: http://docs.inf.mit.bme.hu/ingraph/.

5. Erling, O., et al.: The LDBC social network benchmark: interactive workload. In: SIGMOD, pp. 619–630 (2015)
6. Garcia-Molina, H., Ullman, J.D., Widom, J.: Database Systems - The Complete Book, 2nd edn. Pearson Education, Harlow (2009)
7. Hölsch, J., Grossniklaus, M.: An algebra and equivalences to transform graph patterns in Neo4j. In: GraphQ at EDBT/ICDT (2016)
8. Junghanns, M., et al.: Cypher-based graph pattern matching in GRADOOP. In: GRADES at SIGMOD (2017)
9. Kolovos, D.S., Paige, R.F., Polack, F.A.C.: The epsilon transformation language. In: Vallecillo, A., Gray, J., Pierantonio, A. (eds.) ICMT 2008. LNCS, vol. 5063, pp. 46–60. Springer, Heidelberg (2008). doi:10.1007/978-3-540-69927-9_4
10. Krause, C., Johannsen, D., Deeb, R., Sattler, K.-U., Knacker, D., Niadzelka, A.: An SQL-based query language and engine for graph pattern matching. In: Echahed, R., Minas, M. (eds.) ICGT 2016. LNCS, vol. 9761, pp. 153–169. Springer, Cham (2016). doi:10.1007/978-3-319-40530-8_10
11. Li, C., Chang, K.C., Ilyas, I.F., Song, S.: RankSQL: query algebra and optimization for relational top-k queries. In: SIGMOD, pp. 131–142 (2005)
12. Libkin, L., et al.: Querying graphs with data. J. ACM 63(2), 14:1–14:53 (2016)
13. Neo Technology. openCypher project (2017). http://www.opencypher.org/
14. Pérez, J., et al.: Semantics and complexity of SPARQL. ACM TODS 34(3), 1–45 (2009)
15. Rodriguez, M.A.: A collectively generated model of the world. In: Collective Intelligence: Creating a Prosperous World at Peace, pp. 261–264 (2008)
16. Rodriguez, M.A.: The gremlin graph traversal machine and language (invited talk). In: DBPL, pp. 1–10 (2015)
17. Rodriguez, M.A., Neubauer, P.: The graph traversal pattern. In: Graph Data Management: Techniques and Applications, pp. 29–46 (2011)
18. Rudolf, M., et al.: The graph story of the SAP HANA database. In: BTW (2013)
19. Sakr, S., Elnikety, S., He, Y.: G-SPARQL: a hybrid engine for querying large attributed graphs. In: CIKM, pp. 335–344 (2012)
20. Szárnyas, G., Izsó, B., Ráth, I., Harmath, D., Bergmann, G., Varró, D.: IncQuery-D: a distributed incremental model query framework in the cloud. In: Dingel, J., Schulte, W., Ramos, I., Abrahão, S., Insfran, E. (eds.) MODELS 2014. LNCS, vol. 8767, pp. 653–669. Springer, Cham (2014). doi:10.1007/978-3-319-11653-2_40
21. Szárnyas, G., et al.: The train benchmark: cross-technology performance evaluation of continuous model validation. Softw. Syst. Model. 1–29 (2017)
22. Szárnyas, G., Maginecz, J., Varró, D.: Evaluation of optimization strategies for incremental graph queries. Periodica Polytechnica, EECS (2017)
23. Szárnyas, G., Marton, J.: Formalisation of openCypher queries in relational algebra. Technical report, Budapest University of Technology and Economics (2017). http://hdl.handle.net/10890/5395
24. W3C. Resource Description Framework (2014). https://www.w3.org/RDF/

Spatial Data Management

SliceNBound: Solving Closest Pairs and Distance Join Queries in Apache Spark

George Mavrommatis[1]([✉]), Panagiotis Moutafis[1],
Michael Vassilakopoulos[1], Francisco García-García[2],
and Antonio Corral[2]

[1] Data Structuring & Engineering Lab,
Department of Electrical and Computer Engineering,
University of Thessaly, Volos, Greece
{gmav, pmoutafis, mvasilako}@uth.gr
[2] Department of Informatics, University of Almeria, Almeria, Spain
{paco.garcia, acorral}@ual.es

Abstract. The (K) Closest-Pair(s) Query, KCPQ, consists in finding the (K) closest pair(s) of objects between two spatial datasets. Recently, several systems that enhance Apache Spark with spatial-awareness have been presented, providing a variety of queries for spatial computation, but not the KCPQ. Since queries are of different nature and one processing technique does not fit all cases, we need specialized algorithms for specific queries that exploit the power provided by parallel systems such as Apache Spark. This paper addresses the problem of answering the KCPQ in Apache Spark, by presenting such a specialized, fast algorithm that can easily be imported in any, spatial-oriented or general, Spark-based system. Furthermore, it presents a variant of this algorithm that solves the Distance Join Query. Experiments and comparison to other solutions indicate that our method is fast and efficient.

Keywords: Spatial Query Processing · Closest-pairs query · Distance Join Query · Data partitioning · Apache Spark

1 Introduction

The (K) Closest-Pair(s) Query, KCPQ, finds the (K) closest pair(s) of object(s) (usually ordered by distance), between two spatial datasets. The KCPQ has received considerable attention from the database community, due to its importance in various applications [1, 2] and has been actively studied in centralized environments, when both [3], one [4] or none [5] of the two spatial datasets are indexed.

During the last decades we have witnessed a huge increase of data in various fields. The term Big Data refers to large volumes of unstructured data that appear in a vast variety of contemporary applications and need real-time analysis [6]. The ubiquity of mobile devices, among others, is resulting to a fast increase in the scale of the big input datasets and particularly datasets containing location data. Spatial computing [7] is becoming more and more significant. Running queries on spatial databases is a core operation of spatial computing. Although most commercial and open source spatial

© Springer International Publishing AG 2017
M. Kirikova et al. (Eds.): ADBIS 2017, LNCS 10509, pp. 199–213, 2017.
DOI: 10.1007/978-3-319-66917-5_14

databases provide solutions to answer a variety of spatial queries, processing these queries can become very demanding if the volume of data on which such a query is applied is big (e.g. tens or hundreds of millions of items), or if the number of the combinations of data objects that need to be examined for answering such a query are big. One option for answering spatial queries in reasonable response time is exploiting the scalability provided by Hadoop MapReduce based systems [8, 9], as already done in several specialized, spatial-oriented systems, like SpatialHadoop [10] and Hadoop-GIS [11]. Although these systems outperform spatial extensions of relational database systems, they do not take advantage of the power of distributed memory, nor are they able to reuse intermediate data [12].

Apache Spark (henceforth Spark), the distributed in-memory computation framework was developed to overcome limitations of the MapReduce paradigm [13]. Many researchers have recently used or extended Spark to handle big spatial data [14–19]. Within these software artifacts, several parallel algorithms for spatial operations and queries have been designed and implemented in Spark (KNN joins, spatial joins, range queries, etc.). But, to best of the authors' knowledge, there are only two research works on parallel and distributed KCPQ processing on large spatial data, a challenging task becoming increasingly essential as datasets continue growing. In [20] SpatialHadoop was utilized to perform efficient parallel KCPQ computation and recently an algorithm for processing the KCPQ in Spark [21] was presented. Experimental results in the latter had been quite promising, showing efficiency and potentiality of improvement.

Motivated by these observations, we utilize, elaborate and adapt ideas presented in [5, 20, 21] to propose a new algorithm for the KCPQ in Spark, called SliceNBound (SnB). The contributions of this paper are the following:

- A novel, four-phased, iterative algorithm in "plain" Apache Spark to perform efficient parallel KCPQ processing on big spatial datasets. It is based on a simple and, therefore, not very computationally demanding partitioning scheme that enables the two datasets to share a common partitioning. Additionally, it only exploits built-in functions of Spark, thus making it easy to be imported in any spatial-oriented, or general, Spark-based parallel system.
- A couple of fast heuristic methods that use a two-stage sampling technique to compute good upper bounds of the distance value of the K-th closest pair. These methods can be used as a preprocessing phase for any technique that uses preprocessing.
- A version of SnB algorithm for Distance Join Queries (DJQs). As already mentioned, apart from [21], there is no other research paper on KCPQ processing in Spark, so the SnB technique is being compared to this work. However, in order to compare the performance of our method against other solutions, we have implemented an extension of our algorithm to answer DJQs. In [19], it has been reported that Simba yields superior performance compared against other spatial analytics system (GeoSpark, SpatialSpark, SpatialHadoop, Hadoop GIS etc.). Simba does not support KCPQs, but does support DJQs. Results indicate that, in the case of the DJQ, our method is faster than the one embedded in Simba.
- The execution of an extensive set of experiments using big real-world points datasets that demonstrate the efficiency and scalability of our proposal.

The rest of the paper is organized as follows. In Sect. 2 we present preliminaries concerning Spark, DJQs, KCPQs and we review related work on Spark systems that support spatial operations. In Sect. 3, the new parallel algorithms for processing the KCPQ and DJQ in Spark are being presented. In Sect. 4, we present representative results of the experimentation we have performed, using real-world datasets and taking into account different performance parameters. Finally, in Sect. 5 we discuss the differences between SnB and the method in [21], we present conclusions and plans for related future work directions.

2 Preliminaries and Related Work

In this section, we define KCPQs and DJQs, introduce the basic characteristics of Spark and review Spark extensions supporting spatial query processing.

2.1 Distance Join and K Closest Pairs Query

The Spatial Join Query, one of the most frequent queries in spatial database systems, finds all pairs of objects from two spatial datasets that satisfy a predicate ∂ (i.e. distance, intersect, overlap, etc.). In the special case that ∂ is distance we are dealing with distance join queries (DJQ), according to the following definition.

Definition 1 (Distance Join): Given two datasets, P and Q, and a distance threshold *epsilon* > 0, the distance join between P and Q, denoted as DJQ(P, Q, *epsilon*), finds all pairs (p, q), $p \in P$ and $q \in Q$, within distance *epsilon*:

$$\text{DJQ(P, Q, epsilon)} = \{(p,q) \in P \times Q : dist(p,q) \leq epsilon\} \tag{1}$$

One of the most prominent and studied DJQs in the spatial database field is the K Closest Pairs Query (KCPQ), which is defined as follows.

Definition 2 (K Closest Pairs): Given two datasets P and Q where $P = \{p_1, p_2, .., p_n\}$ and $Q = \{q_1, q_2, .., q_m\}$ and a number $K > 0, K \in N$, the K Closest Pairs Query is an ordered subset of $P \times Q$ denoted by $KCPQ(P, Q, K)$, where $KCPQ(P,Q,K) = \{(p_1,q_1),\ldots,(p_K,q_K) : (p_i,q_i) \in P \times Q, (p_i,q_i) \neq (p_j,q_j), \forall 1 \leq i,j \leq K\}$ and $dist(p_1,q_1) \leq \ldots \leq dist(p_K,q_K) \leq dist(p,q) \forall (p,q) \in P \times Q - KCPQ(P,Q,K)$.

The distance function *dist* is a distance metric defined on points in the data space. A very commonly used such distance function is the Euclidean distance, but depending on the application, other functions may be more appropriate.

The Spatial Join Query is a very expensive query type and is often being used for benchmarking purposes [22]. The KCPQ is a combination of two expensive queries, spatial join and K Nearest Neighbor (KNN), and therefore is even more costly [20].

2.2 Apache Spark

Spark allows a user application to cache data in memory in a flexible manner that lets the application decide what data should be cached and at what point in the processing flow [23].

It uses an advanced job execution scheme based on creation of a directed acyclic graph (DAG) of stages and a lazy evaluation that allows previous knowledge of the full processing path, thus making it easier to optimize the execution. This functionality makes Spark ideal for iterative algorithms implementation.

The Spark API relies, among others, on the abstraction of *Resilient Distributed Dataset* (RDD). Data are being represented as RDDs in the Spark context, and are distributed among Workers of the cluster. Spark provides several methods that operate on the RDDs and finally on the underlying data. Although Spark uses a shared-nothing architecture, it also supports the concept of shared variables that are being materialized as broadcasts and accumulators. By using broadcast variables, Spark sends data to each node, enabling all Workers to share a piece of information.

Broadcast variables may be useful in cases of problems in the field of combinatorial optimization. For example, in NP-hard graph problems such as the maximum clique number, knowing a good lower bound of the maximum clique helps pruning the search space and speeding up the computation [24]. A similar notion holds for the KCPQ. If we know a good upper bound for the k-th smaller distance and transmit it to all Workers, this will lead to discarding a large volume of computation among pairs of points that their distance is greater than the upper bound.

2.3 Spatial Computing in Apache Spark

During the recent few years, several systems and algorithms have been presented that support spatial queries over distributed spatial data using a cluster of commodity machines and Spark. Most systems are either using the on-top or the built-in implementation approach [8]. This section presents a brief overview.

GeoSpark [16] uses Spark as its base layer and adds two more layers, the Spatial RDD (SRDD) Layer and Spatial Query Processing Layer. The SRDD layer consists of three newly defined RDDs, PointRDD, RectangleRDD and PolygonRDD and supports geometrical operations, like Overlap and Minimum Bounding Rectangle. SRDDs are automatically partitioned by using the uniform grid technique, where the global grid file is split into a number of equal geographical size grid cells. GeoSpark provides spatial indexes like Quadtree and R-Tree on a per partition base. The Spatial Query Processing Layer includes spatial range query, spatial join query, and spatial KNN query. Experiments, reported in [16], show that GeoSpark outperforms its Hadoop-based counterparts (e.g., SpatialHadoop).

SpatialSpark [17] supports indexed spatial joins and range queries. Authors report that in some cases of data intensive applications SpatialSpark performs worse on multiple computing nodes than on a single node, thus showing low scalability. This fact is attributed to possible bottlenecks due to communication overheads among computing nodes a factor that is related to the number of partitions.

LocationSpark [18] is built as a library on top of Spark. It provides Dynamic Spatial Query Execution and operations (Range, kNN, Insert, Delete, Update, Spatial-Join, kNN-Join, Spatio-Textual). The system builds two indexes, a global (grid, quadtree and a Spatial-Bloom Filter) and a local per-worker, user-decided index (grid, rtree, etc.). Global index is constructed by sampling the data. Spatial indexes are aiming to tackle unbalanced data partitioning. Additionally, the system contains a query scheduler, aiming to tackle query skew. As reported in [18], LocationSpark outperforms GeoSpark by one order of magnitude.

Simba [19] extends the Spark SQL engine to support spatial queries and analytics through SQL and the DataFrame API. Simba partitions data in a manner such that the resulting partitions are of proper and balanced size and records that are located closely in space are contained in the same partition. It builds a local index per partition and a global index by aggregating information from local indexes. It supports range and kNN queries, kNN and distance joins. As being reported in [19], Simba outperforms SpatialHadoop, HadoopGIS, SpatialSpark and GeoSpark by a few or more orders of magnitude. In the case of distance join queries, Simba is reported to run about 1.2–1.5 times faster than its closest counterparts, GeoSpark and SpatialSpark.

In [20], the KCPQ is being answered by partitioning datasets into blocks and assigning pairs to workers for local computation. Afterwards, all results are being collected and the top-k ones (with the smallest distances) form the solution. Pruning is being done by sampling and computing an upper bound for the solution.

Finally, in [21] we first compute an upper bound *bnd* for the K distances we are seeking, by sampling the datasets and selecting the K smaller distances among points of the sample. Then, each dataset is being independently divided into strips. All combinations of strips from the two datasets are subject to evaluation with respect to their x-axis distance compared to *bnd*. Pairs that may contribute to the solution (overlapping or having their x-distance less than *bnd*) are being processed in the cluster, by using a plane-sweep algorithm. Local results are gathered by the driver program and taking the first (sorted on increasing distance) K tuples with the smaller distances, yields the final, exact solution.

3   SliceNBound in Apache Spark

In this Section, we describe SliceNBound, our four phased approach to KCPQ algorithm on top of Spark. As it can be inferred from its name, our method *slices* the plane into strips and uses sampling to *bound* the solution space. The bound is used as a pruning criterion in several phases of the computation. The algorithm, iteratively and incrementally, approximates the solution in three major phases, until the exact solution is derived in Phase four. Before presenting the phases of the algorithm as a whole, several details need first to be clarified.

3.1   Local Computation of the KCPQ

Regardless of any other detail, one dealing with the KCPQ eventually needs a method to actually compute it. This also holds for parallel systems such as Spark. Some of the most important techniques for the KCPQ computation are the Plane-Sweep (PS) family of algorithms. In our implementation we use the *Classic PS* algorithm [5]:

1. sort in increasing order the entries of the two point sets, P and Q, based on the coordinates of one of the axes (e.g. Y)

2. initialize two pointers p and q to point to the first entry for processing of each sorted array of points. Set the *reference* point be the one with the smallest y-value pointed by one of these two pointers, for example, suppose *reference* $= p$
3. pair up the *reference* point with all the points stored in the other sorted array of points from smaller to larger y-value, satisfying the following condition (δ is the distance of the k-th closest pair found so far):

$$q \cdot y - reference \cdot y < \delta \qquad (2)$$

5. increase to the next entry the pointer of the array that *reference* points to (in our example p is increased) and update the *reference* point with the point of the next smallest y-value pointed by one of the two pointers
6. repeat steps 3–4 until one of the sorted array of points is completely processed.

We have implemented two slight variations of the above algorithm, namely *fy_bounded* and *fy*, depending on whether we already know an upper bound (e.g. from previous computation) or not. Both are being passed as functions to the workers via Spark engine. In both, we maintain a local *maxHeap* of type *Tuple3[Distance, Point, Point]* and with fixed size K. At any point of the computation, the *maxHeap* stores the best K pairs that have been locally found so far. Initially, *maxHeap* is empty.

In the case of *fy*, workers directly store the first K pairs they examine and their corresponding distances. Then, for each consecutive pair it is tested whether condition (2) holds or not. The value of δ is h, the head of *maxHeap*.

In the case of *fy_bounded*, the first K pairs are also stored into *maxHeap*. Condition (2) is tested for every consecutive pair of points, where $\delta = min(h, bnd)$

Only for pairs of points that (2) is true, their distance is being fully computed and compared to the distance value stored at the head of *maxHeap*. If the new distance is smaller, the head is extracted (discarded) and the new pair is inserted, otherwise the pair is discarded.

After computation in all workers finishes, all partial results are collected on the driver side and by taking the K ordered with smaller distance pairs, we have the K closest pairs between the datasets that were submitted.

3.2 Partitioning a Single Dataset

One of the first things one has to tackle when dealing with spatial queries, especially in parallel systems, is selecting a proper data partitioning strategy that will avoid data skew. In most real-world cases, data are not uniformly distributed and using certain partitioning techniques, such as uniform grid, very often leads to partitions containing much more objects than others. In a parallel system, as partitions are assigned to workers, some of them require much more processing time, thus leading to straggler processes that delay the whole computation. One can find many alternative strategies in the literature aiming to tackle the data skew. Most of them require the construction of sophisticated data structures such as Quadtrees or R-trees. These methods have been extensively used in centralized environments with excellent results and in the recent years are being incorporated in spatial-aware parallel systems based on Spark. However, not all queries are of the same nature.

Splitting into strips [22] is a data partitioning technique used in several applications of parallel systems on various disciplines [25]. Partitioning into strips is lightweight, since it does not need extra computing time to construct and maintain complex data structures that may not be necessary in the case of the query we are dealing with. Most spatial-aware systems based on Spark use a twofold indexing scheme. First a global index is created, then data is shuffled to workers and finally each worker creates its own local index. Albeit this is time consuming, it usually rewards when used, but obviously this depends, among others, on the query itself. In our case, for the KCPQ (like other spatial join queries), it is enough to utilize all cluster resources and achieve a high degree of parallelism, in contrast to other queries, like the range query [26]. This is due to the fact that, in the KCPQ, pairs of the solution are usually dispersed all over the search space and indexing would not accelerate computation.

But, still, we need to slice a dataset into N parts (strips) in a way that they will contain roughly equal number of points. This probably means strips of unequal width corresponding to ascending intervals along one of the dimensions (x axis dimension is assumed, w.l.o.g.) and is achieved by the following procedure (also used in [21] and presented here in detail): first, we take a sample of size M, then we extract all x-axis values of the sample into an array *xCoord* and perform quicksort on it. We calculate *step* as the quotient M/N and finally the split (partitioning) points are extracted as certain x-values of *xCoord*, as it can be seen in the following pseudo code snippet. This partitioning is then applied to the whole dataset. Our experiments have shown that this technique, based on function *takeSample()* of Spark works quite well, and produces strips of variable width, containing number of points that vary within less than 10% between them.

---

createSplitPoints(dataSet, N): find split points for a single dataset

```
1: sampledData = dataSet.takeSample(noReplace, M)
2: xCoord = sampledData.map(point=>point.x)
3: quicksort(xCoord)
4: step = M / N
5: splits = Array(xCoord(step), xCoord(step*2),..., xCoord(step*(N-1)))
```

3.3   Pair-Partitioning of Datasets

Upon loading the two datasets, we calculate their max and min values for each dimension, thus creating an *Envelope* = *Env*[*minX*, *maxX*, *minY*, *maxY*] for each one. We use two partioning variants, *parent-child* and *common-merged*, both ending to split both datasets into the same strips, N in the first case and $2N + 1$ in the latter one.

Parent-child partition. We consider one of the two datasets as the *parent* and the other as the *child*. Then *createSplitPoints* is used on parent to partition it into a user defined number of N strips (Fig. 1, left). Afterwards, child dataset is being partitioned by using the split points of the parent. In Fig. 1 parent is P and child is Q. Obviously, some partitions of child Q may be empty, or skewed. The existence of empty partitions hardly affects computing time, since no records are to be examined and skewness of child dataset is not a problem in the case of small datasets (e.g. samples). Note that this partition is applied only on the samples, while the next one is applied on the full datasets.

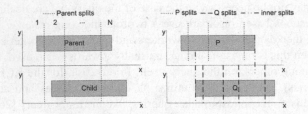

Fig. 1. Pair-partitioning: parent-child (left), common-merged (right)

Common-merged partition. *createSplitPoints* is used on each dataset to compute the split x-coordinates. Then we merge them into an array *allPoints*, together with the *minX*, *maxX* values of both (Fig. 1, right). The array of length $2N + 2$ is being sorted. The first and the last element of this array contain the min and max x-coordinates of all points, and are being removed, thus leaving $2N$ split x-coordinates *allSplits*, from which we create the $2N + 1$ strips, with a minimum width e, for both datasets.

mergeSplitPoints(dataSet1, dataSet2, N, e)	
1: Psplits = createSplitPoints(dataSet1, N)	
2: Qsplits = createSplitPoints(dataSet2, N)	
3: allPoints = Psplits $\cup$ Qsplits $\cup$ {PmaxX, PminX, QmaxX, QminX}	
4: quicksort(allPoints)	
5: left = allPoints(0); right = allPoints(1)	
6: for i = 1 to 2N	
7: if (right - left < e) then allPoints(i) = allPoints(i-1)+e //e=sBound	
8: left = allPoints(i)	//...in l.8 of
9: right = allPoints(i + 1)	//... alg. SnB
10:splits = allPoints.drop(first and last elements)	//...in par. 3.6

In both alternatives, the common Envelopes of the strips from both datasets are formed and broadcasted to the workers. Next, workers assign a proper key to each locally stored point, according to the Envelope it belongs and a shuffling is executed to create the N or $2N + 1$ partitions (according to the alternative used) for each dataset. Thus, points in both datasets that belong to the same strip get the same key.

3.4   Approximating the KCPQ

By cogrouping the two, pair-partitioned datasets, a new RDD is being created. This new, cogrouped, RDD contains items in the form *(stripNo, (Iterable[Point], Iterable[Point]))* where *stripNo* is the number of the strip, the first iterable contains points from P and the second points from Q, that belong to the same strip, therefore are close to each other.

Each partition in this RDD is submitted for computation. Then all results are collected and the top-k among them form an approximation to the solution. Note that this solution is probably not accurate, since there may be points from P and Q in different strips that belong to the result. Still, the larger distance of the pairs found by this approach, is a good upper bound for the final KCPQ.

3.5   Cross-Border Computations

Suppose that we have finished the computation between pairs of strips and have found an upper bound *bnd*, as the distance of the k-th closest pair. Also, suppose that X_i is one splitting point (Fig. 2, left). In order to find the exact solution, we need to check whether points from each dataset belonging to different strips have their distance less than *bnd* as it has been computed in the previous step.

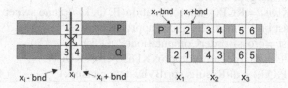

Fig. 2. Cross-border eligible strips (left), cross-border indexing (right)

We have already searched pairs of points in the strip on the left of X_i and pairs on the right of X_i as well. A pair of points (p, q) is a candidate pair only if $dist(p, q) < bnd$ and therefore, one of the following conditions may hold:

- p belongs to Strip 1 and q belongs to Strip 4, or
- p belongs to Strip 2 and q belongs to Strip 3

The bound *bnd* is broadcasted to the workers. Then, on a per-partition basis, data in each dataset is being filtered so that only points between $X_i - bnd$ and $X_i + bnd$ are being included in the newly created RDDs. Additionally, during the procedure, each worker assigns the proper indexes to the points, in a cross-border manner as shown in Fig. 2 (right).

The two newly created RDDs are being joined in order to create all candidate cross-border pairs. In the case that one of the corresponding (sub)partitions is empty, then no pairs will be created. As in previous step, a local *maxHeap* is being utilized during processing. The first K pairs and their corresponding distances are being stored in the *maxHeap*. Then, for each consecutive pair, it is tested whether condition (2) holds or not, where in this case $\delta = h$, the distance of the head of the (local) maxHeap.

3.6   Phases of SliceNBound for the Exact KCPQ

As in many demanding problems, knowing a good bound for the solution in question accelerates the total execution, since this bound helps in pruning the search space. Sampling for upper bound computation has been used in both [20, 21] and reported that this increases efficiency. During the computations, sampling is utilized in two cases, one in order to derive the splitting points and two in order to derive an upper bound of the KCPQ. The algorithm proceeds in four major phases, as being depicted in the following pseudo code snippet.

SliceNBound (P, Q, k, N, fraction)

```
 1: sP = P.sample(noReplace, fraction)
 2: sQ = Q.sample(noReplace, fraction)
 3: parent-childPartition (parent sP, child sQ)
 4: sampledKApproxDist= KCPQ_approximation(sP, sQ, k) //plane sweep
 5: Broadcast sBound = max_distance {sampledKApproxDist}
 6: createSplitPoints(P, N)
 7: createSplitPoints(Q, N)
 8: mergeSplitPoints(P, Q, N, sBound)
 9: common-mergedPartition(P, Q)
10: approxKDist = KCPQ_approximation(P, Q, k) //plane sweep
11: Broadcast bnd = max_distance {approxKDist}
12: crossDist = crossBorderComputation(P, Q, k)
13: allResults = Merge_array(approxKDist, crossKDist)
14: KCPQ(P,Q,k) = allResults.sortByDistance.take(k)
```

Phase one consists of steps 1 to 5 and is used to quickly compute a first upper bound of the KCPQ. Phase two, consisting of steps 6 to 11, computes an approximation of the KCPQ over the whole datasets. Phase three, in step 12, performs the cross-border computation as described in Subsect. 3.5 and, finally, Phase four consisting of steps 13 and 14 collects partial results from Phases two and three and merges them in an array from where the final result is extracted as the top-k (smaller distances). Due to the nature of the partitioning, no pairs are possible to be included in the result twice, so there are no duplicates in the final solution.

Based on our experimental results, we are using parent-child partitioning for computing the KCPQ over the relatively small sample (Step 3 of the above algorithm) and common-merged partitioning for the computation on the whole datasets (Step 9 of the above algorithm), a combination shown to work effectively. Notice that, in case the common-merged partitioning procedure is used as part of the exact KCPQ computation in Phase two, special care has to be taken so that the width of every strip does not fall under the value of $e = bnd$. If such case is found, then the split point is moved rightwards so that width is greater than bnd.

3.7   SliceNBound for the DJQ

In the case of the DJQ, the upper bound is already known. Following this observation, there are three major differences compared to the KCP Query.

First, there is no need for Phase one to compute an initial upper bound via sampling. Second, there is no need for *maxHeap* to store (part of) the K best pairs, since all pairs (p, q) such that $dist(p, q) \leq epsilon$ are part of the solution. Instead, a local list is being utilized. And, third, we don't have to collect results from Phase two before proceeding to Phase three. The algorithm proceeds by performing Phase two and Phase three and each one creates an intermediate RDD. The two RDDs contain all pairs that form the exact result, which can be collected or stored.

4 Experimentation

We have used 2d point datasets to test our KCPQ/DJQ algorithms in Spark derived from OpenStreetMap [10]: WATER resources (5,836,360 line segments), PARKS (11,504,035 polygons) and BUILDINGS of the world consisting (114,736,611 polygons) have been used to create sets of 2D points. Experiments run on a cluster of 5 nodes, each one having 4 vCPUs at 2.1 GHz, with a total of 16 GB of main memory per node (12 GB for Spark), running Ubuntu Linux 16.04 operating system, Spark 2.0.2 on Hadoop 2.7.2 Distributed File System (HDFS with 128 MB block size). Four nodes (4 nodes × 4 vCPUs = 16 vCPUs) were used as HDFS DataNodes and Spark Worker nodes. Java openjdk ver. 1.8.0 and Scala code runner ver. 2.11 were used. We performed each experiment several times and averaged the total response time that expresses the overall CPU, I/O and communication time needed for the execution of each query from startup until count of the results. In all experiments, sample *fraction* for upper bound computation in Phase one was set to 0.01 and sample size for split points computation was set to *noOfRecords/N* (*N* the initial number of partitions).

Our first experiment measures the efficiency of SnB for the KCPQ as compared to the method presented in [21]. It also examines the effect of the number of partitions and the increase of the *K* value. In the case of SnB, each dataset starts with *N* partitions (Phase one) and ends with *2N + 1* (Phases two and three). In Fig. 3 left experiments were conducted for both methods, using the PARKS × WATER datasets combination. The total height of each vertical bar presents time measured for alg. [21] and the bottom part presents time measured for SnB. The upper part of each vertical bar presents the difference (gain) of SnB compared to alg. [21]. It can be seen that the improvement varies around 30–60%, mainly due to the elaborate partitioning schemes. In Fig. 3 right experiments were conducted solely for SnB and are presented versus the best results recorded in [21] (using the same infrastructure and settings) using the

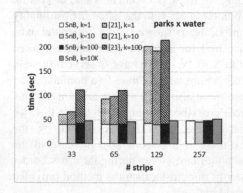
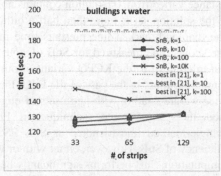

Fig. 3. SnB vs. algorithm in [21] for the KCPQ.

BUILDINGS × WATER datasets combination. There is also a substantial improvement in total response time that exceeds 30%. The total execution time grows moderately as the number of results to be obtained increases. It may be concluded that there is no significant impact on the execution time but, in general, as K increases, pruning gets less effective. Results verify the recommendation of having a number of partitions that equals 2 to 4 times the number of cores. Larger number of partitions overwhelms the computing cluster.

The second experiment (Fig. 4 left) measures the computing time for the case of BUILDINGS x PARKS (line with diamonds, left vertical axis) and the speedup (right vertical axis) of SnB for the KCPQ, varying the number of computing nodes. The dotted line shows the ideal (theoretical) speedup. It could be inferred that the performance of SnB will increase if more computing nodes are added.

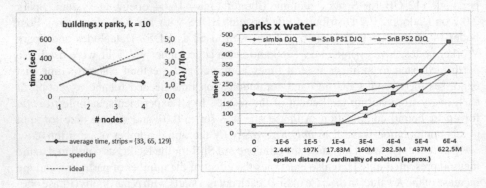

Fig. 4. Total response time and speedup of SnB for the KCPQ (left) and SnB for the DJQ compared to the DJQ in [19] (right).

The third experiment (Fig. 4 right) presents the performance of the SnB algorithm variant for the Distance Join Query, compared to Simba [19], for the case of the PARKS x WATER dataset combination. We have used branch simba-spark-1.6 and the Distance Join example, as provided by the project with very few modifications in order to best fit to our cluster. In both methods, the number of partitions was set to 65, which means starting with 32 for SnB, finally leading to $2N + 1 = 65$. The value of *epsilon* distance varied from 0 (solution contains no points) to 6E-4. In the latter, the cardinality of the solution set is more than half a billion points, as shown on the horizontal axis. The diamond-marked line presents time measured for Simba and the square-marked line shows time measured for SnB (denoted as SnB PS1). We have observed, though, that in contrast to the KCPQ variant of our algorithm where Phase two dominates the computing time, in the case of the DJQ, Phase three also takes a large amount of time, since we are dealing with large sets of eligible cross-border points. To tackle this, we used the plane sweep technique in Phase three as well. As shown by the triangle-marked line (denoted as SnB PS2), there is a considerable increase in the efficiency of SnB. However, either with or without plane sweep during the cross-border computation, SnB performs significantly better in our cluster than the method provided by Simba, in almost all cases taken into consideration.

5 Concluding Remarks

In this paper, we studied the problem of answering the KCPQ and DJQ in Spark and proposed SnB, a family of new parallel algorithms on big spatial datasets, utilizing parent-child and common-merged strip partitioning, local/global bounding and the plane-sweep technique. The performance of the methods has been evaluated with big real-world points datasets. Response times for the KCPQ were compared to the ones in [21]. SnB shares some common parts with the method presented in [21] but also incorporates quite important improvements leading to the substantial reduction of response time and increase in scalability, as observed and recorded in the previous section. Concerning the similarities, both methods use plane sweep to compute the KCPQ. Also, they both utilize sampling to find the partition points for a single dataset and the same technique to strip-partition a single dataset. Both use bounds for the solution, one in the case of [21], where it is computed from a sample and used to find all the eligible pairs of strips, two in SnB.

The major improvements in SnB are as follows: [21] separately partitions each dataset to strips and then performs the computation on the eligible pairs of strips that are extracted by utilizing a bound. SnB proceeds in a completely different manner. It *actually creates* the overlapping pairs by sharing the same partition points between the two datasets. Disjoint pairs are not considered for computation in the first place, but are taken into account in Phase three, now heavily pruned. Initially, sampling is used in combination with parent-child partition to find a good upper bound, since parent-child partition creates pairs of strips with points more or less close to each other. In contrast, upper bound in [21] is being computed with a more naive and time consuming method, leading to the need of selecting of smaller sample fraction and consequently to significant fluctuation in the quality of the bound. Phase two of SnB is preceded by a common-merged partitioning scheme that generates pairs of strips while ensuring that data points are equally shared among them and that all strips have width greater than the bound found so far. In SnB the usage of the bound is twofold: first it serves as a measure for the minimum width of each strip in Phase two and secondly it is used to prune the search space. In Phase three SnB performs the cross border computation on strips that are bounded by the solution found in Phase two, a solution that is already quite accurate and minimizes the need for extra computation.

SnB for the DJQ was also compared to the results derived by using Simba [19]. The presented technique has been proved to be of superior performance.

Future work could include of comparing the performance of the presented technique to the performance of techniques utilizing different partitioning approaches [20]. Since partitioning in strips is well suited for processing the KCPQ and DJQ in parallel and distributed environments, future work could also include of embedding the presented technique in spatially enabled Spark extensions [16–19]. Using of a similar partitioning scheme to answer other queries (e.g. all K Nearest Neighbors, skyline) and adapting the presented technique to data of higher dimensions are also promising research directions.

References

1. Smid, M.: Closest-point problems in computational geometry. In: Sack, J.-R., Urrutia, J. (eds.) Handbook of Computational Geometry, Ch. 20, pp. 877–935. Elsevier (2000)
2. Gao, Y., Chen, L., Li, X., Yao, B., Chen, G.: Efficient k-closest pair queries in general metric spaces. VLDB J. **24**(3), 415–439 (2015)
3. Corral, A., Manolopoulos, Y., Theodoridis, Y., Vassilakopoulos, M.: Algorithms for processing k-closest-pair queries in spatial databases. Data Knowl. Eng. **49**(1), 67–104 (2004)
4. Gutierrez, G., Sáez, P.: The k closest pairs in spatial databases - when only one set is indexed. GeoInformatica **17**(4), 543–565 (2013)
5. Roumelis, G., Corral, A., Vassilakopoulos, M., Manolopoulos, Y.: New plane-sweep algorithms for distance-based join queries in spatial databases. GeoInformatica **20**(4), 571–628 (2016)
6. Chen, M., Mao, S., Liu, Y.: Big data: a survey. Mob. Netw. Appl. **19**, 171–209 (2014)
7. Shekhar, S., Feiner, S.K., Aref, W.G.: Spatial computing. Commun. ACM **59**(1), 72–81 (2016)
8. Eldawy, A., Mokbel, M.F.: The era of big spatial data: a survey. DBSJ J. **13**(1), 25–36 (2015)
9. Dean, J., Ghemawat, S.: MapReduce: simplified data processing on large clusters. In: OSDI 2004, pp. 137–150 (2004)
10. Eldawy, A., Mokbel, M.F.: SpatialHadoop: a MapReduce framework for spatial data. In: ICDE Conference, pp. 1352–1363 (2015)
11. Aji, A., Wang, F., Vo, H., Lee, R., Liu, Q., Zhang, X., Saltz, J.H.: Hadoop-GIS: a high performance spatial data warehousing system over MapReduce. PVLDB **6**(11), 1009–1020 (2013)
12. Zaharia, M., Chowdhury, M., Das, T., Dave, A., Ma, J., McCauly, M., Franklin, M.J., Shenker, S., Stoica, I.: Resilient distributed datasets: a fault-tolerant abstraction for in-memory cluster computing. In: NSDI 2012, pp. 15–28. USENIX (2012)
13. Zaharia, M., Chowdhury, M., Franklin, M.J., Shenker, S., Stoica, I.: Spark: cluster computing with working sets. In: Proceedings of the 2nd USENIX Conference on Hot Topics in Cloud Computing (2010)
14. Chen, D., Shen, C., Feng, J., Le, J.: An efficient parallel top-k similarity join for massive multidimensional data using spark. Int. J. Database Theor. Appl. **8**(3), 57–68 (2015)
15. Dustakar, N.R., Dustakar, S.R.: Computational geometry leveraged by apache spark. J. Innov. Electron. Commun. Eng. **5**(2), 15–31 (2015)
16. Yu, J., Wu, J., Sarwat, M.: GeoSpark: a cluster computing framework for processing large-scale spatial data. In: SIGSPATIAL 2015, Bellevue, WA (2015)
17. You, S., Zhang, J., Gruenwald, L.: Large-scale spatial join query processing in cloud. In: CloudDM Workshop (2015)
18. Tang, M., Yu, Y., Malluhi, Q.M., Ouzzani, M., Aref, W.G.: Locationspark: a distributed in-memory data management system for big spatial data. Proc. VLDB Endowment **9**, 1565–1568 (2016)
19. Xie, D., Li, F., Yao, B., Li, G., Zhou, L., Guo, M.: Simba: efficient in-memory spatial analytics. In: SIGMOD 2016, San Francisco (2016)
20. García-García, F., Corral, A., Iribarne, L., Vassilakopoulos, M., Manolopoulos, Y.: Enhancing SpatialHadoop with closest pair queries. In: Pokorný, J., Ivanović, M., Thalheim, B., Šaloun, P. (eds.) ADBIS 2016. LNCS, vol. 9809, pp. 212–225. Springer, Cham (2016). doi:10.1007/978-3-319-44039-2_15

21. Mavrommatis, G., Moutafis, P., Vassilakopoulos, M.: Closest-pairs query processing in apache spark. In: Proceedings of the Eighth International Conference on Cloud Computing, GRIDs, and Virtualization, pp. 26–31. IARIA (2017)
22. Aji, A., Vo, H, Wang, F.: Effective Spatial Data Partitioning for Scalable Query Processing. arXiv:1509.00910v1 [cs.DB]. Downloaded from https://arxiv.org/pdf/1509.00910v1. 21 December 2016
23. Guller, M.: Big Data Analytics with Spark. Apress, distributed by Springer Science +Business Media, New York (2015)
24. Carraghan, R., Pardalos, P.M.: An exact algorithm for the maximum clique problem. Oper. Res. Lett. **9**, 375–382 (1990)
25. Borges, F., Gutierrez-Milla, A., Suppi, R., Luque, E.: Strip partitioning for ant colony parallel and distributed discrete-event simulation. Procedia Comput. Sci. **51**, 483–492 (2015)
26. Eldawy, A., Alarabi, L., Mokbel, M.F.: Spatial partitioning techniques in SpatialHadoop. Proc. VLDB Endowment **8**(12), 1602–1605 (2015)

A Comparison of Distributed Spatial Data Management Systems for Processing Distance Join Queries

Francisco García-García[1], Antonio Corral[1(✉)], Luis Iribarne[1],
George Mavrommatis[2], and Michael Vassilakopoulos[2]

[1] Department of Informatics, University of Almeria, Almeria, Spain
{paco.garcia,acorral,liribarn}@ual.es
[2] DaSE Lab, Department of Electrical and Computer Engineering,
University of Thessaly, Volos, Greece
{gmav,mvasilako}@uth.gr

Abstract. Due to the ubiquitous use of spatial data applications and the large amounts of spatial data that these applications generate, the processing of large-scale distance joins in distributed systems is becoming increasingly popular. Two of the most studied distance join queries are the K Closest Pair Query (KCPQ) and the ε Distance Join Query (εDJQ). The KCPQ finds the K closest pairs of points from two datasets and the εDJQ finds all the possible pairs of points from two datasets, that are within a distance threshold ε of each other. Distributed cluster-based computing systems can be classified in Hadoop-based and Spark-based systems. Based on this classification, in this paper, we compare two of the most current and leading distributed spatial data management systems, namely SpatialHadoop and LocationSpark, by evaluating the performance of existing and newly proposed parallel and distributed distance join query algorithms in different situations with big real-world datasets. As a general conclusion, while SpatialHadoop is more mature and robust system, LocationSpark is the winner with respect to the total execution time.

Keywords: Spatial data processing · Distance joins · SpatialHadoop · LocationSpark

1 Introduction

Nowadays, the volume of available spatial data (e.g. location, routing, navigation, etc.) is increasing hugely across the world-wide. Recent developments of spatial big data systems have motivated the emergence of novel technologies for processing large-scale spatial data on clusters of computers in a distributed fashion. These Distributed Spatial Data Management Systems (DSDMSs) can be

F. García-García, A. Corral, L. Iribarne and M. Vassilakopoulos — Work funded by the MINECO research project [TIN2013-41576-R].

M. Kirikova et al. (Eds.): ADBIS 2017, LNCS 10509, pp. 214–228, 2017.
DOI: 10.1007/978-3-319-66917-5_15

classified in disk-based [8] and in-memory [18] ones. The disk-based DSDMSs are characterized by being Hadoop-based systems and the most representative ones are Hadoop-GIS [1] and SpatialHadoop [5]. The Hadoop-based systems enable to execute spatial queries using predefined high-level spatial operators without having to worry about fault tolerance and computation distribution. On the other hand, the in-memory DSDMSs are characterized as Spark-based systems and the most representative ones are SpatialSpark [14], GeoSpark [16], Simba [13] and LocationSpark [11,12]. The Spark-based systems allow users to work on distributed in-memory data without worrying about the data distribution mechanism and fault-tolerance.

Distance join queries (DJQs) have received considerable attention from the database community, due to their importance in numerous applications, such as spatial databases and GIS, data mining, multimedia databases, etc. DJQs are costly queries because they combine two datasets taking into account a distance metric. Two of the most representative ones are the K Closest Pair Query ($KCPQ$) and the ε Distance Join Query (εDJQ). Given two point datasets $\mathbb{P}$ and $\mathbb{Q}$, the $KCPQ$ finds the K closest pairs of points from $\mathbb{P} \times \mathbb{Q}$ according to a certain distance function (e.g., Manhattan, Euclidean, Chebyshev, etc.). The εDJQ finds all the possible pairs of points from $\mathbb{P} \times \mathbb{Q}$, that are within a distance threshold ε of each other. Several research works have been devoted to improve the performance of these queries by proposing efficient algorithms in centralized environments [2,9]. But, with the fast increase in the scale of the big input datasets, processing large data in parallel and distributed fashions is becoming a popular practice. For this reason, a number of parallel algorithms for DJQs in MapReduce [3] and Spark [17] have been designed and implemented [6,12].

Apache Hadoop[1] is a reliable, scalable, and efficient cloud computing framework allowing for distributed processing of large datasets using MapReduce programming model. However, it is a kind of disk-based computing framework, which writes all intermediate data to disk between *map* and *reduce* tasks. MapReduce [3] is a framework for processing and managing large-scale datasets in a distributed cluster. It was introduced with the goal of providing a simple yet powerful parallel and distributed computing paradigm, providing good scalability and fault tolerance mechanisms. Apache Spark[2] is a fast, reliable and distributed in-memory large-scale data processing framework. It takes advantage of the Resilient Distributed Dataset (RDD), which allows transparently storing data in memory and persisting it to disk only if it is needed [17]. Hence, it can reduce a huge number of disk writes and reads to outperform the Hadoop platform. Since Spark maintains the status of assigned resources until a job is completed, it reduces time consumption in resource preparation and collection.

Both Hadoop and Spark have weaknesses related to efficiency when applied to spatial data. A main shortcoming is the lack of any indexing mechanism that would allow selective access to specific regions of spatial data, which would in turn yield more efficient query processing algorithms. A solution to this problem

---

[1] Available at https://hadoop.apache.org/.
[2] Available at https://spark.apache.org/.

is an extension of Hadoop, called SpatialHadoop [5], which is a framework that supports spatial indexing on top of Hadoop, i.e. it adopts two-level index structure (global and local) to organize the stored spatial data. And other possible solution is LocationSpark [11,12], which is a spatial data processing system built on top of Spark and it employs various spatial indexes for in-memory data.

In the literature, up to now, there are only few comparative studies between Hadoop-based and Spark-based systems. The most representative one is [10], for a general perspective. But, for comparing DSDMSs, we can find [7,15,16]. Motivated by this fact, in this paper we compare SpatialHadoop and Location-Spark for distance-based join query processing, in particular for KCPQ and εDJQ, in order to provide criteria for adopting one or the other DSDMS. The contributions of this paper are the following:

– Novel algorithms in LocationSpark (the first ones in the literature) to perform efficient parallel and distributed KCPQ and εDJQ, on big real-world spatial datasets
– The execution of a set of experiments for comparing the performance of the two DSDMSs (SpatialHadoop and LocationSpark).
– The execution of a set of experiments for examining the efficiency and the scalability of the existing and new DJQ algorithms.

This paper is organized as follows. In Sect. 2, we review related work on Hadoop-based and Spark-based systems that support spatial operations and provide the motivation for this paper. In Sect. 3, we present preliminary concepts related to DJQs, SpatialHadoop and LocationSpark. In Sect. 4, the parallel algorithms for processing KCPQ and εDJQ in LocationSpark are proposed. In Sect. 5, we present representative results of the extensive experimentation that we have performed, using real-world datasets, for comparing these two cloud computing frameworks. Finally, in Sect. 6, we provide the conclusions arising from our work and discuss related future work directions.

2   Related Work and Motivation

Researchers, developers and practitioners worldwide have started to take advantage of the cluster-based systems to support large-scale data processing. There exist several cluster-based systems that support spatial queries over distributed spatial datasets and they can be classified in Hadoop-based and Spark-based systems. The most important contributions in the context of Hadoop-based systems are the following research prototypes:

– *Hadoop-GIS* [1] extends Hive and adopts Hadoop Streaming framework and integrates several open source software packages for spatial indexing and geometry computation. Hadoop-GIS only supports data up to two dimensions and two query types: rectangle range query and spatial joins.
– *SpatialHadoop* [5] is an extension of the MapReduce framework [3], based on Hadoop, with native support for spatial 2d data (see Sect. 3.2).

On the other hand, the most remarkable contributions in the context of Spark-based systems are the following prototypes:

- *SpatialSpark* [14] is a lightweight implementation of several spatial operations on top of the Spark in-memory big data system. It targets at in-memory processing for higher performance. SpatialSpark adopts data partition strategies like fixed grid or kd-tree on data files in HDFS and builds an index to accelerate spatial operations. It supports range queries and spatial joins over geometric objects using conditions like *intersect* and *within*.
- *GeoSpark* [16] extends Spark for processing spatial data. It provides a new abstraction called Spatial Resilient Distributed Datasets (SRDDs) and a few spatial operations. It allows an index (e.g. Quadtree and R-tree) to be the object inside each local RDD partition. For the query processing point of view, GeoSpark supports range query, KNNQ and spatial joins over SRDDs.
- *Simba* (Spatial In-Memory Big data Analytics) [13] offers scalable and efficient in-memory spatial query processing and analytics for big spatial data. Simba is based on Spark and runs over a cluster of commodity machines. In particular, Simba extends the Spark SQL engine to support rich spatial queries and analytics through both SQL and the DataFrame API. It introduces partitioning techniques (e.g. STR), indexes (global and local) based on R-trees over RDDs in order to work with big spatial data and complex spatial operations (e.g. range query, KNNQ, distance join and KNNJQ).
- *LocationSpark* [11,12] is an efficient in-memory distributed spatial query processing system (see Sect. 3.3 for more details).

As we have seen, there are several distributed systems based on Hadoop or Spark for managing spatial data, but there are not many articles comparing them with respect to spatial query processing. The only contributions in this regard are [7,15,16]. In [15,16], SpatialHadoop is compared with SpatialSpark and GeoSpark, respectively, for spatial join query processing. In [7], Spatial-Hadoop is compared with GeoSpark with respect to the architectural point of view. Motivated by these observations, and since KCPQ [6] is implemented in SpatialHadoop (its adaptation to εDJQ is straightforward), and in Location-Spark neither KCPQ nor εDJQ have been implemented yet, we design and implement both DJQs in LocationSpark. Moreover, we develop a comparative performance study between SpatialHadoop and LocationSpark for KCPQ and εDJQ.

3   Preliminaries and Background

In this section, we first present the basic definitions of the KCPQ and εDJQ, followed by a brief introduction of the preliminary concepts about SpatialHadoop and LocationSpark, the DSDMSs to be compared.

3.1   The K Closest Pairs and ε Distance Join Queries

The $KCPQ$ discovers the K pairs of data formed from the elements of two datasets having the K smallest distances between them (i.e. it reports only the top K pairs). The formal definition of the $KCPQ$ for point datasets (the extension of this definition to other, more complex spatial objects – e.g. line-segments, objects with extents, etc. – is straightforward) is the following:

Definition 1 (K **Closest Pairs Query, KCPQ**). *Let* $\mathbb{P} = \{p_0, p_1, \cdots, p_{n-1}\}$ *and* $\mathbb{Q} = \{q_0, q_1, \cdots, q_{m-1}\}$ *be two set of points, and a number* $K \in \mathbb{N}^+$. *Then, the result of the K Closest Pairs Query is an ordered collection,* $KCPQ(\mathbb{P}, \mathbb{Q}, K)$, *containing K different pairs of points from* $\mathbb{P} \times \mathbb{Q}$, *ordered by distance, with the K smallest distances between all possible pairs:*
$KCPQ(\mathbb{P}, \mathbb{Q}, K) = ((p_1, q_1), (p_2, q_2), \cdots, (p_K, q_K)), (p_i, q_i) \in \mathbb{P} \times \mathbb{Q}, 1 \le i \le K,$
such that for any $(p, q) \in \mathbb{P} \times \mathbb{Q} \setminus KCPQ(\mathbb{P}, \mathbb{Q}, K)$ *we have* $dist(p_1, q_1) \le dist(p_2, q_2) \le \cdots \le dist(p_K, q_K) \le dist(p, q).$

Note that if multiple pairs of points have the same K-th distance value, more than one collection of K different pairs of points are suitable as a result of the query. Recall that $KCPQ$ is implemented in SpatialHadoop [6] using plane-sweep algorithms [9], but not in LocationSpark.

On the other hand, the εDJQ reports all the possible pairs of spatial objects from two different spatial objects datasets, $\mathbb{P}$ and $\mathbb{Q}$, having a distance smaller than a distance threshold ε of each other [9]. The formal definition of εDJQ for point datasets is the following:

Definition 2 (ε **Distance Join Query, εDJQ**). *Let* $\mathbb{P} = \{p_0, p_1, \cdots, p_{n-1}\}$ *and* $\mathbb{Q} = \{q_0, q_1, \cdots, q_{m-1}\}$ *be two set of points, and a distance threshold* $\varepsilon \in \mathbb{R}_{\ge 0}$. *Then, the result of the εDJQ is the set,* $\varepsilon DJQ(\mathbb{P}, \mathbb{Q}, \varepsilon) \subseteq \mathbb{P} \times \mathbb{Q}$, *containing all the possible different pairs of points from* $\mathbb{P} \times \mathbb{Q}$ *that have a distance of each other smaller than, or equal to* ε:
$\varepsilon DJQ(\mathbb{P}, \mathbb{Q}, \varepsilon) = \{(p_i, q_j) \in P \times Q \ : \ dist(p_i, q_j) \le \varepsilon\}$

The εDJQ can be considered as an extension of the $KCPQ$, where the distance threshold of the pairs is known beforehand and the processing strategy (e.g. plane-sweep technique) can be the same as in the $KCPQ$ for generating the candidate pairs of the final result. For this reason, its adaptation to Spatial-Hadoop from $KCPQ$ is straightforward. Note that εDJQ is not implemented in LocationSpark.

3.2   SpatialHadoop

SpatialHadoop [5] is a full-fledged MapReduce framework with native support for spatial data. It is an efficient disk-based distributed spatial query process-ing system. Note that MapReduce [3] is a scalable, flexible and fault-tolerant programming framework for distributed large-scale data analysis. A task to be performed using the MapReduce framework has to be specified as two phases:

the *map* phase is specified by a *map function* takes input (typically from Hadoop Distributed File System (HDFS) files), possibly performs some computations on this input, and distributes it to worker nodes; and the *reduce* phase which processes these results as specified by a *reduce function*. Additionally, a *combiner function* can be used to run on the output of *map* phase and perform some filtering or aggregation to reduce the number of keys passed to the *reducer*.

SpatialHadoop is a comprehensive extension to Hadoop that injects spatial data awareness in each Hadoop layer, namely, the language, storage, MapReduce, and operations layers. *MapReduce* layer is the query processing layer that runs MapReduce programs, taking into account that SpatialHadoop supports spatially indexed input files. The *Operation* layer enables the efficient implementation of spatial operations, considering the combination of the spatial indexing in the storage layer with the new spatial functionality in the *MapReduce* layer. In general, a spatial query processing in SpatialHadoop consists of four steps [5,6] (see Fig. 1): (1) *Preprocessing*, where the data is partitioned according to a specific spatial index, generating a set of partitions or cells. (2) *Pruning*, when the query is issued, where the master node examines all partitions and prunes by a *filter* function those ones that are guaranteed not to include any possible result of the spatial query. (3) *Local Spatial Query Processing*, where a local spatial query processing is performed on each non-pruned partition in parallel on different machines (*map* tasks). Finally, (4) *Global Processing*, where the results are collected from all machines in the previous step and the final result of the concerned spatial query is computed. A *combine* function can be applied in order to decrease the volume of data that is sent from the *map* task. The *reduce* function can be omitted when the results from the *map* phase are final.

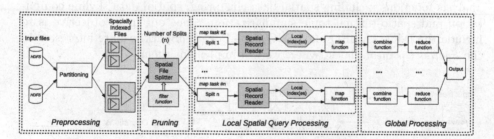

Fig. 1. Spatial query processing in SpatialHadoop [5,6].

3.3   LocationSpark

LocationSpark [11,12] is a library in Spark that provides an API for spatial query processing and optimization based on Spark's standard dataflow operators. It is an efficient in-memory distributed spatial query processing system. LocationSpark provides several optimizations to enhance Spark for managing spatial data and they are organized by layers: memory management, spatial index, query executor, query scheduler, spatial operators and spatial analytical. In the *Memory Management* layer for spatial data, LocationSpark dynamically caches frequently accessed data into memory, and stores the less frequently used data into disk. For the *Spatial Index* layer, LocationSpark builds two levels of spatial indexes (global and local). To build a global index, LocationSpark samples the underlying data to learn the data distribution in space and provides a grid and a region Quadtree. In addition, each data partition has a local index (e.g., a grid local index, an R-tree, a variant of the Quadtree, or an IR-tree). Finally, LocationSpark adopts a new *Spatial Bloom Filter* to reduce the communication cost when dispatching queries to their overlapping data partitions, termed *sFilter*, that can speed up query processing by avoiding needless communication with data partitions that do not contribute to the query answer. In the *Query Executor* layer, LocationSpark evaluates the runtime and memory usage trade-offs for the various alternatives, and then, it chooses and executes the better execution plan on each slave node. LocationSpark has a new layer, termed *Query Scheduler*, with an automatic skew analyzer and a plan optimizer to mitigate query skew. The query scheduler uses a cost model to analyze the skew to be used by the spatial operators, and a plan generation algorithm to construct a load-balanced query execution plan. After plan generation, local computation nodes select the proper algorithms to improve their local performance based on the available spatial indexes and the registered queries on each node. For the *Spatial Operators* layer, LocationSpark supports spatial querying and spatial data updates. It provides a rich set of spatial queries including spatial range query, KNNQ, spatial-join, and KNNJQ. Moreover, it supports data updates and spatio-textual operations. Finally, for the *Spatial Analytical* layer, and due to the importance of spatial data analysis, LocationSpark provides spatial data analysis functions including spatial data clustering, spatial data skyline computation and spatio-textual topic summarization. Since our main objective is to include the DJQs (KCPQ and εDJQ) into LocationSpark, we are interested in the *Spatial Operators* layer, where we will implement them.

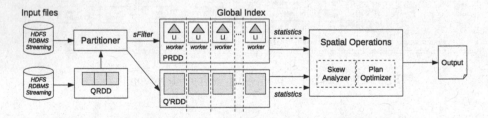

Fig. 2. Spatial query processing for DJQs in LocationSpark, based on [12].

To process spatial queries, LocationSpark builds a distributed spatial index structure for in-memory spatial data. As we can see in Fig. 2, for DJQs, given two datasets $\mathbb{P}$ and $\mathbb{Q}$, $\mathbb{P}$ is partitioned into N partitions based on a spatial index criteria (e.g. N leaves of a R-tree) by the *Partitioner* leading to the

PRDD (Global Index). The *sFilter* determines whether a point is contained inside a spatial range or not. Next, each *worker* has a local data partition $\mathbb{P}_i$ $(1 \leq i \leq N)$ and builds a Local Index (**LI**). *QRDD* is generated from $\mathbb{Q}$ by a member function of RDD (Resilient Distributed Dataset) natively supported by Spark, that forwards such point to the partitions that spatially overlap it. Now, each point of $\mathbb{Q}$ is replicated to the partitions that are identified using the *PRDD* (Global Index), leading to the *Q'RDD*. Then a post-processing step (using the Skew Analyzer and the Plan Optimizer) is performed to combine the local results to generate the final output.

4 DJQ Algorithms in SpatialHadoop and LocationSpark

Since KCPQ is already implemented in SpatialHadoop [6], in this section, we will present how we can adapt KCPQ to εDJQ in SpatialHadoop and how KCPQ and εDJQ can be implemented in LocationSpark.

4.1 KCPQ and εDJQ in SpatialHadoop

In general, the KCPQ algorithm in SpatialHadoop [6] consists of a MapReduce job. The *map* function aims to find the KCP between each local pair of partitions from $\mathbb{P}$ and $\mathbb{Q}$ with a particular plane-sweep KCPQ algorithm [9] and the result is stored in a binary max heap (called *LocalKMaxHeap*). The *reduce* function aims to examine the candidate pairs of points from each *LocalKMax-Heap* and return the final set of the K closest pairs in another binary max heap (called *GlobalKMaxHeap*). To improve this approach, for reducing the number of possible combinations of pairs of partitions, we need to find in advance an upper bound of the distance value of the K-th closest pair of the joined datasets, called β. This β computation can be carried out by sampling globally from both datasets or by sampling locally for an appropriate pair of partitions and, then executing a plane-sweep KCPQ algorithm over both samples.

The method for the εDJQ in MapReduce, adapting from KCPQ in Spatial-Hadoop [6], adopts the *map* phase of the join MapReduce methodology. The idea is to have $\mathbb{P}$ and $\mathbb{Q}$ partitioned by some method (e.g., Grid) into two sets of cells, with n and m cells of points, respectively. Then, every possible pair of cells is sent as input for the *filter* function. This function takes as input, combinations of pairs of cells in which the input set of points are partitioned and a distance threshold ε, and it prunes pairs of cells which have minimum distances larger than ε. By using SpatialHadoop built-in function *MinDistance* we can calculate the minimum distance between two cells (i.e. this function computes the minimum distance between the two MBRs, Minimum Bounding Rectangles, of the two cells). On the *map* phase, each *mapper* reads the points of a pair of filtered cells and performs a plane-sweep εDJQ algorithm [9] (variation of the plane-sweep-based KCPQ algorithm) between the points inside that pair of cells. The results from all *mappers* are just combined in the *reduce* phase and written into HDFS files, storing only the pairs of points with distance up to ε.

4.2 KCPQ and εDJQ in LocationSpark

Assuming that $\mathbb{P}$ is the largest dataset to be combined and $\mathbb{Q}$ is the smallest one, and following the ideas presented in [12], we can describe the *Execution Plan* for KCPQ in LocationSpark as follows. In Stage 1, the two datasets are partitioned according to a given spatial index schema. In Stage 2, statistic data is added to each partition, $S_{\mathbb{P}}$ and $S_{\mathbb{Q}}$, and they are combined by pairs, $S_{\mathbb{PQ}}$. In Stage 3, the partitions from $\mathbb{P}$ and $\mathbb{Q}$ with the largest density of points, $\mathbb{P}_\beta$ and $\mathbb{Q}_\beta$, are selected to be combined by using a plane-sweep KCPQ algorithm [9] to compute an upper bound of the distance value of the K-th closest pair (β). In Stage 4, the combination of all possible pairs of partitions from $\mathbb{P}$ and $\mathbb{Q}$, $S_{\mathbb{PQ}}$, is filtered according to the β value (i.e. only the pairs of partitions with minimum distance between the MBRs of the partitions is smaller than or equal to β are selected), giving rise to $FS_{\mathbb{PQ}}$, and all pairs of filtered partitions are processed by using a plane-sweep KCPQ algorithm. Finally, the results are merged to get the final output.

With the previous *Execution Plan* and increasing the size of the datasets, the execution time increases considerably due to skew and shuffle problems. To solve it, we modify Stage 4 with the query plan that is used for the algorithms shown in [12], leaving the plan as shown in Fig. 3.

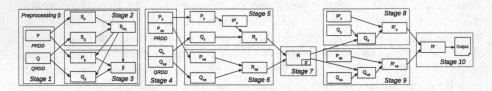

Fig. 3. Execution plan for KCPQ in LocationSpark, based on [12].

Stages 1, 2 and 3 are still used to calculate the β value which will serve to accelerate the local pruning phase on each partition. In Stage 4, using the *Query Plan Scheduler*, $\mathbb{P}$ is partitioned into $\mathbb{P}_{\mathrm{S}}$ and $\mathbb{P}_{\mathrm{NS}}$ being the partitions that present and do not present skew, respectively. The same partitioning is used to $\mathbb{Q}$. In Stage 5, a KCPQ algorithm [9] is applied between points of $\mathbb{P}_{\mathrm{S}}$ and $\mathbb{Q}_{\mathrm{S}}$ that are in the same partition and likewise for $\mathbb{P}_{\mathrm{NS}}$ and $\mathbb{Q}_{\mathrm{NS}}$ in Stage 6. These two stages are executed independently and the results are combined in Stage 7. Finally, it is still necessary to calculate if there is any present candidate for each partition that is on the boundaries of that same partition in the other dataset. To do this, we use β' which is the maximum distance from the current set of candidates as a radius of a range filter with center in each partition to obtain possible new candidates on those boundaries. The calculation of KCPQ of each partition with its candidates is executed in Stages 8 and 9 and these results are combined in Stage 10 to obtain the final answer.

The *Execution Plan* for εDJQ in LocationSpark is a variation of the *K*CPQ one, where the filtering stages are removed, since $S_{\mathbb{PQ}}$ is filtered by ε (i.e. $\beta = \beta' = \varepsilon$), which is the threshold distance known beforehand.

5   Experimentation

In this section we present the results of our experimental evaluation. We have used real 2d point datasets to test our DJQ algorithms in SpatialHadoop and LocationSpark. We have used three datasets from OpenStreetMap[3]: *BUILD-INGS* which contains 115M records of buildings, *LAKES* which contains 8.4M points of water areas, and *PARKS* which contains 10M records of parks and green areas [5]. Moreover, to experiment with the biggest real dataset (*BUILD-INGS*), we have created a new big quasi-real dataset from *LAKES* (8.4M), with a similar quantity of points. The creation process is as follows: taking one point of *LAKES*, p, we generate 15 new points gathered around p (i.e. the center of the cluster), according to a Gaussian distribution with mean = 0.0 and standard deviation = 0.2, resulting in a new quasi-real dataset, called *CLUS_LAKES*, with around 126M of points. The main performance measure that we have used in our experiments has been the total execution time (i.e. total response time). All experiments are conducted on a cluster of 12 nodes on an OpenStack environment. Each node has 4 vCPU with 8 GB of main memory running Linux operating systems and Hadoop 2.7.1.2.3. Each node has a capacity of 3 vCores for MapReduce2/YARN use. The version of Spark used is 1.6.2. Finally, we used the latest code available in the repositories of SpatialHadoop[4] and Location-Spark[5].

Table 1. Configuration parameters used in the experiments.

Parameter	Values (default)
K	1, 10, (10^2), 10^3, 10^4, 10^5
ε $(\times 10^{-4})$	2.5, 5, 7.5, 12.5, (25), 50
Number of nodes	1, 2, 4, 6, 8, 10, (12)
Type of partition	Quadtree

Table 1 summarizes the configuration parameters used in our experiments. Default values (in parentheses) are used unless otherwise mentioned. Spatial-Hadoop needs the datasets to be partitioned and indexed before invoking the spatial operations. The times needed for that pre-processing phase are 94 s for *LAKES*, 103 s for *PARKS*, 175 s for *BUILDINGS* and 200 s for

---

[3] Available at http://spatialhadoop.cs.umn.edu/datasets.html.
[4] Available at https://github.com/aseldawy/spatialhadoop2.
[5] Available at https://github.com/merlintang/SpatialSpark.

CLUS_LAKES. We have shown the time of this pre-processing phase in Spatial-Hadoop (disk-based DSDMS), since it would be the full execution time, at least in the first running of the query. Note that, data are indexed and the index is stored on HDFS and for subsequent spatial queries, data and index are already available (this can be considered as an advantage of SpatialHadoop). On the other hand, LocationSpark (in-memory-based DSDMS) always partitions and indexes the data for every operation. The partitions/indexes are not stored on any persistent file system and cannot be reused in subsequent operations.

Our first experiment aims to measure the scalability of the KCPQ and εDJQ algorithms, varying the dataset sizes. As shown in the left chart of Fig. 4 for the KCPQ of real datasets (*LAKES* × *PARKS*, *BUILDINGS* × *PARKS* and *BUILDINGS* × *CLUS_LAKES*), the execution times in both DSDMSs increase linearly as the size of the datasets increase. Moreover, LocationSpark is faster for all the datasets combinations except for the largest one (e.g. it is 29 s slower for the biggest datasets, *BUILDINGS* × *CLUS_LAKES* (BxC_L)). However, it should be noted that SpatialHadoop needs a pre-indexing time of 175 and 200 s for each dataset (depicted by vertical lines in the charts) and that difference can be caused by memory constraints on the cluster.

As we have just seen for KCPQ, the behavior of the execution times when varying the size of the datasets is very similar for εDJQ. For instance, for the combination of large datasets (see the right chart of Fig. 4), *BUILDINGS* × *CLUS_LAKES* (BxC_L), SpatialHadoop is 32 s faster than LocationSpark. However, for smaller sets, LocationSpark shows better performance (e.g. it is 96 s faster for the middle size datasets, *BUILDINGS* × *PARKS* (BxP)). From these results with real data, we can conclude that both DSDMSs have similar performance, in terms of execution time, even showing LocationSpark better values in most of the cases, despite the fact that neither pre-partitioning nor pre-indexing are done.

The second experiment studies the effect of the increasing both K and ε value for the combination of the biggest datasets (*BUILDINGS* × *CLUS_LAKES*). The left chart of Fig. 5 shows that the total execution time for real datasets grows

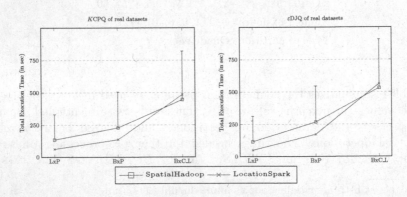

Fig. 4. KCPQ (left) and εDJQ (right) execution times considering different datasets.

slowly as the number of results to be obtained (K) increases. Both DSDMSs, employing *Quadtree*, report stable execution times, even for large K values (e.g. $K = 10^5$). This means that the Quadtree is less affected by the increment of K, because Quadtree employs regular space partitioning depending on the concentration of the points. As shown in the right chart of Fig. 5, the total execution time grows as the ε value increases. Both DSDMSs (SpatialHadoop and LocationSpark) have similar relative performance for all ε values, with SpatialHadoop being faster, except for $\varepsilon = 50 \times 10^{-4}$, where LocationSpark outperforms it (i.e. LocationSpark is 377 s faster). This difference is due to the way in which εDJQ is calculated in the latter, where fewer points are used as candidates and skew cells are dealt with its *Query Plan Scheduler*. For smaller ε values SpatialHadoop preindexing phase reduces time considerably for very large datasets.

The main conclusions that we can extract for this experiment are: (1) the higher K or ε values, the greater the possibility that pairs of candidates are not pruned, more tasks would be needed and more total execution time is consumed and, (2) LocationSpark shows better performance especially for higher values of K and ε thanks to its *Query Plan Scheduler* and the reduction of the number of candidates.

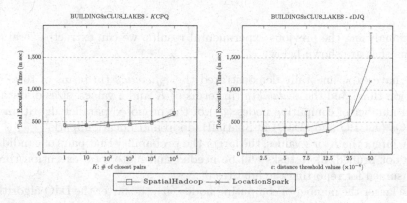

Fig. 5. KCPQ cost (execution time) vs. K values (left) and εDJQ cost (execution time) vs. ε values (right).

The third experiment aims to measure the speedup of the DJQ MapReduce algorithms (KCPQ and εDJQ), varying the number of computing nodes (n). The left chart of Fig. 6 shows the impact of different number of computing nodes on the performance of parallel KCPQ algorithm, for $BUILDINGS \times PARKS$ with the default configuration values. From this chart, it could be concluded that the performance of our approach has direct relationship with the number of computing nodes. It could also be deduced that better performance would be obtained if more computing nodes are added. LocationSpark is still showing a better behavior than SpatialHadoop. In the right chart of Fig. 6, we can observe a similar trend for εDJQ MapReduce algorithm with less execution time, but in this case LocationSpark shows worse performance for a smaller number of nodes. This is due to the fact that LocationSpark and εDJQ depends more on the available memory and when the number of nodes decreases, this memory also decreases considerably.

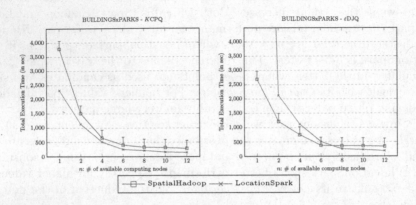

Fig. 6. Query cost with respect to the number of computing nodes n.

By analyzing the previous experimental results, we can extract several conclusions that are shown below:

- We have experimentally demonstrated the *efficiency* (in terms of total execution time) and the *scalability* (in terms of K and ε values, sizes of datasets and number of computing nodes (n)) of the proposed parallel algorithms for DJQs (KCPQ and εDJQ) in SpatialHadoop and LocationSpark.
- The larger the K or ε values, the larger the probability that pairs of candidates are not pruned, more tasks will be needed and more total execution time is consumed for reporting the final result.
- The larger the number of computing nodes (n), the faster the DJQ algorithms are.
- Both DSDMSs have similar performance, in terms of execution time, although LocationSpark shows better values in most of the cases (if an adequate number of processing nodes with adequate memory resources are provided), despite the fact that neither pre-partitioning nor pre-indexing are done.

6   Conclusions and Future Work

The KCPQ and εDJQ are spatial operations widely adopted by many spatial and GIS applications. These spatial queries have been actively studied in centralized environments, however, for parallel and distributed frameworks has not attracted similar attention. For this reason, in this paper, we compare two of the most current and leading DSDMSs, namely SpatialHadoop and Location-Spark. To do this, we have proposed novel algorithms in LocationSpark, the first ones in literature, to perform efficient parallel and distributed KCPQ and εDJQ algorithms on big spatial real-world datasets, adopting the plane-sweep technique. The execution of a set of experiments has demonstrated that LocationSpark is the overall winner for the execution time, due to the efficiency of in-memory processing provided by Spark and additional improvements as the *Query Plan Scheduler*. However, SpatialHadoop is a more mature and robust DSDMS because of time dedicated to investigate and develop it (several years) and, it provides more spatial operations and spatial partitioning techniques. Future work might cover studying other Spark-based DSDMSs like *Simba* [13], implement other spatial partitioning techniques [4] in LocationSpark and, design and implement other DJQs in these DSDMSs for further comparison.

References

1. Aji, A., Wang, F., Vo, H., Lee, R., Liu, Q., Zhang, X., Saltz, J.H.: Hadoop-GIS: a high performance spatial data warehousing system over MapReduce. PVLDB **6**(11), 1009–1020 (2013)
2. Corral, A., Manolopoulos, Y., Theodoridis, Y., Vassilakopoulos, M.: Algorithms for processing K-closest-pair queries in spatial databases. Data Knowl. Eng. **49**(1), 67–104 (2004)
3. Dean, J., Ghemawat, S.: MapReduce: simplified data processing on large clusters. In: OSDI Conference, pp. 137–150 (2004)
4. Eldawy, A., Alarabi, L., Mokbel, M.F.: Spatial partitioning techniques in SpatialHadoop. PVLDB **8**(12), 1602–1613 (2015)
5. Eldawy, A., Mokbel, M.F.: SpatialHadoop: a MapReduce framework for spatial data. In: ICDE Conference, pp. 1352–1363 (2015)
6. García-García, F., Corral, A., Iribarne, L., Vassilakopoulos, M., Manolopoulos, Y.: Enhancing SpatialHadoop with closest pair queries. In: Pokorný, J., Ivanović, M., Thalheim, B., Šaloun, P. (eds.) ADBIS 2016. LNCS, vol. 9809, pp. 212–225. Springer, Cham (2016). doi:10.1007/978-3-319-44039-2_15
7. Lenka, R.K., Barik, R.K., Gupta, N., Ali, S.M., Rath, A., Dubey, H.: Comparative analysis of SpatialHadoop and GeoSpark for geospatial big data analytics, CoRR abs/1612.07433 (2016)
8. Li, F., Ooi, B.C., Özsu, M.T., Wu, S.: Distributed data management using MapReduce. ACM Comput. Surv. **46**(3), 31:1–31:42 (2014)
9. Roumelis, G., Corral, A., Vassilakopoulos, M., Manolopoulos, Y.: New plane-sweep algorithms for distance-based join queries in spatial databases. GeoInformatica **20**(4), 571–628 (2016)
10. Shi, J., Qiu, Y., Minhas, U.F., Jiao, L., Wang, C., Reinwald, B., Özcan, F.: Clash of the titans: mapreduce vs. spark for large scale data analytics. PVLDB **8**(13), 2110–2121 (2015)
11. Tang, M., Yu, Y., Malluhi, Q.M., Ouzzani, M., Aref, W.G.: Locationspark: a distributed in-memory data management system for big spatial data. PVLDB **9**(13), 1565–1568 (2016)
12. Tang, M., Yu, Y., Aref, W.G., Mahmood, A.R., Malluhi, Q.M., Ouzzani, M.: In-memory distributed spatial query processing and optimization, April 2017. http://merlintang.github.io/paper/memory-distributed-spatial.pdf
13. Xie, D., Li, F., Yao, B., Li, G., Zhou, L., Guo, M.: Simba: efficient in-memory spatial analytics. In: SIGMOD Conference, pp. 1071–1085 (2016)

14. You, S., Zhang, J., Gruenwald, L.: Large-scale spatial join query processing in cloud. In: ICDE Workshops, pp. 34–41 (2015)
15. You, S., Zhang, J., Gruenwald, L.: Spatial join query processing in cloud: Analyzing design choices and performance comparisons. In: ICPPW Conference, pp. 90–97 (2015)
16. Yu, J., Wu, J., Sarwat, M.: GeoSpark: a cluster computing framework for processing large-scale spatial data. In: SIGSPATIAL Conference, pp. 70:1–70:4 (2015)
17. Zaharia, M., Chowdhury, M., Das, T., Dave, A., Ma, J., McCauly, M., Franklin, M.J., Shenker, S., Stoica, I.: Resilient distributed datasets: a fault-tolerant abstraction for in-memory cluster computing. In: NSDI Conference, pp. 15–28 (2012)
18. Zhang, H., Chen, G., Ooi, B.C., Tan, K.-L., Zhang, M.: In-memory big data management and processing: a survey. TKDE **27**(7), 1920–1948 (2015)

A Generic and Efficient Framework for Spatial Indexing on Flash-Based Solid State Drives

Anderson Chaves Carniel[1]([✉]), Ricardo Rodrigues Ciferri[2],
and Cristina Dutra de Aguiar Ciferri[1]

[1] Department of Computer Science, University of São Paulo,
São Carlos, SP 13566-590, Brazil
accarniel@gmail.com, cdac@icmc.usp.br
[2] Department of Computer Science, Federal University of São Carlos,
São Carlos, SP 13565-905, Brazil
ricardo@dc.ufscar.br

Abstract. Speeding up the spatial query processing on *flash-based Solid State Drives* (SSDs) has become a core problem in spatial database applications, and has been carried out aided by *flash-aware spatial indices*. Although there are some existing flash-aware spatial indices, they do not exploit all the benefits of SSDs, leading to loss of efficiency. In this paper, we propose a new generic and efficient *F*ramework for spatial *IND*exing on SSDs, called *eFIND*. It takes into account all the intrinsic characteristics of SSDs by employing (i) a *write buffer* to avoid random writes; (ii) a *read buffer* to decrease the overhead of random reads; (iii) a *temporal control* to avoid interleaved reads and writes; (iv) a *flushing policy* based on the characteristics of the indexed spatial objects; and (v) a log-structured approach to provide *data durability*. Performance tests showed that eFIND is very efficient. Compared to existing indices, eFIND improved the construction of spatial indices from 22% up to 68% and the spatial query processing from 3% up to 50%.

1 Introduction

Spatial indices are largely used to improve spatial query processing since they reduce the search space of spatial data, discarding portions of the dataset where the answer cannot be found [4]. Nowadays, there is an increase number of spatial database applications requiring the use of spatial indices to retrieve efficiently spatial objects stored in *flash-based Solid State Drives* (SSDs) [2]. In fact, SSDs have been widely used as secondary storage in notebooks, desktops, and database servers because of their improved characteristics compared to Hard Disk Drives (HDDs) [9]. These characteristics include smaller size, lighter weight, lower power consumption, better shock resistance, and faster reads and writes.

On the other hand, SSDs have intrinsic characteristics that introduce several system implications [3,7]. A well-known characteristic is that a write requires more time and power consumption than a read. In addition, random writes can lead to erase-before-update operations and thus, sequential writes are preferable.

© Springer International Publishing AG 2017
M. Kirikova et al. (Eds.): ADBIS 2017, LNCS 10509, pp. 229–243, 2017.
DOI: 10.1007/978-3-319-66917-5_16

To deal with these characteristics, some *flash-aware spatial indices* have been proposed [6,8,10,11], which extend indices originally designed for HDDs like the R-tree [5]. We term these indices for HDDs as *disk-based spatial indices*.

In general, existing flash-aware spatial indices employ buffers in the main memory to store index modifications, instead of directly performing random writes on the SSD. When the buffer is full, these indices perform a flushing operation, which can encompass a flushing policy to choose a set of nodes to be written sequentially to the SSD. Further, the structure of the index can also be changed to reduce even more the number of writes [6].

However, current flash-aware spatial indices do not exploit all the benefits of SSDs. First, these indices execute an excessive number of random reads, which can degenerate the performance of SSDs [7]. Second, they perform interleaved reads and writes, also impacting negatively the performance [3,7]. Third, they employ flushing policies that do not take into account characteristics of the spatial index, leading to unnecessary writes.

In this paper, we go one step forward to existing work by proposing the *efficient Framework for spatial INDexing on SSDs* (eFIND). It is a generic framework that transforms a disk-based spatial index into a flash-aware spatial index without requiring modifications in the structure and algorithms of the underlying index. Instead, eFIND efficiently changes the way in which reads and writes are performed on the SSD. Hence, eFIND can be incorporated into existing spatial database systems without high implementation costs. To achieve its efficiency, eFIND employs a set of design goals, which are introduced in this paper and specifically defined to take into account all the intrinsic characteristics of SSDs.

Our experiments showed that eFIND is very efficient since it provides a consonance among the following elements:

- *write buffer* to avoid random writes to the SSD;
- *read buffer* to decrease the overhead of random reads;
- *temporal control* to avoid interleaved reads and writes;
- *flushing algorithm* to sequentially write modifications according to a *flushing policy* that is based on the characteristics of the indexed spatial objects;
- *log-structured approach* to guarantee data durability.

This paper is organized as follows. Section 2 introduces our design goals. Section 3 surveys related work. Section 4 proposes eFIND. Section 5 details the experiments. Section 6 concludes the paper and presents future work.

2   Design Goals for Flash-Aware Spatial Indices

Despite the fact that the intrinsic characteristics of SSDs have been studied in the literature [3,7,9], it remains unclear how to deal with them to achieve good spatial indexing performance. Indeed, we will see in Sect. 3 that existing flash-aware spatial indices do not consider important intrinsic characteristics of SSDs. In this section, we introduce a set of design goals that correlate the intrinsic characteristics of SSDs with the characteristics of the spatial indexing.

These design goals should be employed as a basis to create efficient and robust flash-aware spatial indices.

Goal 1 - Avoid excessive random reads in frequent locations. There is a common assumption that the random read is the fastest operation of SSDs. However, because of the reliability management on reads and resource conflicts [7], reads can become the worst operation of SSDs if they are performed always in the same location or interleaved with writes [3]. To achieve Goal 1, a flash-aware spatial index should use a buffer for reads in the main memory, called *read buffer*.

Goal 2 - Prioritize sequential writes instead of random writes. Random writes are the most expensive operations of SSDs and can lead to erase-before-updates [3,7]. To achieve Goal 2, a flash-aware spatial index should employ a buffer in the main memory to store the most recent modifications of the index, called *write buffer*. Further, flushing operations should be made by applying in batch a set of modifications on SSDs.

Goal 3 - Avoid interleaved reads and writes. The performance of SSDs can be negatively affected by interleaved reads and writes. This is due to the internal buffer and parallelism of SSDs, which may require that a read operation must wait for a previous write request that is still internally buffered [3,7]. To achieve Goal 3, a flash-aware spatial index should use read and write buffers together with a *temporal control of reads and writes* to take into account the order of these reads and writes in future operations.

Goal 4 - Choose dynamically modifications to be flushed. A flushing operation that writes all the modifications in the write buffer is inefficient since it leads to a big write to the SSD, degenerating its performance. Further, it writes nodes that would be potentially modified soon [10]. To achieve Goal 4, a flash-aware spatial index should use a *flushing unit creator* that picks a set of modified nodes, according to a *flushing policy*, to be written in batch to the SSD. The flushing operation should also consider Goals 2 and 3.

Goal 5 - Provide data durability. System crashes and power failures impact the write buffer since buffered modifications would be lost. To achieve Goal 5, a flash-aware spatial index should use a *log-structured approach* that sequentially writes to the SSD the modifications stored in the write buffer. Hence, it is possible to rebuild the write buffer after a system crash.

3   Related Work

There are few flash-aware spatial indices that have been proposed in the literature. The first indices, i.e., the *RFTL* [11] and the *LCR-tree* [8], are straightforward extensions of the R-tree. They do not change the structure of the R-tree and only employ a write buffer tò deal with the well-known poor performance of random writes of SSDs.

Subsequently, *FAST* [10] has been proposed to generalize the write buffer for any hierarchical index, and to guarantee data durability. It also provides a flushing algorithm that writes a flushing unit chosen from a flushing policy. But, FAST can write a flushing unit containing a node without modification and thus, leading to unnecessary writes to the SSD. This is due to the static creation of flushing units as soon as nodes are created in the index.

Posteriorly, the *FOR-tree* [6] was introduced. It improves the flushing algorithm by dynamically creating flushing units considering only the modifications stored in the write buffer. Further, it allows overflowed nodes by eliminating splits in the R-tree. According to a read counter, a merge back operation groups the overflowed nodes to be stored in the next level of the tree, growing up the tree if needed. However, the counter for the root node is never incremented in an index construction. As a consequence, the spatial objects are stored in this node with overflowed nodes in a sequential form instead of a hierarchical form.

Table 1 compares the existing flash-aware spatial indices taking as a basis our design goals. In this table, we note that there is a lack of indices aimed to improve the performance of reads, and to avoid interleaved reads and writes.

On the other hand, eFIND fulfills all the design goals. As a result, it improves the performance of index construction and spatial query processing on SSDs, as reported in Sect. 5.

Table 1. Comparison of flash-aware spatial indices according to our proposed design goals (Sect. 2).

		RFTL	LCR-tree	FAST	FOR-tree	eFIND
(1)	Read buffer					✓
(2)	Write buffer	✓	✓	✓	✓	✓
(3)	Temporal control of reads/writes					✓
(4)	Flushing units and flushing policy			✓	✓	✓
(5)	Data durability			✓		✓

4   The Efficient Framework for Spatial Indexing on SSDs

This section details eFIND, a generic framework that transforms any hierarchical spatial index to an efficient flash-aware spatial index by following the design goals of Sect. 2. Figure 1 depicts its architecture, which includes three components: (i) *Buffer Manager*, (ii) *Flushing Manager*, and (iii) *Log Manager*. These components are detailed as follows.

Buffer Manager. This component contains two in-memory buffers: (i) a *read buffer*, which stores frequently accessed nodes (Goal 1); and (ii) a *write buffer*, which stores the modifications of nodes generated from *insertions*, *updates*, and *deletions* over the spatial index (Goal 2).

Flushing Manager. This component contains three interacting elements: (i) *flushing unit creator*, which creates flushing units by grouping sequential

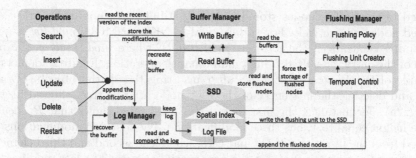

Fig. 1. The architecture of eFIND.

nodes; (ii) *flushing policy*, which ranks the flushing units according to the quantity of modifications, recency of modifications, and structure of the index (Goal 4); and (iii) *temporal control*, which stores frequently accessed nodes to the read buffer and avoids interleaved reads and writes (Goal 3).

Log Manager. This component is responsible to keep log of all writes performed on the SSD, thus providing data durability (Goal 5). Modifications lost after a system crash can be recovered by dispatching a *restart operation*.

The three components interact in a flushing operation as follows. When the write buffer is full, the Flushing Manager performs this operation by writing only some modifications to the SSD, while the Log Manager keeps log of the flushed nodes. Thus, the spatial index can be partially stored in three different locations: (i) the SSD, (ii) the read buffer, and (iii) the write buffer. For *searching purposes*, a specific algorithm that considers all these locations is applied to retrieve nodes from the index (see Algorithm 2 in Sect. 4.1).

4.1   Maintenance Operations

Here, we detail how the eFIND's components are employed to perform maintenance operations, which insert, update, or delete objects in the spatial index.

Data Structures. We use Fig. 2 to exemplify the use of the data structures on an R-tree index (Fig. 2a). The temporal control employs two arrays, named TCR and TCW, to store the node identifiers of the last performed reads and writes, respectively. Figure 2b shows that the last read nodes are R, B, B, and C, and the last written nodes are A, B, C, and R.

The write buffer stores, in a hash table named *Write Buffer Table*, the modifications of nodes that were not applied to the SSD yet. The key of this hash table is the unique node identifier ($node\_id$) and its value stores modifications in the format (h, mod_count, $timestamp$, $status$, mod_list). Here, h refers to the height of the modified node; mod_count is the quantity of in-memory modifications; $timestamp$ informs when the last modification was made; $status$ is the type of modification that can be NEW, MOD, or DEL to represent nodes newly created in the buffer, nodes stored in the SSD but with modified entries, and deleted nodes, respectively. For *status* equal to NEW or MOD, *mod_list* is a list containing elements in the form (*e*, *mod_result*), where *mod_result* is the new value of the entry *e* resulted from a maintenance operation. Figure 2c depicts an example of the *Write Buffer Table* after inserting the spatial objects *O9* and *O10*, and deleting the last entry of the root node in an R-tree (showed in gray in Fig. 2a). For instance, the first line of this hash table shows that *R*, located in the height equal to 1, has three modifications. The first two modifications are derived from the insertions of *O9* and *O10*, where the second and third elements of *R* were adjusted to encompass the new spatial objects. The last modification is related to the deletion of the last entry of *R*, which was performed in the time equal to 43460. Note that the same format of an entry of the underlying index of eFIND is used for each value of *mod_result*. In Fig. 2c, *MBR* considers the current modifications to calculate a minimum bounding rectangle (MBR). For instance, (*B*, *MBR(B)*) corresponds to an entry of *R* taking into account current modifications of *B* in the write buffer.

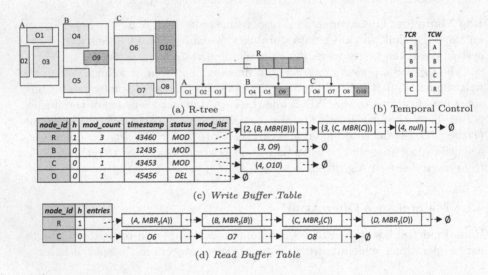

Fig. 2. Graphical representation of the eFIND's data structures using an R-tree index.

The read buffer maintains, in another hash table named *Read Buffer Table*, the nodes that are already stored in the SSD by prioritizing nodes located in the highest levels of the index and frequently accessed by searching routines. The key of this hash table is the unique node identifier (*node_id*) and its value stores (*h*, *entries*), where *h* is its height and *entries* is a list of node entries of the underlying index. Figure 2d depicts that *R* and *C* are stored in the *Read Buffer Table*. The entries of *R* reflect what is stored in the SSD and thus, MBR_S considers only the stored MBR. That is, the modifications of *B* and *C*, and the remotion of *D* are not taken into account. Further, the node *C* is cached since it is the last accessed node before inserting the new objects into the index.

Algorithm 1. Execution of a maintenance operation OP using eFIND

---

1 $N \leftarrow$ a list of modified nodes from the in-memory execution of OP;
2 **foreach** N_i **in** N **do**
3      $HashWEntry \leftarrow$ the entry of node N_i in the $WriteBufferTable$;
4   **if** $HashWEntry$ *is not NULL* **then**
5   **if** N_i *is a deleted node* **then**
6   free all modifications contained in the $HashWEntry$;
7   set the *status* of the $HashWEntry$ to DEL;
8   **else**
9   append the changes of N_i to the *mod_list* of $HashWEntry$;
10   **else**
11         $HashWEntry \leftarrow$ a new entry in the $WriteBufferTable$ with key $= N_i$;
12   **if** N_i *is a newly created node* **then**
13   set the *status* of the $HashWEntry$ to NEW;
14   **else**
15   set the *status* of the $HashWEntry$ to MOD;
16   append the changes of N_i to the *mod_list* of $HashWEntry$;
17   call *Log Manager* to append the changes of N_i to the log file;
18 **if** $WriteBufferTable$ *is full* **then**
19   call *Flushing Manager* to execute a flushing operation (Algorithm 3);

To guarantee data durability, eFIND also keeps the modifications in a log file. The format (*node_id, height, mod_count, timestamp, status, entry, mod_result*) is employed to store each modification in the log file. It has a very similar format to the format employed in the *Write Buffer Table*.

General Algorithm. Algorithm 1 shows how eFIND executes a maintenance operation. Since eFIND only changes the way in which reads and writes are performed on the SSD, the first step executes the corresponding maintenance operation as an *in-memory operation* (line 1). It returns a list of modifications made on the index, which can include the creation of new nodes resulted from splits, the deletion of underflowed nodes, and the adjustment of MBRs. After that, these modifications are stored in the *Write Buffer Table* (lines 2 to 16). If a modified node has an entry in this hash table, the corresponding entry is modified accordingly (lines 4 to 9). Otherwise, the algorithm creates a new hash entry (line 11), which can be a newly created node (line 13) or a node with in-memory modifications (line 15). Note that we call the Log Manager to keep log of all modifications of a node (line 17). This is a low latency operation since involves only sequential writes. The final step of the algorithm is the execution of a flushing operation, if the buffer is full (lines 18 and 19).

The execution of an in-memory maintenance operation needs a new algorithm for retrieving nodes. For instance, to choose a leaf node to accommodate or delete a spatial object. The reason is that a node can be cached in the *Read Buffer Table*

Algorithm 2. Retrieving a node N using eFIND

---

1   $HashWEntry \leftarrow$ the hash entry of N in the $WriteBufferTable$;
2   **if** $HashWEntry$ *has status equal to NEW or DEL* **then**
3 | return the node pointed by $HashWEntry$;

4   $R \leftarrow$ empty;
5   $HashREntry \leftarrow$ the hash entry of N in the $ReadBufferTable$;
6   **if** $HashREntry$ *is not NULL* **then**
7 | $R \leftarrow HashREntry$;
8 | update the recency of access of N in the $ReadBufferTable$;
9   **else**
10 | $R \leftarrow$ read N from the SSD;
11 | possibly store R in $ReadBufferTable$ according to a replacement policy;
12 | append to TCR that the node N was read from the SSD;

13 **if** $HashWEntry$ *has status equal to MOD* **then**
14 | unify the modifications contained in $HashWEntry$ to R;

15 return R;

---

and/or contain modifications in the *Write Buffer Table*. Algorithm 2 details how to retrieve a node considering all these aspects. If the node is a newly created node or a deleted node, the algorithm returns the node pointed by its corresponding entry in the *Write Buffer Table* (lines 2 and 3). Otherwise, it gets the last stored version of the node that is either buffered in the *Read Buffer Table* (lines 6 to 8) or stored in the SSD (lines 9 to 12). In the latter case, the node is read from the SSD and stored in the *Read Buffer Table* (lines 10 and 11) if it has available space, or if the height of the least recently accessed node is smaller than or equal to the height of the read node. Thus, the *Read Buffer Table* only stores nodes near to the root node, which are frequently accessed in the index. Further, the read operation is appended to TCR for the temporal control of reads performed on the SSD (line 12). The last step of the algorithm is to apply any change contained in the *Write Buffer Table* to the node (lines 13 and 14) to be returned (line 15). For instance, to retrieve the node C from the index of Fig. 2a, Algorithm 2 gets its cached version from the *Read Buffer Table* and then applies its modification contained in the *Write Buffer Table*, without requiring reads from the SSD.

Flushing Operation. Algorithm 3 details the flushing operation (executed at line 19 in Algorithm 1). Instead of writing all the modifications to the SSD, eFIND smartly selects only some of them to be written to the SSD. Firstly, the algorithm forms a list F composed of nodes with the oldest modifications contained in the *Write Buffer Table* according to the *timestamp* (line 1). By using a parameter value p, the size of F is the $p\%$ of the total number of modified nodes stored in the *Write Buffer Table*.

Algorithm 3. Execution of the eFIND's flushing operation

---

1   $F \leftarrow$ a list of the first $p\%$ oldest modified nodes of the *WriteBufferTable*;
2   sort in ascending order the list F by the node identifiers;
3   $FU \leftarrow$ a list of flushing units created from the list F;
4   **foreach** FU_i **in** FU **do**
5      $\left\lfloor FU_i.d \leftarrow \sum_{j=1}^{FU_i.n} FU_i[j].mod\_count * (FU_i[j].h + 1) \right.$

6   sort in descending order the list FU by the d value;
7   *chosenFU* $\leftarrow$ a flushing unit picked by the *Temporal Control*;
8   *nodesToFlush* $\leftarrow$ an empty list of nodes to be flushed;
9   **foreach** N_i **in** *chosenFU* **do**
10     $RN \leftarrow$ retrieve node N_i (Algorithm 2);
11   append RN to *nodesToFlush*;
12   **if** RN *is a frequently accessed node* **then**
13   store RN in the *ReadBufferTable*;
14   **else**
15   update RN in the *ReadBufferTable* if it is stored there;

16   write in batch *nodesToFlush* to the SSD;
17   call *Log Manager* to append the flushed nodes to the log file;
18   append to *TCW* that the nodes in *nodesToFlush* were written to the SSD;
19   free the modifications of the nodes in *nodesToFlush* from the *WriteBufferTable*;

Assuming a flushing unit size equal to s, the next s nodes of F, previously sorted by the node identifiers (line 2), define a flushing unit (line 3). Thus, flushing units are formed by sequential nodes. For each created flushing unit, the algorithm computes a degree (lines 4 and 5) that is used to sort the flushing units in descending order (line 6). The degree of a flushing unit is the sum of the number of modifications of each node times its height. Hence, the height works as a weight to give higher degrees for nodes located in the highest levels of the index, which are not so frequently modified. After this ranking, the temporal control uses TCW to choose a flushing unit that leads to a sequential write or possibly to a stride write considering previous performed writes (line 7).

Next, the nodes that compose the chosen flushing unit are retrieved (lines 8 to 11). Here, the temporal control plays an important role (line 12). By using TCR, the temporal control can store a node in the *Read Buffer Table* based on its read frequency (line 13). Thus, it avoids a possible read after a write operation. Otherwise, the algorithm updates the content of a node if necessary (line 15). It then sequentially writes the flushing unit to the SSD (line 16), which was built considering a balance between recency of the modifications in the write buffer, quantity of modifications, and the structure of the index. Finally, the algorithm keeps log of the flushed nodes, appends the made writes to TCW, and frees them from the *Write Buffer Table* (lines 17 to 19).

4.2 Search Operations

eFIND does not change the search algorithm of the underlying index. But, the algorithm should use Algorithm 2 to retrieve nodes. For instance, the classical recursive searching operation of the R-tree should use Algorithm 2 to retrieve a node when the entry that points to this node satisfies a determined topological relationship. When the node is a leaf node, the searching operation returns the most recent version of the entries to answer a spatial query.

4.3 System Restart

Since eFIND guarantees data durability, it permits to reconstruct the *Write Buffer Table* after a system crash, fatal error, or failure power. Thus, after a system restart, eFIND recovers all the modifications that were not effectively applied to the index. This is possible by reading the log file in reverse order since the modifications are written to the log as append-only operations (line 17 in Algorithm 1). Further, the log file possibly can store flushed nodes (line 17 in Algorithm 3). This means that modifications of these nodes prior to the flushing operation can be ignored since they were already written to the SSD. Optionally, after the log file reaches a specific size, eFIND may use a compactor to remove modifications already flushed to the SSD.

5 Experimental Evaluation

5.1 Experimental Setup

Dataset. We extracted a real spatial dataset from the OpenStreetMap[1], which consisted of 1,486,557 complex regions possibly with holes representing the buildings of Brazil like hospitals, universities, schools, houses, and stadiums.

Configurations. We compared the following configurations: (i) the *R-tree* with a standard LRU buffer of 512 KB caching nodes in the highest levels of the tree; (ii) the *FAST R-tree* with a buffer of 512 KB, log capacity of 10 MB, and FAST* flushing policy; and (iii) the *eFIND R-tree* with a buffer of 512 KB where 80% was dedicated for the write buffer and 20% for the read buffer, log capacity of 10 MB, flushing percentage (p) of 40%, and read frequency of 40% for the temporal control. We applied the quadratic split algorithm of the R-tree for all the configurations, and the best parameters values of the FAST R-tree according to [10]. We also employed the best parameters values of eFIND based on an experimental evaluation that studied the impact of the design goals on our proposed framework. We considered FAST as the state of art since, as reported in Sect. 3, it is the most complete related work. We did not compare the FOR-tree because it failed to construct indices over our dataset.

---

[1] http://www.openstreetmap.org/.

Varied Parameters. We employed page sizes (i.e., node sizes) from 2 KB to 16 KB for all the configurations. Further, we used flushing unit sizes from 1 to 5 for the FAST R-tree and the eFIND R-tree.

Workloads. We executed two workloads: (i) index construction, and (ii) processing of intersection range queries (IRQ) [4], considering three different sets of query windows. Each set was composed of 100 query windows corresponding respectively to 0.001%, 0.01%, and 0.1% of area of the total extent of Brazil. Considering that the selectivity of a query is the ratio between the number of returned objects and the total objects, these sets of query windows form spatial queries with low, medium, and high selectivity, respectively. For each configuration, we executed these workloads as a sequence, i.e., the index construction followed by the processing of the three sets of IRQs. Each sequence was executed 10 times. We flushed the system cache after the execution of each sequence and calculated the elapsed time as follows. For the first workload, we collected the average elapsed time. For the second workload, we collected the average elapsed time of the execution of each IRQ, and then calculated the sum of the average elapsed times of the 100 IRQs for each set of query windows. Hence, we analyzed the performance of each index in specific workloads.

Running Environment. We employed FESTIval [1], an open-source PostgreSQL extension to benchmark disk-based and flash-aware spatial indices[2]. We performed the tests locally to avoid network latency. The experiments were conducted on a local server equipped with an Intel® Core™ i7-4770 with frequency of 3.40 GHz, 32 GB of main memory, and an SSD 480 GB Kingston V300. We used Ubuntu Server 14.04 64 bits, PostgreSQL 9.5, and PostGIS 2.2.0.

5.2 Index Construction

Figure 3 depicts the obtained results for index construction. In this figure, the R-tree only has elapsed times for the flushing unit size equal to 1 in all the page sizes since it only writes one node by time to the SSD. Clearly, the eFIND R-tree greatly overcame its competitors for all the page sizes, followed by the R-tree, which in turn outperformed the FAST R-tree. Compared to the R-tree, the eFIND R-tree introduced performance gains varying from 22% to 43%. If compared to the FAST R-tree, the performance gains of the eFIND R-tree were yet more expressive, ranging from 62% to 68%.

The eFIND R-tree exploited the benefits of SSDs because eFIND takes into account the intrinsic characteristics of these memories. For instance, compared to its closest competitor, the R-tree, the eFIND R-tree decreased the number of writes by up to 99% and the number of reads by up to 96%. The contribution of the read buffer to obtain these results was very important. Even using a relative small percentage of the buffer size dedicated for the read buffer, eFIND provided an expressive reduction in the number of reads without impairing the number of writes. Further, the use of the temporal control guaranteed that frequently

---

[2] http://gbd.dc.ufscar.br/festival/.

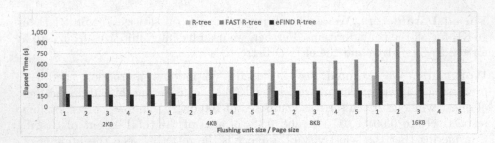

Fig. 3. Performance results to construct spatial indices on the SSD.

accessed nodes were stored beforehand in the read buffer. Moreover, eFIND did not show significant overhead in flushing operations.

Regarding the FAST R-tree, it reduced the number of writes by up to 99% compared to the R-tree, but at the same time increased the number of reads from 9% to 123%. This means that the FAST R-tree more than doubled the number of reads in some situations. Hence, the reduction of the number of writes was important but not sufficient to achieve good performance on SSDs.

In general, the construction of the indices by using the page sizes equal to 2 KB and 4 KB required less time than the construction by using other page sizes. The page size equal to 16 KB provided the worst results since the indices had to process several entries in the main memory. For the FAST R-tree and the eFIND R-tree, the flushing unit size equal to 2 provided the best performance results. This indicates that writing two nodes achieved a good granularity for writes.

5.3 Spatial Query Processing

Figure 4 depicts the obtained results for processing of IRQs. Considering the best performance results, the eFIND R-tree overcame its competitors, followed by the R-tree, which in turn outperformed the FAST R-tree in most cases.

Regarding query windows with 0.001% (Fig. 4a), all the configurations showed faster elapsed times using page sizes equal to 2 KB and 4 KB. The main reason is that these IRQs have low selectivity, where each IRQ returned up to 82 objects from the dataset. Thus, smaller page sizes required the processing of less entries in the main memory, reducing the processing time. For the page size equal to 4 KB, the eFIND R-tree provided performance gains of 32% over the R-tree and of 50% over the FAST R-tree. Here, the read buffer of eFIND improved the retrieval of nodes from the index. Comparing the R-tree to the FAST R-tree, while the former avoided some reads by accessing nodes from its LRU buffer, the latter had to read the majority of the accessed pages from the SSD. This impaired the performance of the FAST R-tree.

With respect to query windows with 0.01% (Fig. 4b), the page size equal to 16 KB showed the best performance results for the majority of the indices, although these results were very close to the page size equal to 8 KB.

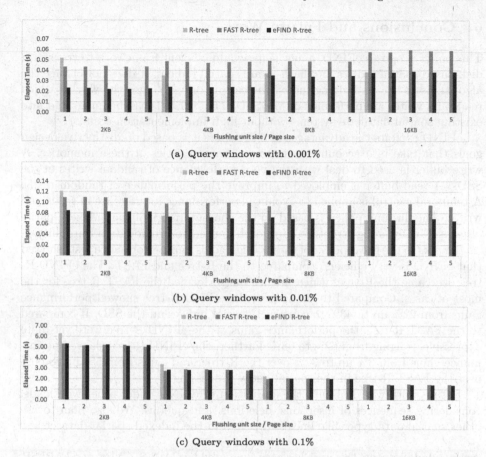

(a) Query windows with 0.001%

(b) Query windows with 0.01%

(c) Query windows with 0.1%

Fig. 4. Performance results to process IRQs on the SSD using query windows with 0.001% (a), 0.01% (b), and 0.1% (c) of area of the total extent of Brazil.

Since these IRQs have a medium selectivity, returning up to 11,771 objects, the use of smaller page sizes decreased the performance by requiring more accesses to entries. For the page size equal to 16 KB, the eFIND R-tree showed reductions of 3% compared to the R-tree and of 29% compared to the FAST R-tree.

Regarding query windows with 0.1% (Fig. 4c), the largest employed page size, i.e. 16 KB, showed the fastest elapsed times for all the configurations. The high selectivity of these IRQs, which returned up to 515,117 objects, required the traversal of several nodes of the index to find all the objects that answer the spatial query. Due to this, the page size containing the greatest number of elements improved the performance of these spatial queries. For the page size equal to 16 KB, the eFIND R-tree obtained performance gains of 9% over the R-tree and of 5% over the FAST R-tree.

6   Conclusions and Future Work

This paper proposes eFIND, a new generic and efficient framework to transform disk-based spatial indices into flash-aware spatial indices. It does not change the algorithms of the underlying index. Instead, it only changes the way in which reads and writes are performed on the SSD. Thus, eFIND can be integrated into existing spatial database systems without high implementation costs.

eFIND exploits the advantages of SSDs since it is based on distinctive design goals that take into account the intrinsic characteristics of these memories. A write buffer is used to deal with the poor performance of random writes of the SSDs. A read buffer is employed to improve the performance of random reads. A temporal control is applied to avoid the performance interference of reads and writes. The flushing policy is based on the characteristics of the indexed spatial objects. Finally, a log-structured approach is used to provide data durability.

Thanks to the great power of these design goals, our performance tests showed that eFIND is very efficient, compared to the R-tree and FAST. The eFIND R-tree overcame the R-tree, which in turn outperformed the FAST R-tree for the most of cases. Compared to the R-tree, the eFIND R-tree showed performance gains from 22% up to 43% to construct spatial indices on the SSD. If compared to the FAST R-tree, the performance gains of the eFIND R-tree were yet more expressive, ranging from 62% to 68%. Further, the eFIND R-tree showed performance gains from 3% up to 32% over the R-tree, and gains from 5% up to 50% over the FAST R-tree in the spatial query processing.

Future work will deal with the evaluation of eFIND under other workloads, considering sequences of insertions, deletions, and queries. We also plan to use other spatial data types like lines and points in the indexed spatial dataset.

Acknowledgments. This work has been supported by CAPES, CNPq, and FAPESP. A. C. Carniel has been supported by the grant #2015/26687-8, FAPESP. R. R. Ciferri has been supported by the grant #311868/2015-0, CNPq.

References

1. Carniel, A.C., Ciferri, R.R., Ciferri, C.D.A.: Experimental evaluation of spatial indices with FESTIval. In: Brazilian Symposium on Databases - Demonstration Track, pp. 123–128 (2016)
2. Carniel, A.C., Ciferri, R.R., Ciferri, C.D.A.: The performance relation of spatial indexing on hard disk drives and solid state drives. In: Brazilian Symposium on GeoInformatics, pp. 263–274 (2016)
3. Chen, F., Koufaty, D.A., Zhang, X.: Understanding intrinsic characteristics and system implications of flash memory based solid state drives. In: International Conference on Measurement and Modeling of Computer Systems, pp. 181–192 (2009)
4. Gaede, V., Günther, O.: Multidimensional access methods. ACM Comput. Surv. **30**(2), 170–231 (1998)
5. Guttman, A.: R-trees: A dynamic index structure for spatial searching. In: ACM SIGMOD International Conference on Management of Data, pp. 47–57 (1984)

6. Jin, P., Xie, X., Wang, N., Yue, L.: Optimizing R-tree for flash memory. Expert Syst. Appl. **42**(10), 4676–4686 (2015)
7. Jung, M., Kandemir, M.: Revisiting widely held SSD expectations and rethinking system-level implications. In: International Conference on Measurement and Modeling of Computer Systems, pp. 203–216 (2013)
8. Lv, Y., Li, J., Cui, B., Chen, X.: Log-compact R-tree: an efficient spatial index for SSD. In: International Conference on Database Systems for Advanced Applications, pp. 202–213 (2011)
9. Mittal, S., Vetter, J.S.: A survey of software techniques for using non-volatile memories for storage and main memory systems. IEEE Trans. Parallel Distrib. Syst. **27**(5), 1537–1550 (2016)
10. Sarwat, M., Mokbel, M.F., Zhou, X., Nath, S.: Generic and efficient framework for search trees on flash memory storage systems. GeoInformatica **17**(3), 417–448 (2013)
11. Wu, C.H., Chang, L.P., Kuo, T.W.: An efficient R-tree implementation over flash-memory storage systems. In: ACM SIGSPATIAL International Conference on Advances in Geographic Information Systems, pp. 17–24 (2003)

Parallel and Distributed Data Processing

Incremental Frequent Itemsets Mining
with MapReduce

Kirill Kandalov[1]([✉]) and Ehud Gudes[1,2]

[1] Open University, Ra'anana, Israel
kirill.kandalov@gmail.com
[2] Ben-Gurion University, Beer-Sheva, Israel
ehud@cs.bgu.ac.il

Abstract. Frequent itemsets mining is a common task in data mining. Since sizes of today's databases go far beyond capabilities of a single machine, recent studies show how to adopt classical algorithms for frequent itemsets mining for parallel frameworks such as MapReduce. Even then, in case of a slight database update a re-run of the MapReduce mining algorithm from the beginning on the whole data set is required and could be far from optimal. Thus, a variation of these algorithms for incremental database update is desired.

The current paper presents a general algorithm for incremental frequent itemsets mining and shows how to adapt it to the parallel paradigm. It also provides optimizations that are unique to a constrained model of MapReduce for an effective algorithm.

1 Introduction

The amount of information generated in our world has grown in the last few decades at an exponential rate. This process resulted with a new term called "Big Data". Classical databases (DB) are unable to handle such size and velocity of data. So, special tools were developed for this task. One of them is the MapReduce (MR) framework [11]. It was originally developed by Google, but currently, the most researched version is an open source project called Hadoop [12]. MR provides parallel distributed model and framework that scales to thousands of machines.

Frequent Itemsets (FI) Mining (FIM) is the most computational intensive part of association rules mining [3, 5, 13]. Solving FIM efficiently allows efficiently finding the association rules. The association rules mining has been heavily researched and few solutions were proposed for running classical FIM algorithms in the MR framework [15, 16, 20, 21]. These algorithms find frequent itemsets for a given static database. But in the world of constantly aggregated data, databases are dynamic. So, there's a need for an algorithm that will be able to update the FI effectively when the database is updated, instead of rerunning the full FIM algorithm on the whole DB.

There exist incremental versions of FIM algorithm [9, 19]. Some of these algorithms even suit a distributed environment [10] but not for the MR model. Because the MR model is more limited than general distributed or parallel computation models, these algorithms cannot be used as is. So, the algorithm should be carefully designed to

© Springer International Publishing AG 2017
M. Kirikova et al. (Eds.): ADBIS 2017, LNCS 10509, pp. 247–261, 2017.
DOI: 10.1007/978-3-319-66917-5_17

be efficient. In this paper, we investigate and develop algorithms for incremental FIM for the MR model. Specifically, our contributions are:

1. Defining a general scheme for incremental FIM which is agnostic to the underlined FIM algorithm and is suitable for the MR framework.
2. Evaluation of the general incremental scheme with one of the best known so far FIM algorithm for MR.
3. Usage of additional optimizations for general distributed DB.
4. Optimizing the algorithm to overcome MR weak points.

This paper is structured as follows. Section 2 discusses the background and related work. Section 3 defines the problem. Section 4 presents several algorithms and optimizations to solve it. Section 5 shows the experimental evaluation of our algorithms and comparison to a full scan algorithm and we conclude in Sect. 6.

2 Background and Related Work

2.1 Association Rules, Frequent Itemsets and Incremental Update

Association Rule mining was introduced in [3, 5] as a market basket analysis. Finding items that were bought together - if a customer bought item X, then he has a high probability to buy also item Y ($X \Rightarrow Y$). A pre-requisite to finding association rules is the mining of frequent itemsets, itemsets that appear in at least some percent of transactions.

One of the most well-known algorithms for association rules is the Apriori algorithm [3, 5]. This algorithm uses a pruning rule called Apriori - an itemset may be frequent iff all its subsets are also frequent. Apriori algorithm iteratively generates candidates for frequent itemsets and prunes them. Once FI are found it generates all association rules.

There exist several versions of implementing Apriori in the distributed environment [4, 22]. The most important idea is that if an itemset is frequent in a union of distributed databases, it must be frequent in at least one of them.

The idea not to generate candidates but calculate frequent itemsets directly was used in FP-growth [23]. That algorithm uses the original DB to build a compact FP-Tree. This tree is used to find all frequent itemset. The algorithm is reported to be faster than Apriori based algorithms. In our tests in the cloud the algorithm wasn't significant faster than Apriori based. But the higher memory requirements limited the use of many types of machines in the cloud. As this work isn't about comparison between different FIM algorithms we chose to use Apriori based algorithm for our tests (see IMRApriori in Sect. 2.3).

The idea of maintenance of association rules and frequent itemsets during database update has been discussed shortly after the first algorithms for FIM appeared. The reason for it is that the updated part is usually much smaller than the full DB and this fact could be used for faster algorithms. A well-known efficient algorithm for it is called "Fast Update" (FUP) [9]. It is based on the fact that for an item to be frequent in the updated

Table 1. Cases for item to be frequent and the outcome

	Frequent in DB	Not frequent in DB
Frequent in deltaDB	Frequent in DB+	Unknown
Not frequent in deltaDB	Unknown	Not frequent in DB+

database (DB+) it must be frequent in the DB or the new added transactions (deltaDB). Table 1 describes the options for an itemset to become frequent in DB+.

FUP is working iteratively by mining only the new delta DB by a method similar to Apriori. At the end of each iteration, the algorithm decides if the itemset is frequent, not frequent or needs to be counted in the old DB. The last kind of transactions being recounted in old DB. The survivors are used to create candidates for next iteration.

ARMIDB [13] is based on FUP but tries to minimize scans over the original DB. It uses data from the original FI, updates them and then uses a technique called "Look Ahead for Promising Items" (LAPI). Then it scans for candidates in deltaDB only for those items that may be frequent in DB+. It's not a distributed algorithm like ours.

2.2   MapReduce Model and Incremental Computation

MapReduce is a parallel model and framework introduced in [10]. The abstract model requires defining of two functions (algorithms):

Map	(k1,v1)	→	list(k2,v2)
reduce	(k2,list(v2))	→	list(v2)

The framework takes care of everything else - reading input, splitting it to distributed nodes, running the map function tasks, sending results (partitioning) to the reduce function tasks and stores the final results from the reducers. Each Map and Reduce combination called a Job. Some algorithms may require multiple consecutive MR jobs. E.g. Apriori may require K jobs. Data in MR is saved in a distributed file system (HDFS in Hadoop) as blocks of constant size with common sizes of 32, 64 and 128 MB.

In MapReduce, each chunk of split input is called simply a "Split". The standard way of MapReduce to split the input is by using a constant size of HDFS blocks or its multiplier. Some of the MR steps could be customized to support advanced scenarios (e.g. instead of reading input line by line, the whole file could be read as a single chunk).

Most of the time, when data is changed or added the result of the algorithm also changes. MR doesn't provide built-in tool in the model that supports an update of the result. There're researches which attempt to enhance the MR model to support this.

One of the attempts is the Incoop system [8]. This paper proposes a way, that is almost transparent to the user of MR, to keep the results of the algorithm updated as new data is added. The system treats the computations as a Directed Acyclic Graph (DAG) of data that flows from input to output and on the way, is transformed by user functions. When data is updated the system reruns only the part of the graph that has some new input. This system uses a Memoization technique to keep the data-algorithm-result dependencies. This works well for the Mapper (map job) as few data records mostly affect small number of Mappers. In case a new key-value pair is generated for the Reducer, it will need to rerun its function on the whole input (new and old values).

A similar approach can be found in DryadInc [18]. The main difference with MR is that Dryad allows any DAG of computations and not only Map and Reduce. In the incremental version, there's also a Cache server that keeps input/output relations and result is being reused if it wasn't changed from a previous run.

2.3   Apriori MapReduce Algorithm

There are several known MapReduce Apriori algorithms. PApriori algorithm [16] uses classical algorithm inside the main program in a sequential way except for the frequency count, which is done in parallel with the MR algorithm. Apriori-Map/Reduce [20] is similar to PApriori but also does the candidate generation in parallel by using a MR job. Both algorithms require K steps to find the frequent itemsets of length K.

MRApriori [21] is different and presents a two steps algorithm. The first step is to run Apriori inside each Mapper to find "locally" frequent itemsets. Reducer joins the results of all Mappers and they become candidates for global frequent itemsets set. The second job/step is just counting of all candidates' appearance in each split and filtering only the frequent itemsets that pass the minimum support.

IMRApriori [15] works similar to MRApriori with one add-on/observation that an itemset may become candidate itemset only if it appeared locally frequent in "enough" splits. More precisely, let $S_1, \ldots, S_m$ be splits of DB in step 1. Denote their sizes to be $|S_i|$. If itemset X is locally frequent in k (k $\leq$ m) splits. Without loss of generality, let's say $S_1, \ldots, S_k$. Let Ci be the count of occurrence of X in split Si. Let minsup be the minimum support. Then:

$$Count(X, DB) \leq \sum_{1}^{k} C_i + \sum_{k+1}^{m}(|S_i| * minsup - 1) \tag{1}$$

If this number is less than minsup*|DB| then there is no need to calculate X's occurrences in DB in step 2 as it doesn't have a chance to be frequent anymore (early pruning). This observation is applied in the first Reducer of IMRApriori. This algorithm is shown to outperform all previous ones [15] and a variation of it will be used by us.

2.4   Join Operation and MapReduce

The Join operation of two or more datasets is one of the standard operations in relational DB. The operation combines records from input datasets by some rule (predicate). Several Join algorithms for MR were proposed [1, 2, 7]. One of them is a repartition join: data from all datasets sent to Mappers, that identify which dataset is used and outputs the input together with a dataset's "tag". The key is the predicate value (in equi-join it's the value for the join). Reducer groups records by the input key, extracts dataset tags and generates the combinations. It is used in our algorithm and therefore relevant here.

3   Problem Definition

Following is a formal definition of frequent itemsets mining (similar to [1, 2]): Let $I = \{i_1, i_2, ..., i_m\}$ be set of all unique items. Transaction T is a subset of items I ($T{\subseteq}I$). It is said to contain itemset X, if $X{\subseteq}T$. Database DB is a set of transactions. Number of occurrences of X in DB is defined as: $Count(X,DB) = |\{T|T{\in}DB, X{\in}T\}|$.

Support of itemset X is s and it's a fraction of transactions in DB that contain X:

$$s = support(X, DB) = Count(X, DB)/|DB|$$

Frequent itemsets is a set of all itemsets X such that support(X,DB) $\geq$ minsup. We'll denote this set as FI. We'll also denote frequent itemsets of specific DB with minimum support minsup as FI(DB,minsup). We may omit minsup from the function if the support is clear from the context. We'll call $X \in FI$ frequent itemset or just "frequent". The process of finding the FI set is called Frequent Itemsets Mining (FIM).

Definition of incremental frequent itemsets mining: Let DB be a database, PK be some previous knowledge that we acquired during the FIM over DB, deltaDB be the set of additional transactions and DB+ be a new database defined as DB+ = DB $\cup$ deltaDB. Given also minsup and FI(DB,minsup) the problem is to find FI (DB+, minsup). We'll refer to FI(deltaDB) as deltaFI.

4   Algorithms

4.1   General Scheme

We first propose a general algorithm (Algorithm 1) for incremental frequent itemsets mining. It can be used for any distributed or parallel framework, but it also suits the model of MapReduce. The algorithm is loosely based on the FUP [9] algorithm and shares similarity to ARMIDB [13] (see Sect. 2.1). The idea is first to find all frequent itemsets in the new database FI(deltaDB,minsup), unite (join) the new frequent itemsets with the old frequent itemsets and, lastly, revalidate itemsets which are in the "unknown" state in each database. This algorithm is general as it doesn't set any constraints on the FIM algorithm. We can use Apriori or non Apriori based algorithm (e.g. FP-growth) for FIM, as long as these algorithms generates full and correct FI.

Brief description of the algorithm: Steps 1–2 find frequent itemsets of deltaDB. Steps 3 just checks if this is an incremental run or not. If it is an incremental run, we proceed mining FI(DB + ,minsup). As mentioned in Sect. 2.1, adding new transactions may generate locally frequent itemsets that have 3 options. To determine to which option each itemset applies we propose using MR Join job (steps 4–6). The key of the "Join"'s reducer would be the itemset itself and the list of values would be the occurrences of the itemset in the different DB parts together with their count. Step 5 determines the further processing for each itemset and is composed of three cases:

1. Frequent in deltaDB and DB - it's frequent in DB+ and outputted immediately (6).
2. Frequent only in deltaDB - we need to count it in DB (7–8 by using count MR job).
3. Frequent only in DB - we need to count it in deltaDB (9–10 by using same MR job).

All 3 outputs of steps 6, 8 and 10 represent together FI(DB + ,minsup).

The scheme contains 3 different kinds of MR jobs: A. FIM of deltaFI by using any MR algorithm for it. B. Join MR job. Any join algorithm may be used (our evaluation used Repartition Join). The only modification is that the Reducer should have 3 output files (instead of just one). C. Count of itemsets inside the database. It may be applied twice, once for DB and once for deltaDB.

General Incremental Frequent Itemsets Mining

   Input: minsup, deltaDB, DB, FI(DB), Optional: Previous knowledge (PK).
1: Mine Frequent Itemsets with MapReduce **Job** on deltaDB with minsup
2: Save frequent itemset for deltaDB as deltaFI
3: **if** no (old) DB exists **then** report deltaFI as total frequent itemsets and **end**.
4: Join **Job**: Join Mapper loads all frequent itemsets and send them to reducer
5: Join Reducer joins itemset I from FI(DB) and deltaFI and categories I into
   one of 3 sets: Appears in both FIs, appears only in deltaFI and only in FI(DB)
6: All itemsets that appear in FI of both DBs are outputted as frequent itemsets
7: Count **Job**: Itemsets that are frequent only in deltaDB are counted in old DB
8: Count   Reducer: outputs   only   itemset   I   that   satisfies
   Count(I,DB)+Count(I,deltaDB)≥minsup
9: Count **Job**: Itemsets that're frequent only in DB are counted in deltaDB
10: Count   Reducer: outputs   only   itemset   I   that   satisfies
   Count(I,DB)+Count(I,deltaDB)≥minsup
11: **end** /* All outputs together generate FIM(DB+) */

Algorithm 1: General Incremental Frequent Itemsets Mining

There's at most one pass over old DB for counting. There's no requirement for any additional input for general FIM (i.e. Previous Knowledge, PK).

4.2   Early Pruning Optimizations

In Algorithm 1 only one step requires accessing the old DB, whose size may be huge compared to deltaDB. This is the step of recounting local FI(deltaDB) which didn't appear in FI(DB). To minimize access to the old DB we suggest using early pruning techniques which consider the relation between the old DB size and the deltaDB size. These are additions to the early pruning of IMRApriori technique but are not unique for MR algorithms and could be used in every incremental FIM algorithm:

Let inc be size of deltaDB relatively to DB size and n the size of DB (n = |DB|) then the size of deltaDB is inc*n or inc*|DB|.

$$I\varepsilon FI(DB+,minsup) \to Count(I,DB) + Count(I,deltaDB) = Count(I,DB+)$$
$$\geq minsup|DB+| = minsup(|DB| + inc * |DB|) \qquad (2)$$

Lemma 1 (Absolute Count):

$$I\varepsilon FI(DB, minsup(1+inc)) \to I\varepsilon FI(DB+, minsup) \qquad (3)$$

Proof of Lemma 1:

$$I\varepsilon FI(DB, minsup * (1+inc)) \to$$
$$Count(I,DB+) \geq Count(I,DB) \geq minsup * (1+inc)|DB| = minsup * |DB+|$$

I.e. $Count(I,DB) \geq minsup(1+inc)n \to I\varepsilon FI(DB+,minsup)$;

Lemma 1 ensures that if I is "very" frequent in old DB (support is at least minsup * (1 + inc)) then it will be frequent in DB+ even if I didn't appear in deltaDB at all.

Lemma 2 (Minimum Count):

$$I\varepsilon FI(DB+,minsup) \to Count(I,DB) \geq n(minsup + minsup * inc - inc) \qquad (4)$$

Proof of Lemma 2:

$$Count(I,deltaDB) \leq |deltaDB| = inc * n;$$
$$|deltaDB| + Count(I,DB) \geq Count(I,deltaDB) + Count(I,DB)$$
$$\geq minsup * |DB+| = minsup * (1+inc) * n;$$

$$Count(I,DB) \geq minsup * (1+inc) * n - |deltaDB|$$
$$= minsup * (1+inc) * n - inc * n$$
$$= n * (minsup + minsup * inc - inc);$$

For answering if I can be in FI(DB + ,minsup) without even looking at deltaDB we need to know if IεFI(DB,minsup + minsup*inc - inc). Lemma 2 puts a lower bound of occurrences of I in old DB, so it'd still have a potential to appear in FI(DB+).

Lemma 3:

$$Count(I,deltaDB) \geq minsup * (1+inc) * n \to I\varepsilon FI(DB+,minsup); \qquad (5)$$

Proof of Lemma 3: Similar to Lemma 1 conclusion:

$$Count(I, DB+) = Count(I, DB) + Count(I, deltaDB) \geq Count(I, deltaDB)$$
$$\geq minsup * (1 + inc) * n = minsup * |DB + |;$$

Lemma 3 tells us that if I is "very" frequent in deltaDB (deltaDB is large enough or minsup is small enough), then I will appear in FI of DB+ no matter what. So, it's a pruning condition. If itemset I satisfies it then there's no need to count I in old DB.

Observation from Lemma 3 (Absolute Count Delta):

$$I \varepsilon FI(deltaDB, minsup * (1 + 1/inc)) \rightarrow I \varepsilon FI(DB+, minsup); \qquad (6)$$

Proof:

$$I \varepsilon FI(deltaDB, minsup * (1 + 1/inc)) \rightarrow Count(I, deltaDB)$$
$$\geq minsup * (1 + 1/inc) * |deltaDB|$$
$$= minsup * (1 + 1/inc) * inc * n = minsup(1 + inc) * n;$$

To use the above lemmas in our algorithm we modify the FIM algorithm to keep the itemset together with its potential "minimum count" and "maximal count" (for each Split). If there's no information from some split about some itemset I, we're using observation from IMRApriori and set "maximal count" to be Ceil(|Split Size|* minsup)−1 and "minimum count" set to 0. This is done in the Reducer of Stage 1 of IMRApriori. Let $\chi_i(I)$ be an indicator function that is 1 iff I was locally frequent in split S_i. The Reducer would output a triple <I, mincount, maxcount>:

$$mincount = \sum_{i=1}^{|Splits|} \chi_i(I) * Count(I, S_i) \qquad (7)$$

$$maxcount = \sum_{i=1}^{|Splits|} \chi_i(I) * Count(I, S_i) + (1 - \chi_i(I)) * (\lceil |S_i| * minsup \rceil - 1) \qquad (8)$$

Note that when exact count is known then mincount equals to maxcount. If maxcount < minsup*|DB| then I is pruned. If mincount ≥ minsup*DB then I is globally frequent and should not need to be recounted in missed splits. Lemmas are applied in the algorithm during the Join step. We know the sizes of DB and deltaDB (n and inc). So, we compare potential counts of itemset directly to sizes of databases. Algorithm 2 determines the total potential counts and takes the corresponding decision.

Split size information and total DB size is being passed as "previous knowledge" (PK) input into FIM incremental algorithm in "General Scheme", Algorithm 1.

By using early pruning optimization we're able to reduce the number of candidates. Which reduces the output of MR and saves CPU in future jobs that otherwise would require counting of the non-potential "candidates". The above optimization is valid for any distributed framework. Next section shows optimizations specific to MapReduce.

4.3   Optimizing Algorithm for MapReduce

There are few known drawbacks of the MapReduce framework [12, 14] that can harm the performance of any algorithm. We'll concentrate on the overhead of establishing a new computational job - creation of physical process for Mapper and IO. Our performance evaluation (see Sect. 5) of the General Scheme, showed that CPU time of the algorithms is lower compared to full process of DB+ from scratch but parallel run time could be the same. It happens for databases that are small and whose delta is also small. The first reason for this is that the incremental scheme has many more MR jobs compared to non-incremental. The overhead of job creation overrides the benefit of the incremental algorithm run time (although each machine still consumes less energy). Another reason is that the general scheme needs to read deltaDB from remote location several times for each job execution. In the non-incremental FIM algorithm the amount of IO reads of whole DB+ is dependent on the underlying FIM algorithm and later output of FI(DB+). In the general scheme the same FIM is executed only on deltaDB with output of FI(deltaDB), but there's also a requirement to read all FI(deltaDB) back from a network disk to join with FI(DB). Moreover, it's required to read deltaDB again for the recounting step.

Join's Reducer Phase with Lemmas Applied

Input: I, List<DBPartMarker, MinCount, MaxCount>, |DB|. |deltaDB|, minsup
1:    init values: MinDB = 0, MaxDB = |DB|, MinDelta = 0, MaxDelta = |deltaDB|
2:    **foreach** T in List<DBPartMarker,MinCount, MaxCount>
3:        **if** DBPartMarker = DB **then** MinDB=MinCount, MaxDB=MaxCount
4:        **else** MinDelta = MinCount, MaxDelta = MaxCount
5:    **end foreach**
6:    Min = MinDB + MinDelta
7:    Max = MaxDB + MaxDelta
8:    **if** Min >=minsup*(|DB|+|deltaDB|) **then** //Lemma 1 or 3
9:        **Output** <I,Min,Max> in frequent itemset file and **end**
10:    **if** Max < minsup*(|DB|+|deltaDB|) **then end** //by Lemma 2
11:    **if** MinDB<>MaxDB **then Output**<I,MinDelta,MaxDelta> in "Count in DB" file
12:    **if** MinDelta<>MaxDelta **then Output**<I,MinDB,MaxDB> in "Count in deltaDB"

Algorithm 2: Optimized Join's Reducer Phase with Lemmas Applied

---

To work around these limitations of the incremental scheme we suggest to reduce the number of jobs that are used in the general scheme. The Join job (steps 6–7) is required to read the output of the previous FIM job (step 4) immediately. We suggest to merge this step with the Join job. It should receive additional input of FI(DB) and instead of writing only FI(deltaDB) it will do the "join" of the results. It will still have 3 outputs. All optimizations discussed in Sect. 4.2 should also be applied.

Next job that would be removed is the recounting step in deltaDB (steps 11–12). The only itemsets that could be qualified for this output are itemsets that weren't frequent in deltaDB and were frequent in DB. We suggest counting all itemsets from FI(DB) during step 3 of FIM in deltaDB, there's already a pass over deltaDB anyway. Enhance FIM algorithm of candidate validation (recounting) to count also FI(DB).

Our performance evaluation also revealed additional condition when the incremental algorithm performs worse than non-incremental one. It happens when the input for some Split is very small and the minimum support level is also small. Under such conditions the minimum required occurrences for an itemset to become candidate for frequent itemset is very low, it may be as low as a single transaction and then almost all combinations of items would become candidates. Such small job may run longer than mining whole DB+. To overcome this problem, we propose few techniques:

1. In case deltaDB is being split by the MR framework to Splits, it is being split to chunks of equal predefined size. We need to make sure that last chunk is larger than previous (instead of being smaller). MR systems, like Hadoop, append the last smaller input part to a previous chunk. So, the last Split actually becomes largest.
2. If the total input divided by minimum Split amounts is still too small, it's preferably to manually control the number of Splits. In most of the cases it's better to sacrifice parallel computation for gaining a speed up with less workers or even use a single worker. Once again, it's possible to control the splitting process in Hadoop.
3. If deltaDB is still very small for a single worker to process it effectively, it's better to use non-incremental algorithm for calculation of DB+.

Merge of MR jobs into one allowed us to achieve an algorithm that has only two steps. First step is MapReduce FIM step for deltaDB only. The second step is the optional step of counting candidates in old DB (at most one pass).

5 Experimental Results

5.1 Algorithms and Datasets

We compare the performance of three algorithms described in Sect. 4. We denote the algorithm from Sect. 4.2 - incremental algorithm with IMRApriori and early pruning optimization as "Delta". The algorithm from Sect. 4.3 with minimum steps is called "DeltaMin". The baseline for the comparison of the algorithms is running previously published non-incremental FIM algorithm - IMRApriori [15] on DB+. It is called "Full" (justification for comparing only to "full" is discussed in Sect. 5.4).

The first tested dataset is synthetically generated T20I10D100000 K (will be referenced as T20) [6] it contains almost 100 M (D100000 K) transactions of average length 20 (T20) and average length of maximal potential itemset is 10 (I10), of size 13.7 Gb. The second dataset is "WebDocs" [17]. It is based on real world information collected by web spiders. Its size is 1.48 Gb with almost 1.7 M transactions (5 M + unique items).

The datasets were cut to two equal size halves. The first half of each dataset is used as the baseline of 100% size (DB). The other part was used to generate deltaDB. E.g. Webdocs was cut to a file of 740 Mb. Its delta of 5% was cut from the part that was left out and its size would be 37 Mb. The running time of the incremental algorithms on the

5% delta was compared to the full process of 777 Mb file (joined 740 Mb base file with 5% delta of 37 Mb to a single file). Similarly, T20 baseline of 100% have size of 6.7 GB with 10% increments of 700 Mb. We used different minsup values for each dataset. For T20 we tested minsup 0.1% and 0.2%. Webdocs was tested with 15%, 20%, 25%, 30% (although we show graphs only of 15% and 20%).

More datasets were tested but not presented in this paper, like synthetic T5I2D10000K, T40I10D100K, T10I5D1000K. Previous works used them for benchmarks. We omit their results as they're too small to consider them as large databases.

5.2   Evaluation of Algorithms

The executions were done in Google Compute Engine Cloud (GCE) by directly spanning VMs with Hadoop with "bdutil" script. We tried different cluster sizes with 4, 5, 10 and 20 cores (we used instance types of n1-standard-1 or n1-standard-2). We measured "Run time" (measured by the "Driver" program from the start of the run until output is ready) and "CPU time" (time that all cluster machines consumed together as measured by the MR framework). In our experiments, we were changing deltaDB sizes, minimal support level and the cluster size.

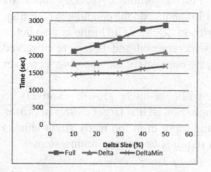

Fig. 1. T20 Run time minsup 0.1% Cluster 5

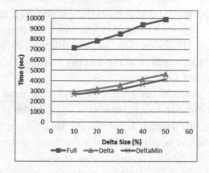

Fig. 2. T20 Cpu time minsup 0.1% Cluster 5

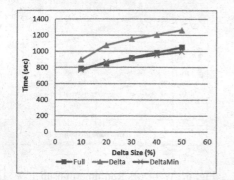

Fig. 3. T20 Run time minsup 0.2% Cluster 5

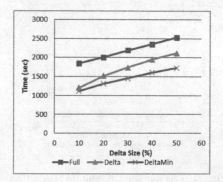

Fig. 4. T20 Cpu time minsup 0.2% Cluster 5

5.3   Results and Analysis

Figures 1 and 2 demonstrate run and cpu times of each algorithm for dataset T20 on GCE cluster of size 5 and minsup 0.1%. It shows that incremental algorithms behave better than full in both parameters. The increase in delta increases computation time. Figures 3 and 4 are similar to 1 & 2 but the minsup is 0.2%. In this case the run time of "Delta" is higher than "Full", "DeltaMin" behaves same or better than "Full". Cpu time of "Delta" and "DeltaMin" is still lower than of "Full". Figures 5 and 6 show algorithms run and cpu times behavior for T20, minsup 0.1% and inc 10% as GCE cluster size changes from 5 to 10 and to 20 nodes/cores. We can see that the algorithms scale well with more cores added to the system. Figures 7 and 8 show WebDocs with minsup 20% run and cpu time as (inc)rement size varies. It shows that run time of "Delta" is no better than "Full" but its cpu time is better. Figures 9 and 10 show same for minsup 15%. As frequent itemsets mining for minsup of 15% is more computationally extensive than 20% the times are higher. In this case, incremental algorithms behave much better than the full algorithm.

Our evaluation showed that incremental algorithms do similar or better than full recalculation on smaller size datasets with larger support in terms of Cpu time. While the Run time wasn't always better for less optimized "Delta" algorithm.

As the support threshold decreased all incremental algorithms had better Run and Cpu time than "Full". They outperformed the "Full" algorithm by several times. Its explanation is that more time is required to mine FI than just do the candidates counting.

The Run time of "DeltaMin" is always superior to "Delta". CPU time is also better than "Delta"'s. The largest differences between incremental and full were observed when incremental algorithms managed to eliminate completely the counting step over the old DB.

Algorithms' run time and cpu time showed almost linear growth relatively to increase in input size (deltaDB). Cluster size change showed that larger cluster improves the run time. It's not always linear as splitting the input in to too many small chunks generate too many locally frequent itemsets and requires longer recounting steps.

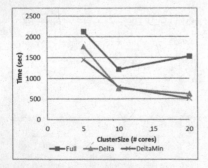

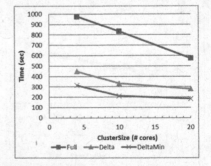

Fig. 5. T20 Run time minsup 0.1%, inc 10%

Fig. 6. WebDocs Run time minsup 15% inc 10%

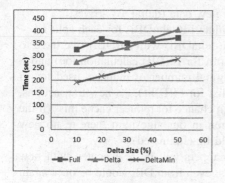

Fig. 7. WebDocs minsup 20% Run Time

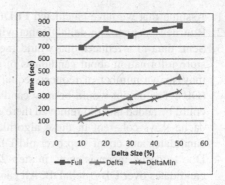

Fig. 8. WebDocs minsup 20% Cpu Time

5.4 Comparison to Previous Work

FUP algorithm [9] was the first to provide incremental scheme which is based on mining the deltaDB. The algorithm is not distributed or parallel. It mines deltaDB by iterative steps from candidates of size 1 to K and stop when no more candidates available. At each step, this algorithm scans old DB to check validity of its candidates. Implementing this algorithm as is in MR would require K scans over the old DB. Which would generate K-times more IO than our algorithms and would be less effective.

ARM IDB [13] provides optimizations on incremental mining via using TID-lists intersection and its "LAPI" optimization. The algorithm doesn't deal with distributed environment (and of course MR) so it has no way to scale out.

Incoop and DryadInc don't support more than one input for DAG (and we need to be able to get DB and candidate set as an input). As there's no known way to overcome this difference it doesn't allow a direct comparison. These systems can't extract useful information from the knowledge of the algorithm goal or specific implementation and therefore improve their runtime. Let's assume that there's a way to compare these systems and go over the following execution example (we use 2-steps IMRApriori):

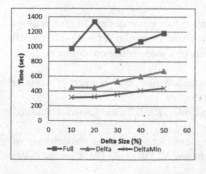

Fig. 9. WebDocs minsup 15% Run time

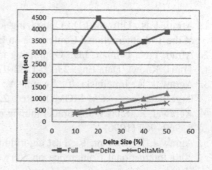

Fig. 10. WebDocs minsup 15% Cpu time

1. Assume that we have calculated FI(DB) and new data (deltaDB) arrives.
2. Run new mappers on deltaDB that will calculate FI(deltaDBi) for each split i.
3. Send data to a reducer that would merge FI(deltaDBi) with FI(DBi) (we have to store information about candidates of each split for FI(DB) calculation too) and it may spawn number of reducers equal to original number of reducers used for FI (DB). The output would be new candidates list for FI(DB+). As new transactions, could be added and old removed there's a high chance for reducer cache miss (if the cache stays coherent in these algorithms, then in our algorithm there're no new candidates and count step over old DB would be eliminated).
4. Due to cache inconsistency, in step 2, all mappers will be re-executed with new candidates over DB ∪ deltaDB. All candidates, even from DB, would be once again recalculated against DB. In our algorithm, we would never count FI(DB) in DB.
5. Output would be sent to reducers to join data of DB and deltaDB to generate FI.

The above example demonstrates additional computation that generic incremental system has compared to our solution. If we change the FIM algorithm to some K steps MR algorithm, then fan-out of DAG would only grow (and may require K passes over old DB) and general incremental frameworks would become even less effective.

6 Conclusions

This paper presented a generic way of performing incremental frequent itemset mining in the MapReduce framework. We also proposed several optimizations for the algorithms to run more effectively in MR. Our experimental evaluation showed that "DeltaMin" algorithm perform better than the non-incremental computation. The lower is the support rate, the harder the computations are, the more benefit could be achieved by incremental algorithms.

Future work may examine additional optimizations for MapReduce based algorithms. They may also test the effect of the Split size and the number of generated candidates on these algorithms. Another interesting topic is to check how the split size can be combined with the cluster size and find the optimal point for the Run time.

References

1. Afrati, F.N., Ullman, J.D.: Optimizing joins in a map-reduce environment. In: Proceedings of the 13th International Conference on Extending Database Technology, pp. 99–110 (2010)
2. Afrati, F.N., Ullman, J.D.: Optimizing multiway joins in a map-reduce environment. IEEE Trans. Knowl. Data Eng. 23(9), 1282–1298 (2011)
3. Agrawal, R., Imieliński, T., Swami, A.: Mining association rules between sets of items in large databases. ACM SIGMOD Rec. 22(2), 207–216 (1993)
4. Agrawal, R., Shafer, J.: Parallel mining of association rules. IEEE Trans. Knowl. Data Eng. 8, 962–969 (1996)
5. Agrawal, R., Srikant, R.: Fast algorithms for mining association rules. In: Proceedings of 20th International Conference on Very large data bases, VLDB, vol. 1215, pp. 487–499 (1994)

6. Agrawal, R., Srikant, R.: Quest Synthetic Data Generator. IBM Almaden Research Center, San Jose, California. http://www.almaden.ibm.com/cs/quest/syndata.html
7. Blanas, S., Patel, J.M., Ercegovac, V., Rao, J., Shekita, E.J., Tian, Y.: A comparison of join algorithms for log processing in mapreduce. In: Proceedings of the 2010 ACM SIGMOD International Conference on Management of data, pp. 975–986 (2010)
8. Bhatotia, P., Wieder, A., Rodrigues, R., Acar, U.A., Pasquin, R.: Incoop: MapReduce for incremental computations. In: Proceedings of the 2nd ACM Symposium on Cloud Computing, p. 7. ACM (2011)
9. Cheung, D.W., Han, J., Ng, V.T., Wong, C.Y.: Maintenance of discovered association rules in large databases: an incremental updating technique. In: Proceedings of the Twelfth International Conference on Data Engineering, 1996, pp. 106–114. IEEE (1996)
10. Das, A., Bhattacharyya, D.K.: Rule mining for dynamic databases. In: Sen, A., Das, N., Das, S.K., Sinha, B.P. (eds.) IWDC 2004. LNCS, vol. 3326, pp. 46–51. Springer, Heidelberg (2004). doi:10.1007/978-3-540-30536-1_6
11. Dean, J., Ghemawat, S.: MapReduce: simplified data processing on large clusters. Commun. ACM **51**(1), 107–113 (2008)
12. Doulkeridis, C., Nørvåg, K.: A survey of large-scale analytical query processing in MapReduce. Int. J. Very Large Data Bases **23**(3), 355–380 (2014)
13. Duaimi, I.G., Salman, A.: Association rules mining for incremental database. Int. J. Adv. Res. Comput. Sci. Technol. **2**, 346–352 (2014)
14. Ekanayake, J., Li, H., Zhang, B., Gunarathne, T., Bae, S.H., Qiu, J., Fox, G.: Twister: a runtime for iterative mapreduce. In: Proceedings of the 19th ACM International Symposium on High Performance Distributed Computing, pp. 810–818. ACM (2010)
15. Farzanyar, Z., Cercone, N.: Efficient mining of frequent itemsets in social network data based on MapReduce framework. In: Proceedings of the 2013 IEEE/ACM International Conference on Advances in Social Networks Analysis and Mining, pp. 1183–1188. ACM (2013)
16. Li, N., Zeng, L., He, Q., Shi, Z.: Parallel implementation of apriori algorithm based on MapReduce. In: 2012 13th ACIS International Conference on Software Engineering, Artificial Intelligence, Networking and Parallel & Distributed Computing, pp. 236–241. IEEE (2012)
17. Lucchese, C., Orlando, S., Perego, R., Silvestri, F.: WebDocs: a real-life huge transactional dataset. In: FIMI, vol. 126 (2004)
18. Popa, L., Budiu, M., Yu, Y., Isard, M.: DryadInc: reusing work in large-scale computations. In: USENIX workshop on Hot Topics in Cloud Computing (2009)
19. Thomas, S., Bodagala, S., Alsabti, K., Ranka, S.: An efficient algorithm for the incremental updation of association rules in large databases. In: KDD, pp. 263–266 (1997)
20. Woo, J.: Apriori-map/reduce algorithm. In: The 2012 International Conference on Parallel and Distributed Processing Techniques and Applications (PDPTA 2012)
21. Yahya, O., Hegazy, O., Ezat, E.: An efficient implementation of Apriori algorithm based on Hadoop-Mapreduce model. Int. J. Rev. Comput. **12**, 59–67 (2012)
22. Zaki, M.J., Parthasarathy, S., Ogihara, M., Li, W.: New algorithms for fast discovery of association rules. In: KDD, vol. 97, pp. 283–286 (1997)
23. Han, J., Pei, J., Yin, Y.: Mining frequent patterns without candidate generation. ACM Sigmod Rec. **29**(2), 1–12 (2000)

Towards High Similarity Search Throughput by Dynamic Query Reordering and Parallel Processing

Filip Nalepa[✉], Michal Batko, and Pavel Zezula

Faculty of Informatics, Masaryk University, Brno, Czech Republic
f.nalepa@gmail.com

Abstract. Current era of digital data explosion calls for employment of content-based similarity search techniques since traditional searchable metadata like annotations are not always available. In our work, we focus on a scenario where the similarity search is used in the context of stream processing, which is one of the suitable approaches to deal with huge amounts of data. Our goal is to maximize the throughput of processed queries while a slight delay is acceptable. We extend our previously published technique that dynamically reorders the incoming queries in order to use our caching mechanism more effectively. The extension lies in adoption of a parallel computing environment which allows us to process multiple queries simultaneously.

Keywords: Stream processing · Similarity search · Parallel processing

1 Introduction

Huge amounts of unstructured data are being produced nowadays resulting from the current digital media explosion. Many tasks targeting the processing of such data involve, in some form, searching in the data. Unfortunately, traditional search techniques based on exact match of data attributes often cannot be applied to such data types. Instead, content-based search that treats the data by similarity is a viable option. Such search then usually uses *k-nearest-neighbors queries* (kNN), which retrieve the k objects that are the most similar to a given query object.

Due to the nature of the data and applications which use them, it can be desired to view the data as a potentially infinite stream which is continuously being created. For example, consider a text search-engine crawler that gathers images from the web and needs to continuously annotate them by textual descriptions according to the image content. Another example can be a spam filter that receives incoming emails and compares them to some learned spam knowledge base so that the spam messages can be detected. Finally, consider a news notification system which needs to compare the newly published articles to the profiles of all the subscribed users to find out who should be notified.

A subtask of all these applications is processing the streamed data items by some form of content-based searching. The performance of these applications

© Springer International Publishing AG 2017
M. Kirikova et al. (Eds.): ADBIS 2017, LNCS 10509, pp. 262–277, 2017.
DOI: 10.1007/978-3-319-66917-5_18

is mostly determined by the number of processed data items in a given time interval, i.e., the throughput is the most important metric. The individual query search time can be improved by applying some similarity indexing technique, for which there are efficient algorithms based on the metric model of similarity [20]. As opposed to interactive applications focusing on the single query optimization, in our scenario, we can afford a slight delay of the single query processing if the overall throughput of the system is improved. Performance of such stream processing applications is studied in [12,13].

I/O costs typically have a significant effect on the performance of similarity search techniques. In our work, we exploit the fact that some orderings of the processed queries can result in considerably lower I/O costs and overall processing times than random orderings. This is based on the assumption that two similar queries need to access similar data of the search index. By obtaining an appropriate ordering of queries, the accessed data can be cached in the main memory and reused for evaluation of similar queries thus lowering down the I/O costs. We have previously published [14] a technique which dynamically reorders the incoming queries that allows to achieve a significant improvement of the throughput.

In this paper, we provide an extension of the technique by adopting it to a parallel computing environment where multiple queries can be processed simultaneously by individual query processors. Due to the nature of our approach, it is very important how the streamed query objects are distributed among the query processors, i.e., which query object goes to which query processor. The main contribution of this work is a proposal of effective and efficient ways in which the query objects are spread among the processors so that high throughput is achieved.

The rest of the paper is organized as follows. First, we present related work on caching and query reordering in similarity search, and parallelization in stream processing. In Sect. 3 we formally define our problem. The originally published technique is summarized in Sect. 4. Its adoption to the parallel computing environment is presented in Sect. 5. Experimental evaluation can be found in Sect. 6.

2   Related Work

The usage of a caching mechanism in similarity search has been proposed in several papers to reduce the amount of I/O operations. In [6], the authors propose caching of similarity search results and reusing them to produce approximate results of similar queries. The concept of caching similarity search results is used also in [16]. The paper focuses on caching policies which incrementally reorganize the cache to ensure that the cached items cover the similarity space efficiently. The Static/Dynamic cache presented in [19] consists of a static part to store queries (along with their results) that remain popular over time and a dynamic part to keep queries that are popular for a short period of time. Authors of [4] present a caching system to obtain quick approximate answers. If the cache cannot provide the answer, the distances computed up to that moment are used to query the index so that the computations are not wasted.

Another way to improve the throughput of a stream of kNN queries, is to reorder the queries. In [17], the authors optimize nearest neighbor search for videos. Intersections of candidate sets between every pair of queries need to be computed and updated periodically. This approach is designed for a relatively small number of queries (tens). Since we have tens of thousands of queries to be evaluated (as will be seen in Sect. 6), the overhead of such computations is likely to be very high.

The authors of the paper [18] propose D-cache which stores distances computed during previous queries. The Snake Table [2] uses a cache of distances to improve performance of processing streams of queries with snake distribution (i.e., consecutive query objects are similar). In [1], an inverted cache index stores statistics about usefulness of data partitions in order to modify priority queues.

All the aforementioned techniques are designed for interactive applications when queries are evaluated immediately. We focus on scenarios when delays are affordable and the throughput is the main issue which calls for a different type of a solution.

Speaking generally of stream processing, parallelization is a common technique for throughput enhancement. This is typically referred to as an *operator replication* where an operator is a component for processing the stream. Each data item of the stream is sent to one of the replicas where it is subsequently processed. There is a number of issues which need to be dealt with, e.g., determining optimal number of replicas or creating an appropriate strategy for deciding which data item is processed by which replica. These challenges have been widely explored [7,8,10,11]. In our work, we do not aim at enhancing parallel stream processing of general applications. We rather focus on designing schemas of the parallelization applicable to our specific case which is not covered by the general approaches.

In [3], techniques for parallelization of similarity search are studied. The approaches are based on creating distributed metric index structures and parallelization of a single query evaluation. In our work, we use the parallelization to evaluate multiple queries concurrently. In our case, we avoid the overhead related to the distribution of a single query to multiple processing components and the overhead caused by merging partial results into the final answer.

3   Problem Definition and Objectives

Suppose there is a domain of complex objects D (e.g., images) and a large database containing such objects $X \subseteq D$. Let $s = (d_1, d_2, \ldots)$ be a stream, i.e., a potentially infinite sequence of data items. Each item of the stream is a pair $d_i = (q_i, t_i)$ where $q_i \in D$ is a query object and t_i is the time it was created (entered the application). It holds that $t_i \leq t_{i+1}$ for each i and $t_1 = 0$.

As a universal model of similarity we use the metric space (D, d) [20], where d is a total distance function $d : D \times D \to R$. The distance between two objects corresponds to the level of their dissimilarity ($d(p, p) = 0$, $d(o, p) \geq 0$).

For each query object q_i in the stream s, a k-nearest neighbors query $NN(q_i, k)$ is executed which returns k nearest objects from the database to the query object. It is allowed to change the order of the processed query objects. More precisely, at time t, any query object q_i, where (q_i, t_i) is a data item of the stream and $t_i \leq t$, can be processed.

The goal is to process the query objects of the stream so that the throughput is maximized. Specifically, we want to maximize the number of processed query objects of a given stream until a given time T. Alternatively, the criteria can be defined as minimization of the number of unprocessed query objects at the time T, i.e., the number of (q_i, t_i) where $t_i \leq T$ and q_i is not processed.

Our objective is to extend our previously proposed technique enhancing the throughput of the similarity search. The extension lies in parallel processing of multiple queries. In particular, we focus on the ways to distribute the queries among individual query processors to gain maximum effectiveness.

4 Enhancing Throughput with a Single Query Processor

In this paper, we build upon our previous work [14] in which we proposed a technique for enhancing the throughput of the similarity search by dynamically reordering the queries combined with a caching mechanism to lower down the I/O costs during processing. In this section, we summarize this technique.

We consider a generic metric index which uses data partitioning $P = \{p_1, \ldots, p_n\}$ where $p_i \subseteq D$. When evaluating a query, a subset of the partitions $Q \subseteq P$ needs to be accessed. The partitions are typically stored on a disk [20]. A frequent bottleneck of similarity search techniques is the reading of the partitions from the disk during a query evaluation. Our solution aims at decreasing the number of disk accesses and consequently the time to process the queries.

We make use of the following feature of data partitioning methods. If two query objects are very similar to each other, the sets of accessed data partitions are also very similar. This property can be used to speed up the processing of query objects q_1 and q_2. First, q_1 is evaluated, and the accessed data partitions are kept in the main memory cache. When q_2 is being evaluated, the data partitions stored in the cache can be reused to avoid expensive disk accesses.

However, the caching itself is not typically enough for the speedup. Since huge databases are often used in practice, there is a very low probability that two subsequent queries in the stream are similar enough to access similar sets of data partitions during their evaluation. For the cache to be sufficiently utilized, the query objects in the stream need to be reordered so that sequences of similar query objects are obtained.

To sum it up, the approach consists of two parts. The first one is the in-memory caching of recently loaded data partitions and reusing them for evaluation of subsequent queries. The second one is the query object reordering allowing to process sequences of similar query objects to maximize the cache utilization.

4.1   Query Ordering

The problem of the query object ordering can be viewed as a graph problem. Let $s = ((q_1, t_1), (q_2, t_2), \ldots)$ be the stream to be processed. Let $((q_1, t_1), \ldots, (q_k, t_k))$ be a finite subsequence of s so that $t_k \leq t$ and $t_{k+1} > t$ for a given t, i.e., the query objects which have become available by the time t. We define the *query graph* $G_t = (V, E)$ at the time t. The set of vertices is comprised of the subsequence items $V = \{(q_1, t_1), \ldots, (q_k, t_k))\}$. In other words, each query object of the stream subsequence represents a vertex in the query graph.

The graph is complete, i.e., there is an edge between every pair of vertices (q_i, t_i) and (q_j, t_j) where $i \neq j$. A value is associated with each edge denoting the query time to process q_i right after q_j or vice versa. The assigned time is based on the extent of the cache utilization.

To formally define the throughput maximization, given the time limit t, the task is to find the longest path $((q_{i_1}, t_{i_1}), \ldots, (q_{i_k}, t_{i_k}))$ in G_t so that $start_k < t$. The path represents the ordering of the queries; $start_k$ is the time when the last query object q_{i_k} starts to be evaluated. The length of the path is measured as the number of vertices, i.e., the number of processed query objects. This is in fact a variation of the traveling salesman problem (an NP-hard problem). There is an added difficulty since the query graph evolves throughout the time, i.e., G_t is not completely known before the time t.

Since searching for the optimal solution is unfeasible, we apply a greedy approach trying to minimize average edge values (i.e., the query times). For this we proposed a combined heuristics of a dense subgraph and the nearest-neighbor strategy. The intuition is to find a subgraph containing short edges between the vertices and process the corresponding query objects in a nearest-neighbor manner, thus minimizing query times. To implement the heuristics, we build hierarchical clusters of the query objects using a pivoting technique. In particular, let there be a fixed set of objects in the metric space; we will denote them as pivots. When a new query object q is to be added to the graph, distances of the query object q to all the pivots are computed. The pivots are ordered from the nearest to the farthest one which defines a permutation of the pivots. This pivot permutation identifies the cluster where the query object belongs. By taking just a prefix of the permutation, hierarchical clustering is obtained. The length of a common prefix of two query objects is used to approximate the corresponding edge value in the query graph.

4.2   Architecture

This section describes the architecture of the system using the proposed technique. Its schema is depicted in Fig. 1.

Let us have a stream $((q_1, t_1), (q_2, t_2), \ldots)$. A query object q_i enters the application at the time t_i, and it is inserted into a component called *buffer*. The buffer is used to temporarily store the incoming query objects which are awaiting processing. This is the component where the query reordering takes place.

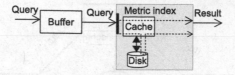

Fig. 1. Architecture

Another part of the architecture is the metric index which takes care of the query object evaluation. It contains a disk where the database of objects is stored and a main memory cache which is used to store the recently loaded data partitions from the disk.

When the metric index is ready for processing another query, a query object is picked from the buffer according to the ordering strategy. During the query processing, the metric index exploits the cache to possibly use any data partitions obtained from evaluating recent queries. If the data are not in the cache, they are loaded from the disk.

5   Parallelization

In general, a way to speed up computer processing is to use parallel computations. In stream processing, the parallelization can be accomplished by creating several instances (replicas) of the processing component. Each data item of the stream is sent to one of the replicas where it is processed. This results in such a scheme where several data items can be processed in parallel. In this section we discuss the applicability of such parallelization method to our case of processing a stream of query objects.

Formally, given the stream $((q_1, t_1), (q_2, t_2), \ldots)$, the number of replicas r and the time limit t, we generate a set of r disjoint paths (i.e., no vertex can be a part of two paths) in the query graph G_t : $\{((q_{i_{11}}, t_{i_{11}}), (q_{i_{12}}, t_{i_{12}}), \ldots, (q_{i_{1k_1}}, t_{i_{1k_1}})), \ldots, ((q_{i_{r1}}, t_{i_{r1}}), (q_{i_{r2}}, t_{i_{r2}}), \ldots, (q_{i_{rk_r}}, t_{i_{rk_r}}))\}$ so that $start_{k_j} < t$ for $1 \leq j \leq r$ where $start_{k_j}$ is the time when the processing of the k_j^{th} query object of the jth replica starts. Each path represents the order of the query objects processed at a particular replica. The goal is to identify such paths which in sum give the largest number of vertices (query objects) to maximize the throughput.

5.1   Parallel Architecture

A generic architecture of the parallel processing system can be seen in Fig. 2. There are several instances (replicas) of the query processor whose task is to evaluate query objects. Each query processor maintains its own cache. During evaluation, the query processor needs to access some data partitions which can be either acquired from the cache or they have to be loaded from a disk. The figure contains just two replicas for simplicity reasons.

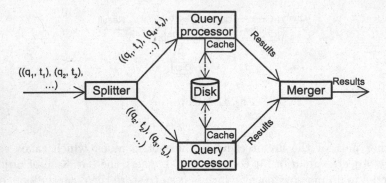

Fig. 2. Generic architecture of parallel processing

In our experiments, we consider the situation when the database is shared among all the query processors, i.e., they access the same storage space (the disk in the figure). Such a setup allows to quickly add or remove query processors without significant overhead which is advantageous in dynamic environments when it is needed to create or remove replicas on the fly to adapt to load changes.

An important component of the architecture is the splitter which serves as the entry point to the rest of the system. It decides for each query object by which query processor it will be processed. The splitting strategy significantly influences the efficiency of the processing as will be seen in the experiment results. We present three approaches that differ in the functionality of the splitter.

5.2   Push Technique

The first approach is based on a push technique. In this scenario every query processor instance possesses its own buffer of waiting query objects. As soon as a query object arrives at the splitter, it is pushed to the instance of the query processor having the least number of query objects waiting in the buffer. This ensures all the buffers are loaded evenly. The schema is depicted in Fig. 3a. Each query processor continuously evaluates query objects of its own buffer; the same technique for the ordering of the query objects is applied as for the case with a single processor. An advantage of this approach is the low overhead of the splitter, so it can scale well with increasing number of query processors. A disadvantage is that the query graph is not considered for the distribution of query objects among the replicas which can result in ineffective distribution.

Consider two query objects q_1, q_2 needing similar data partitions for their evaluation, i.e., there is a short edge between them in the query graph. If the two query objects are processed by the same replica, the cache may be used to speedup the processing. If each of them is processed by a different replica, the similarity of the query objects cannot be benefit from the cache. This implies that the query objects connected by short edges in the query graph should be processed by the same replica whenever possible.

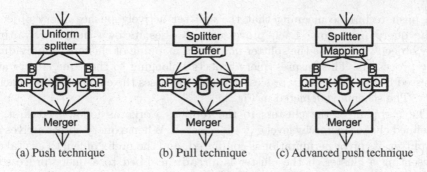

(a) Push technique   (b) Pull technique   (c) Advanced push technique

Fig. 3. Replication schemas; QP – query processor, B – buffer, C – cache, D – disk

The dimension of the time plays its role too. Consider two data items (q_1, t_1), (q_2, t_2) where $t_1 \le t_2$ and a path where q_2 is scheduled for processing at some point after q_1. The bigger the difference between the entry times t_2 and t_1 is, the smaller is the probability that the data partitions needed for processing q_1 are still in the cache when processing q_2 since q_2 may be processed a long time after q_1. This observation is based on the query ordering strategy presented in Sect. 4.1.

5.3 Pull Technique

The disadvantages of the first approach are avoided by the second one which is based on a pull technique. There is one shared buffer for all the query processors which pull query objects from it. The schema can be seen in Fig. 3b. In this case the splitter is in charge of the buffer. A query processor sends a request to the splitter which, according to the ordering strategy, returns a query object from the buffer. The splitter maintains a state for each query processor consisting of the last processed query object by this processor. This is used to enable processing of the query clusters (defined in Sect. 4.1) independently for each query processor. Moreover, the state is used to lock the currently processed cluster. This ensures that a particular cluster cannot be simultaneously processed by multiple query processors. This enables to achieve higher cache utility than in the case of the push technique since all the query objects of the particular cluster which are currently in the buffer are processed by the same replica. Possible disadvantages may occur with a large number of query processors. The splitter then becomes a bottleneck that gets overloaded with requests from the processors. Another possible disadvantage may be when the size of the shared buffer grows over the limits which can fit into the memory.

5.4 Advanced Push Technique

The third approach (advanced push technique) tries to combine advantages of the two previous approaches while mitigating their disadvantages. It is based on a push technique meaning that the splitter actively pushes query objects to the query processors. Each processor possesses its own buffer of waiting query objects. This time the splitter maintains mapping of clusters to individual query processors. This ensures query objects belonging to the same cluster are processed by the same query processor which increases the chances of high cache utility. The schema is depicted in Fig. 3c.

The mapping of the clusters to the replicas works as follows. At first, a predefined cluster hierarchy level $e \geq 1$ is selected. When a query object arrives at the splitter, its pivot permutation is computed, and the prefix of length e is taken representing a cluster. If the cluster is already mapped to a query processor, the query object is sent there for processing. Otherwise the cluster is mapped to the replica having the least number of query objects in the buffer at that moment. The distribution of the streamed query objects among the clusters is not known beforehand and it can change throughout the time. Therefore having such a mapping mechanism may lead to imbalanced load of the replicas. To prevent this, whenever the difference of the maximal and minimal buffer size of all the replicas exceeds a given threshold, remapping of clusters to the replicas is performed to balance the buffer sizes.

The advantage of this technique is that the splitter maintains just the mapping of the clusters to the replicas, and the burden of query ordering is spread among individual replicas. At the same time, the mapping ensures high effectiveness of the distribution of the query objects to the replicas.

6   Experiments

In this section, we provide results of experiments with parallel processing of a stream of query objects using the three approaches presented in the previous section.

6.1   Setup

Let us start with describing the setup of the experiments.

We use the M-Index [15] structure to index the metric-space data. It employs many principles of metric space partitioning, pruning, and filtering, thus reaches very high search performance. The actual data are partitioned into buckets which are stored as separate files on a disk and read into the main memory during query evaluations. To partition the data, M-Index uses a set of pivots. To insert an object into the index, the pivots are sorted based on the distance to the object. In this way, a pivot permutation is obtained which identifies the data partition to insert the object. During a similarity search, mutual distances between the query object and the pivots are used to reduce the set of data partitions which need to be accessed. The M-Index supports executing approximate kNN queries among other operations. One of the stop conditions of a query evaluation is given by the maximum number of accessed objects (the size of a candidate set). Such a stop condition is used in our experiments.

The M-Index uses the same set of pivots as are used for the query graph construction. This is beneficial for the effectiveness of the query ordering since the partitioning schema of the metric space used in the M-Index and used for the query graph construction is synergic. This also improves efficiency since the distances from a query object to the pivots can be computed just once and used both in the query graph and in the M-Index.

For the experiments, we use the Profimedia dataset of images [5]. We created two different subsets of the images and extracted their visual-feature descriptors. The generated datasets are: 1 million Caffe descriptors [9] (4096 dimensional vectors) and 10 million MPEG-7 descriptors (280 dimensional vectors with complex distance function). Separately, we created streams of images represented by corresponding descriptors. During each experiment, images from the respective collection are continuously streamed to the application and being processed as approximate 10-NN queries. For the approximate kNN queries, we used candidate sets of size $1,000$ for the Caffe dataset and size $2,000$ for the MPEG-7 dataset.

The maximum size of the cache is set to 40,000 descriptors for the Caffe dataset (i.e., 4% of the database); up to 90,000 descriptors are stored for the MPEG-7 dataset (i.e., 0.9% of the database). The least recently used policy is used when inserting to the full cache. In particular, the data partitions with the oldest last access time are discarded and replaced with the new partitions of the current query so that the maximum size of the cache is preserved. This strategy is appropriate since there is a high probability that recently needed partitions will be reused for evaluation of subsequent queries.

All the query processors are run on a single machine in multithreaded environment. For the advanced push technique, the remapping of the subclusters occurs when the ratio of the maximal and minimal buffer size of the replicas exceeds 1.2 (i.e., 20%).

6.2   Evaluation

Fixed Input Frequency. In the first group of experiments, the input frequency of query objects was fixed to 10 ms. This simulates the standard stream processing scenario when the application cannot control the rate of incoming data. The experiments were run with different numbers of query processors: 2, 4 and 8. Each experiment was run for 2 hours for the 1 mil. Caffe dataset and for 4 hours for the 10 mil. MPEG-7 dataset.

The results can be seen in Figs. 4 and 5 depicting the evolution of the buffer sizes during the experiments. For the push techniques, the overall buffer size is taken as a sum of all the buffer sizes at individual replicas. The worst throughput was observed for the simple push technique because of its ineffective distribution of the query objects to the replicas. The other two approaches provide comparable results which are significantly better than those obtained for the simple push approach. Although the advanced push technique distributes the query objects much more effectively than the simple push technique, the pull technique manages to keep the buffer size a little lower most of the time.

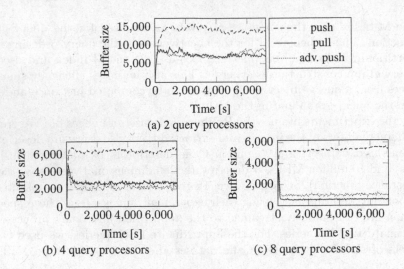

Fig. 4. The buffer size evolution in time during processing of the 1 mil. Caffe dataset

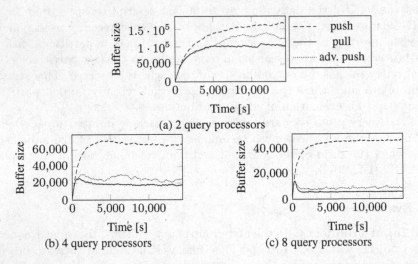

Fig. 5. The buffer size evolution in time during processing of the 10 mil. MPEG-7 dataset

This shows an advantage of the centralized buffer where a better ordering of the query objects can be obtained due to the access to all the buffered query objects rather than just to portions of the data at individual replicas. Another disadvantage of the partial buffers is a fragmentation of clusters among multiple replicas caused by remapping of the clusters to the replicas. On the other hand, if the splitter becomes a bottleneck, the advanced push technique is expected to provide the best performance.

Table 1. Parallelization statistics for the 1 mil. Caffe dataset

# Query processors	Push			Pull			Adv. push		
	2	4	8	2	4	8	2	4	8
Max delay [s]	354	179	90	358	180	90	199	99	81
Median delay [s]	319	140	110	193	100	70	75	23	11
Load difference	1	1	1	0.99	0.99	1	0.98	0.98	0.98

Table 2. Parallelization statistics for the 10 mil. MPEG-7 dataset

# Query processors	Push			Pull			Adv. push		
	2	4	8	2	4	8	2	4	8
Max delay [s]	3356	1440	964	2197	554	327	3131	718	345
Median delay [s]	1300	636	435	917	186	59	1080	255	90
Load difference	0.99	1	0.99	0.99	0.99	0.99	0.99	0.97	0.97

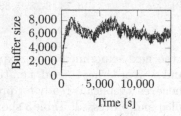

Fig. 6. The buffer size evolution in time per replica for adv. push technique with 4 replicas, 10 mil. MPEG-7 dataset

Tables 1 and 2 present the results considering maximal and median delays (the time since a query object arrives at the application until it is processed). The delays decrease with higher number of query processors. The tables also capture the difference in the number of processed queries by individual query processors computed as $\frac{s}{b}$ where s is the smallest number of queries processed by a single replica; b is the biggest number of queries processed by a single replica. In all the scenarios, the differences in the load of the query processors were negligible.

Due to the nature of the advanced push technique, there can be temporal buffer size imbalances of the individual replicas, and remapping of clusters has to be performed. Figure 6 depicts how the buffer size evolves for individual replicas during the experiment with the MPEG-7 dataset and 4 replicas. Each curve represents the buffer size of one replica. It can be seen the relative difference between the maximal and minimal buffer sizes is kept within the predefined limit of 20%.

Figure 7 shows results of the experiments comparing throughput for high input frequencies (3, 4 and 5 ms). These were conducted for the Caffe dataset using 4 replicas and the pull technique. While the 5 ms input frequency can be coped with, higher frequencies cause the buffer to grow to very large sizes.

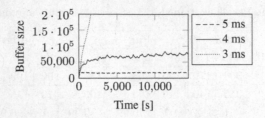

Fig. 7. The buffer size evolution for pull technique with 4 replicas and various input frequencies, 1 mil. Caffe dataset

Table 3. Throughput speedup for parallel processing with no reordering relatively to a single query processor

# Query processors	2	4	8
Caffe	1.90	3.37	5.56
MPEG-7	1.73	3.38	5.83

No Optimizations. In the next experiments, we explore the parallelization impact when no caching and no reordering is used, i.e., the queries are processed in their original order. Whenever a replica is ready for processing, another query object of the stream is pulled and processed. Table 3 shows throughput speedup for different numbers of used query processors. The speedup is computed as the ratio of the average query time (i.e., the time to evaluate a single query object) when just one query processor is used and the average query time when a given number of query processors are used. Since all the replicas access the same database storage, the speedup is not linear. The results serve as the baseline for the following experiments where the optimization techniques are used.

Fixed Buffer Size. Another set of experiments was conducted with a fixed size of the buffer of 10,000 query objects so that we can compare the throughput of the techniques with a steady state of the buffers. For the push approaches, the buffer size constraint was applied as a limit of the sum of the individual buffer sizes. A next query object was loaded from the stream by the splitter after every processed query object to keep the overall buffer size constant. 100,000 query objects were evaluated during each experiment. Table 4 captures the throughput speedup (the first three rows). A single query processor with no reordering and no caching is taken as the baseline. The speedup is computed as the ratio of the time needed to process 100,000 query objects with a single query processor with no reordering and no caching, and the time needed by a given number of query processors with given optimization techniques applied. For the reference, the results contain also the speedup using one query processor with applied reordering and caching (i.e., our original single replica approach). The best speedup can be observed for the pull technique, followed by the advanced push technique. Comparing 8 query processors of the pull technique to a single query

Table 4. Throughput speedup with fixed buffer size relatively to a single query processor with no reordering; the column headers are in the format: # query processors; technique (*PS* – push, *PL* – pull, *AP* – adv. push)

Dataset	Buffer size	1; *PS*	2; *PS*	2; *PL*	2; *AP*	4; *PS*	4; *PL*	4; *AP*	8; *PS*	8; *PL*	8; *AP*
Caffe	10^4	3.7	5.9	6.7	6.5	8	11.2	10.2	8.2	15.6	13
MPEG-7	10^4	2	3.5	4.2	4.2	4.6	6.8	6.4	6.1	10	9
Caffe	$10^4 \cdot$ # query proc.	3.7	6.9	7.8	7.7	11.2	14.1	14.3	15.2	22.8	21
MPEG-7	$10^4 \cdot$ # query proc.	2	4.1	4.9	5.1	6.8	9.7	9.6	9.8	16.3	16.2

processor with applied optimizations, the processing is 4.2 times faster for the Caffe dataset and 5 times faster for the MPEG-7 dataset. One reason of such a rather small factor is sharing the same database storage. Another reason is a small buffer size relatively to the number of query processors which results in small cache utility.

We considered another scenario when, together with the number of query processors, also the buffer size is increased. We used the factor of 10,000 to set the buffer size; in particular, for 2 query processors, 20,000 size of the buffer was used, 40,000 for 4 query processors and 80,000 for 8 replicas. See the last two rows of Table 4 for the results. For the MPEG-7 dataset, the speedup of the pull technique using 8 query processors compared to one query processor with applied optimizations is 8.1, i.e., better than linear speedup. The buffer size can be observed to be an important aspect of the proposed approaches. (This observation was also made in our previous work for the single processor cases.) Also very small differences between the pull and the advanced push techniques can be noted as the partial buffers of the push technique are of larger sizes than in the previous scenario and so the distribution of query objects among the replicas can be more effective.

To sum up, the distribution strategy of the query objects by the splitter to individual replicas was observed to have a significant impact on the overall throughput of the system. The highest effectiveness of the distribution is achieved by the pull technique. However, it is closely followed by the advanced push technique which eliminates possible scalability issues when a large number of replicas are used. We used an HDD for all the experiments, and different numbers can be expected for an SSD. However, the proposed approaches should still bring improvement due to decompression which has to be carried out when loading the data from the disk.

7   Conclusion

We have presented an extension of the technique for enhancing the throughput of similarity search query processing by dynamic query reordering. The extension lies in the adoption to a parallel environment where multiple queries can be processed simultaneously. We described three approaches to such parallel processing and showed the importance of an appropriate distribution strategy of the query objects to individual query processors.

The best results are achieved with the pull technique when the query ordering is centralized. However, it is closely followed by the advanced push approach when the query objects are intelligently distributed to the query processors and the ordering itself is performed at individual query processors by themselves.

Acknowledgements. This work was supported by the Czech national research project GA16-18889S.

References

1. Antol, M., Dohnal, V.: Optimizing query performance with inverted cache in metric spaces. In: Pokorný, J., Ivanović, M., Thalheim, B., Šaloun, P. (eds.) ADBIS 2016. LNCS, vol. 9809, pp. 60–73. Springer, Cham (2016). doi:10.1007/978-3-319-44039-2_5
2. Barrios, J.M., Bustos, B., Skopal, T.: Analyzing and dynamically indexing the query set. Inf. Syst. **45**, 37–47 (2014)
3. Batko, M., Novak, D., Falchi, F., Zezula, P.: Scalability comparison of peer-to-peer similarity search structures. Future Gener. Comput. Syst. **24**(8), 834–848 (2008)
4. Brisaboa, N.R., Cerdeira-Pena, A., Gil-Costa, V., Marin, M., Pedreira, O.: Efficient similarity search by combining indexing and caching strategies. In: Italiano, G.F., Margaria-Steffen, T., Pokorný, J., Quisquater, J.-J., Wattenhofer, R. (eds.) SOFSEM 2015. LNCS, vol. 8939, pp. 486–497. Springer, Heidelberg (2015). doi:10.1007/978-3-662-46078-8_40
5. Budikova, P., Batko, M., Zezula, P.: Evaluation platform for content-based image retrieval systems. In: Gradmann, S., Borri, F., Meghini, C., Schuldt, H. (eds.) TPDL 2011. LNCS, vol. 6966, pp. 130–142. Springer, Heidelberg (2011). doi:10.1007/978-3-642-24469-8_15
6. Falchi, F., Lucchese, C., Orlando, S., Perego, R., Rabitti, F.: Similarity caching in large-scale image retrieval. Inf. Process. Manag. **48**(5), 803–818 (2012)
7. Gedik, B.: Partitioning functions for stateful data parallelism in stream processing. VLDB J. Int. J. Very Large Data Bases **23**(4), 517–539 (2014)
8. Gedik, B., Schneider, S., Hirzel, M., Wu, K.L.: Elastic scaling for data stream processing. IEEE Trans. Parallel Distrib. Syst. **25**(6), 1447–1463 (2014)
9. Jia, Y., Shelhamer, E., Donahue, J., Karayev, S., Long, J., Girshick, R., Guadarrama, S., Darrell, T.: Caffe: convolutional architecture for fast feature embedding. In: Proceedings of the ACM International Conference on Multimedia, pp. 675–678. ACM (2014)
10. Lakshmanan, G.T., Li, Y., Strom, R.: Placement strategies for internet-scale data stream systems. Int. Comput. IEEE **12**(6), 50–60 (2008)
11. Lakshmanan, G.T., Li, Y., Strom, R.: Placement of replicated tasks for distributed stream processing systems. In: Proceedings of the Fourth ACM International Conference on Distributed Event-Based Systems, pp. 128–139. ACM (2010)
12. Mera, D., Batko, M., Zezula, P.: Towards fast multimedia feature extraction: Hadoop or storm. In: 2014 IEEE International Symposium on Multimedia (ISM), pp. 106–109. IEEE (2014)
13. Nalepa, F., Batko, M., Zezula, P.: Performance analysis of distributed stream processing applications through colored petri nets. In: Kofroň, J., Vojnar, T. (eds.) MEMICS 2015. LNCS, vol. 9548, pp. 93–106. Springer, Cham (2016). doi:10.1007/978-3-319-29817-7_9

14. Nalepa, F., Batko, M., Zezula, P.: Enhancing similarity search throughput by dynamic query reordering. In: Hartmann, S., Ma, H. (eds.) DEXA 2016. LNCS, vol. 9828, pp. 185–200. Springer, Cham (2016). doi:10.1007/978-3-319-44406-2_14
15. Novak, D., Batko, M., Zezula, P.: Metric index: an efficient and scalable solution for precise and approximate similarity search. Inf. Syst. **36**(4), 721–733 (2011)
16. Pandey, S., Broder, A., Chierichetti, F., Josifovski, V., Kumar, R., Vassilvitskii, S.: Nearest-neighbor caching for content-match applications. In: Proceedings of the 18th International Conference on World Wide Web, pp. 441–450. ACM (2009)
17. Shao, J., Huang, Z., Shen, H.T., Zhou, X., Lim, E.P., Li, Y.: Batch nearest neighbor search for video retrieval. IEEE Trans. Multimedia **10**(3), 409–420 (2008)
18. Skopal, T., Lokoc, J., Bustos, B.: D-Cache: universal distance cache for metric access methods. IEEE Trans. Knowl. Data Eng. **24**(5), 868–881 (2012)
19. Solar, R., Gil-Costa, V., Marín, M.: Evaluation of static/dynamic cache for similarity search engines. In: Freivalds, R.M., Engels, G., Catania, B. (eds.) SOFSEM 2016. LNCS, vol. 9587, pp. 615–627. Springer, Heidelberg (2016). doi:10.1007/978-3-662-49192-8_50
20. Zezula, P., Amato, G., Dohnal, V., Batko, M.: Similarity Search: The Metric Space Approach, vol. 32. Springer, Berlin (2006)

Comparative Evaluation of Distributed Clustering Schemes for Multi-source Entity Resolution

Alieh Saeedi[(✉)], Eric Peukert, and Erhard Rahm

Database Group, Department of Computer Science,
University of Leipzig, Leipzig, Germany
{saeedi,peukert,rahm}@informatik.uni-leipzig.de

Abstract. Entity resolution identifies semantically equivalent entities, e.g., describing the same product or customer. It is especially challenging for big data applications where large volumes of data from many sources have to be matched and integrated. Entity resolution for multiple data sources is best addressed by clustering schemes that group all matching entities within clusters. While there are many possible clustering schemes for entity resolution, their relative suitability and scalability is still unclear. We therefore implemented and comparatively evaluate distributed versions of six clustering schemes based on Apache Flink within a new entity resolution framework called Famer. Our evaluation for different real-life and synthetically generated datasets considers both the match quality as well as the scalability for different number of machines and data sizes.

1 Introduction

Entity resolution (ER) – also called deduplication, record linkage or object matching - is the task of identifying records that refer to the same real-world entity, such as specific costumers, products or publications. This problem is of key importance for improving data quality and for integrating data from multiple sources. Numerous approaches for entity resolution have been developed and investigated [4, 13]. They derive match decisions typically based on the combined similarity of several attribute values and possibly on the contextual similarity of entities (for example, two publications may match if they have both highly similar titles and co-authors). To achieve high efficiency for large datasets, one has to avoid comparing each entity to all other entities. This is achieved by so-called blocking strategies [4] where only records within the same block (partition) need to be compared with each other, e.g., only publications from the same year. Entity resolution can also be performed in parallel on multiple processors and computing nodes to achieve additional performance improvements [12].

Most previous ER approaches compare pairs of entities and determine binary match mappings consisting of all correspondences or links between two matching entities. This is a natural approach when one has to integrate only a few data

M. Kirikova et al. (Eds.): ADBIS 2017, LNCS 10509, pp. 278–293, 2017.
DOI: 10.1007/978-3-319-66917-5_19

sources but it does not scale well since the number of binary mappings grows quadratically with the number of sources. As a result, integrating data from 200 sources would require the determination (and maintenance) of 19.900 mappings which is not practically feasible with today's ER tools. Grouping all matching entities within single clusters is a better approach for integrating data from multiple data sources as it allows a more compact match representation than with binary links [15,17]. It also simplifies the fusion of the matching entities for data integration by combining and consolidating the attributes values of the different cluster members. Furthermore, it allows an incremental integration of additional entities and data sources by comparing them with the set of previously determined clusters.

In our research, we aim at scalable ER approaches for Big Data that are able to deal with large data volumes and multiple data sources. We therefore need ER approaches that support clustering matching entities and exploit both blocking and distributed (parallel) processing. For this study, we implemented distributed versions of six previously proposed clustering techniques to analyze their quality and scalability. The considered clustering schemes require as input a so-called similarity graph containing all links between matching entities and try to find additional links by considering indirect matches and to eliminate weaker links in favor of more plausible ones. The clustering schemes are part of a new framework called FAMER (FAst Multi-source Entity Resolution system) for distributed multi-source entity resolution. Famer is implemented on top of the distributed dataflow framework Apache Flink to achieve a high scalability to large amounts of data and many machines.

In the next section, we briefly discuss related work. In Sect. 3, we provide an overview about our ER framework Famer. Section 4 describes the considered clustering algorithms and their distributed implementation. In Sect. 5, we evaluate the match quality and scalability of the approaches for different datasets. Section 6 concludes.

2   Related Work

There is a huge literature about ER and there are several books and surveys to provide an overview about the main methods and tools, e.g., [4,13]. The parallel implementation of ER methods has also been studied but mainly for MapReduce (e.g., [12]). Only few studies considered more recent Big Data frameworks such as Apache Flink or Apache Spark [14] but not yet for clustering-based ER schemes. Our distributed ER framework will build on known blocking and matching techniques (see next section) and their parallel implementation using Apache Flink.

Previous clustering approaches for ER [2,3,7,8,16] first determine a pairwise matching between entities and apply clustering within a post-processing step. The most straight-forward clustering approach is computing connected components based on the transitive closure of binary match links. This approach can often improve recall by identifying indirectly matching entities but may lead to poor precision since indirect matches may not be similar enough to really represent the same real-word object. For our evaluation, we use connected components as the base strategy and consider five additional clustering schemes that have proven to be effective in previous studies. In particular, we study parallel versions of correlation clustering [3], Center [8], Merge Center [8], and two versions of Star [1] clustering. Previous evaluations such as in [7] did not consider parallel clustering schemes and focused on clustering within single datasets rather than across multiple data sources.

3   Famer Framework for Multi-source Entity Resolution

Figure 1 illustrates the main components and processing steps of the Famer framework for distributed multi-source entity resolution. The components are similar to the ones in previous entity resolution tools but support more than two sources and are implemented in Apache Flink to achieve a parallel execution for high scalability. The input of Famer are thus multiple data sources with the entities to be matched and clustered. The output is a collection of clusters where all entities within a cluster match with each other and different clusters refer to different real-world objects. In this paper, we assume that all sources are duplicate-free[1] so that we only have to find matching entities between sources. The final match clusters should thus only contain entities from different sources so that the maximal cluster size is limited by the number of sources. All entities of a cluster are assumed to match with each other, so that a cluster of m entities represents $\frac{m \cdot (m-1)}{2}$ match pairs.

Famer consists of two main parts (Fig. 1): generation of a similarity graph based on pairwise matches and clustering. The first component has several steps (blocking, pairwise comparison, match classification), which can be customized according to a configuration input. We provide more details on the different steps below. We also illustrate the workflow of our framework for the person records in Table 1 that originate from four sources and contain erroneous attribute values as typical for real-world data. The table groups already the duplicate records referring to the same person.

In the first phase, we start with a *blocking step* to reduce the number of comparisons compared to a naïve approach where each entity of a data source has to be compared against all entities of any other source. We support different blocking techniques such as Standard Blocking (SB) and Sorted Neighborhood as well as single- and multi-pass blocking [4]. For SB, which we will use in our evaluation, entities are partitioned into blocks by a predefined blocking key (to be provided in the configuration input) on attribute values such that only entities with the same blocking key need to be compared with each other. For the person records in Table 1, we assume that the two initial letters of the surnames form the blocking key. Table 2 shows the resulting blocking key values and blocks sharing the same key value. For this example, blocking reduces the number

---

[1] This is not a main restriction since we could first deduplicate the individual data sources before applying the workflow.

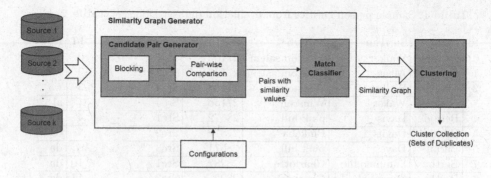

Fig. 1. Overview of the famer approach for multi-source entity resolution

of comparisons from 55 to only 16. On the other hand, we may now miss some matches if similar entities are assigned to different blocks (e.g., entity with id 1 is not paired with entities 0, 2, and 3). Multi-pass blocking can reduce this problem (at the expense of more comparisons) by partitioning the entities according to multiple blocking keys.

After blocking, all entities of a block from any of the input data sources are pairwise compared with each other. For each entity pair we compute the similarity of their attribute values for the attributes and similarity functions specified in the configuration input. These similarity values are used in the following match classification step to decide about whether or not a pair of entities is assumed to match. The classification approach is also specified in the configuration input, e.g., by match rules specifying the required minimal similarity for the considered attributes. The output of this step is the set of matching entity pairs (links) together with a combined similarity value per link. This output is stored as a similarity graph where entities are represented as vertices and match links as edges.

The *clustering* step of Famer aims at grouping together all matching vertices of the similarity graph based on the link structure of the graph and possibly the similarity values. Clustering algorithms typically try to group entities such that the similarity between entities within a cluster is maximized while the similarity between entities of different clusters is minimized. Compared to the similarity graph, the clustering algorithm can ideally add all missing matches (links) and remove all wrong links. The clustering algorithms we implemented and evaluated are described in Sect. 4.

Figure 2 illustrates the results of the described workflow for the sample entities of Table 1 and the blocking using SB as shown in Table 2. The entities are pairwise compared within the blocks and a rule-based match classification is applied resulting in the similarity graph shown in the middle of Fig. 2. Compared to the matches assumed in Table 1, the graph misses some links between matching entities, e.g., between 0 and 2. The final clustering determines five fully connected graphs (clusters) which are meant to represent different entities. For the example, the clusters include links missing in the similarity graph.

Table 1. Sample person entities from evaluation dataset DS3.   **Table 2.** Keys

Id	Name	Surname	Suburb	Post code	SourceId
0	ge0rge	Walker	winston salem	271o6	Src1
1	George	Alker	winstom salem	27106	Src2
2	George	Walker	Winstons	27106	Src3
3	Geoahge	Waker	Winston	271oo	Src4
4	Bernie	Davis	pink hill	28572	Src1
5	Bernie	Daviis	Pinkeba	2787z	Src2
6	Bernii	Davs	pink hill	28571	Src3
7	Bertha	Summercille	Charlotte	28282	Src1
8	Bertha	Summeahville	Charlotte	2822	Src2
9	Brtha	Summerville	Charlotte	28222	Src4
10	Bereni	dan'lel	Pinkeba	27840	Src3
11	Bereni	Dasniel	Pinkeba	2788o	Src4

Id	Key
0	wa
2	wa
3	wa
1	al
4	da
5	da
6	da
10	da
11	da
7	su
8	su
9	su

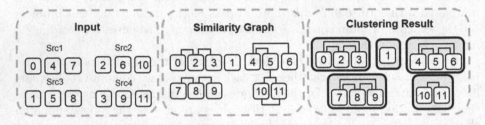

Fig. 2. Applying famer to the data of Table 1.

Compared to Table 1, all clusters are correctly found except for the singleton cluster with entity 1 that was not matched with matching entities 0, 2 and 3 due to the assumed blocking approach.

Famer is implemented using Apache Flink and a new extension for graph analytics called Gradoop [11]. Hence, all match and clustering approaches can be executed in parallel on clusters of variable size. Gradoop supports an extended property graph model so that we store the attribute values of entities as key value properties. Analogously, the similarity values of matching entity pairs are represented as edge properties. For the implementation of the parallel clustering schemes we also use the Gelly library of Flink supporting a so-called vertix-centric programming of graph algorithms (see next section).

4   Clustering Approaches

In this section, we present the considered clustering approaches for entity resolution and their parallel implementation. As described in the previous section, all algorithms use as input a similarity graph with entities from multiple data sources and similarity edges indicating the computed degree of similarity. In addition to the computation of connected components, Famer supports parallel versions of the Center clustering, Merge Center clustering, two versions of Star clustering, and correlation clustering.

The parallel implementations are based on a vertex-centric programming model, also known as 'think like a vertex', to iteratively execute a user-defined program in parallel over all vertices of a graph. In particular, we use the two-step Scatter-Gather model of Gelly that breaks up vertex programming into two functions. In the Scatter step, a value is distributed to all vertex neighbors, and in the Gather step the inputs from the neighbors are collected to update the state of a vertex. The computation proceeds in synchronized iteration steps, called supersteps. Each scatter and each gather execution is performed in a different superstep. Supersteps are executed synchronously, so that messages sent during one superstep are guaranteed to be delivered in the beginning of the next superstep [10]. The vertex functions are executed by a configurable number of worker nodes among which the graph data is partitioned, e.g., according to a hash or range partitioning on the vertex ids. We will explain the vertex-centric implementation in detail for one of the clustering schemes (Center); the other implementations follow similar approaches.

4.1   Connected Components

The subgraphs of a graph that are not connected to each other are called connected components. Having the input similarity graph, the connected components are easy to determine in a vertix-centric way by letting every vertex iteratively add all its direct neighbors to its cluster. The approach is therefore easy to implement with Scatter-Gather (as shown in [10]). In the evaluation, we use this approach as a baseline for the comparison with the other clustering schemes.

4.2   Center Clustering

In contrast to connected components, the Center clustering algorithm [8] utilizes the similarity values (weights) of the edges in the similarity graph. In the sequential algorithm, edges are first sorted based on these weights in descending order and put in a queue. Edges are then removed from the queue and processed one by one. For each edge $e_{(v_i, v_j)}$, if both v_i and v_j are unassigned to any cluster, one of them will be center and the other will belong to the cluster of that center. If one of them is center and the other is unassigned, the unassigned vertex will belong to the cluster of the center vertex. If both vertices are centers or both of them are non-centers, or one of them is non-center and the other is unassigned, that edge is ignored.

We propose and implemented a parallel version of the Center algorithm (see Algorithm 1). In each round of the algorithm for all unassigned vertices, the outgoing edge with the highest weight must be found. The vertices on both sides of this edge are then processed. If one of them is center, the other will belong to the cluster of that vertex (lines 6–8). If one of them is assigned to another cluster (line 9), i.e., both vertices belong to different clusters, the edge between these two vertices is removed (line 10). If both vertices are unassigned and the edge

---

Algorithm 1. Parallel Center

Data: $G = (V, E)$
1 assignVertexPriorities(V)
   /* priority according to a random permutation of vertices   */
2 $Center \leftarrow \{\}$
3 **for** $v_i \in V$ **in Parallel do**
4   **repeat**
5        $v_{nn} \leftarrow \underset{j}{\mathrm{argmax}}(e_{(v_i, v_j)})$
6   **if** $(v_{nn} \in Center)$ **then**
7           $v_i.SetClusterId(nn)$
8           $V \leftarrow V - \{v_i\}$
9   **else if** $(v_{nn} \notin V)$ **then**
10          $E \leftarrow E - \{e_{(v_i, v_{nn})}\}$
11   **else**
12          $v_k \leftarrow \underset{j}{\mathrm{argmax}}(e_{(v_{nn}, v_j)})$
13   **if** $((i = k \wedge i > nn) \vee (v_{nn} = Null))$ **then**
14             $Center \leftarrow Center \cup \{v_i\}$
15             $v_i.SetClusterId(i)$
16             $V \leftarrow V - \{v_i\}$
17   **until** $(v_i \in V)$

---

between them is for both the outgoing edge with the highest weight (line 13, $i = k$), then one of them is assumed as center (line 14) and the other will belong to the same cluster in the next round. For selecting the center in this case we make use of initially assigned (line 1) vertex priorities as done in the sequential algorithm. Hence, the vertex with higher priority is considered as center (line 16, $i > nn$). If a vertex is not connected to any other vertex (line 13, $v_{nn} = Null$), it is a singleton. The algorithm iterates until all vertices are assigned to a cluster (line 17).

We implemented parallel Center using the Scatter-Gather model (see Algorithm 2). The algorithm applies two phases that are iteratively executed for all vertices. Phase 1 (Scatter1, Gather1) finds for each vertex v_i its neighboring vertex with the currently highest edge weight, and phase 2 (Scatter2, Gather2) processes the status of the found vertex and assigns v_i to an existing cluster or considers it as a center. Again, we initially assign a priority per vertex (line 3). In phase 1, for each vertex v_i the neighbor with the K-highest edge weight (nearest neighbor NN) is found (lines 13–21). K is a helper variable. It helps to prevent that already assigned vertices are chosen again as neighbor. It is attached to each vertex and initialized with 1 (lines 5–7). It will be incremented in phase 2 when a vertex neighbor has been assigned to a cluster (lines 39–41). In phase 2, all neighbors of a vertex v_i are sorted and processed in descending order of the

---

Algorithm 2. Parallel Center with Scatter-Gather

Data: $G = (V, E)$

```
1  Algorithm Center
2      assignVertexPriorities(V)
       /* set priority according to a
          random permutation of vertices    */
3      for (v_i ∈ V) do
4       |  v_i.K ← 1
5      end
6      repeat
7          |    Phase1: Scatter1 (Vertex)
8          |            Gather1 (Vertex, MessageIterator)
           |    Phase2: Scatter2 (Vertex)
9          |            Gather2 (Vertex, MessageIterator)
10     until (V ≠ {})
11 Procedure Scatter1 (Vertex v)
12     for (e ∈ getOutEdges()) do
13      |  msg.Src ← v.getId()
14      |  msg.Weight ← e.getWeight()
15      |  sendMessageTo(edge.target(), msg)
16     end
17 Procedure Gather1 (Vertex v, MessageIterator messages)
18     Array ← messages.Sort()
       /* Messages are sorted based on
          their weights descendingly         */
19     v.NN ← Array[v.K].getSrc()
```

```
20 Procedure Scatter2 (Vertex v)
21     msg.Src ← v.getId()
22     msg.NN ← v.getNN()
23     msg.Priority ← v.getPriority()
24     for (e ∈ getOutEdges()) do
25      |  msg.Weight ← e.getWeight()
26      |  sendMessageTo(edge.target(), msg)
27     end
28 Procedure Gather2 (Vertex v, MessageIterator messages)
29     Array ← messages.Sort()
       /* sorted based on weights
          descendingly                       */
30     for (i : v.K → Array.Size()) do
31      |  m ← Array[i]
32      |  if (m.getSrc().isCenter()) then
33      |   |  v.ClusterId ← m.getId()
34      |   |  v.assigned ← true
35      |   |  break
36      |  end
37      |  else if (m.getSrc().isAssigned()) then
38      |   |  v.K + +
39      |  end
40      |  else if (v.NN= Null ∨ (v.NN = m.getSrc() ∧
           v.Priority > m.getPriority()) ) then
41      |   |  v.ClusterId ← m.getSrc()
42      |   |  v.center ← true
43      |   |  v.assigned ← true
44      |   |  break
45      |  end
46     end
```

---

edge weights (for the edge to v_i) (lines 32–38). Then vertex v_i is set as center similar to Algorithm 1 (lines 42–47).

4.3 Merge Center

The Merge Center clustering algorithm [8] is a modified version of Center. In contrast to Center, it merges two clusters if a vertex in one cluster is similar to the center of another cluster. Our parallel implementation for Merge Center is very similar to parallel Center but applies an extra iteration for merging clusters. This iteration is initiated right after all vertices are assigned to a cluster. The merge processing is repeated until there are no further cluster changes.

4.4 Star Clustering

The Star clustering algorithm [1] initially computes the degree for each vertex of the similarity graph. Then in each iteration, the unassigned vertex with the highest degree becomes center and all its direct neighbors are assigned to its cluster. The algorithm terminates when all vertices are assigned to a cluster. In contrast to all other clustering approaches, Star clustering can result in overlapping clusters. As a consequence, it introduces the need of a post-processing to select the best cluster for entities that have been assigned to several clusters.

Our parallel version of the Star algorithm is described in Algorithm 3. Initially, the degree of all vertices is computed and if the degree of a vertex is greater than the degree of all its neighbors, that vertex becomes a center (lines 4–7). If the degree of two adjacent vertices is equal, the one with higher priority is

---

Algorithm 3. Parallel Star

Data: $G = (V, E)$

1 $V \leftarrow \{v_1, ..., v_n\}$
        `/* A random permutation of vertices`   `*/`
2 $Center \leftarrow \{\}$
3 **repeat**
4   **for** $(v_i \in V)$ *in Parallel* **do**
5        $v_{max} \leftarrow \underset{v_j \in \{v_j | e(v_i, v_j) \in E\} \cup \{v_i\}}{\mathrm{argmax}} (ComputeDegree(v_j))$
6   **if** $(v_i = v_{max})$ **then**
7           $Center \leftarrow Center \cup \{v\}$

8   **for** $(v_i \in V)$ *in Parallel* **do**
9   **for** $(e(v_i, v_j) \in E)$ **do**
10   **if** $(v_j \in Center)$ **then**
11             $v_i.addClusterId(v_j.getId())$
12             $V \leftarrow V - \{v_i\}$

13 **until** $(V \neq \{\})$

---

assumed as center. Similar to the previous parallel algorithms, vertex priority is initially determined by generating a random permutation of vertices (line 1). Then each center and all its neighbors are considered as a cluster. (lines 8–12). The Scatter-Gather version of Algorithm 3 uses three phases. In the first phase the degree of each vertex is computed. In the second phase, centers are selected, and in the final phase, clusters are grown around centers.

We use two methods for computing the degree of vertices resulting into algorithms Star-1 and Star-2. For Star-1, we count the number of outgoing edges of a vertex, while Star-2 is based on the average similarity degrees of the outgoing edges of a vertex.

4.5 Correlation Clustering

The original correlation clustering approach [2] uses a graph with positive and negative edge weights to indicate whether two vertices are similar (positive edge weight) or dissimilar (negative edge weight). The goal is to find a clustering that either maximizes agreements (sum of positive edge weights within a cluster plus the absolute value of the sum of negative edge weights between clusters) or minimizes disagreements (absolute value of the sum of negative edge weights within a cluster plus the sum of positive edge weights across clusters). Gionis et al. propose an approximate and iterative solution for this optimization problem [6] that randomly selects an unassigned vertex as a cluster center in each round. Then all unassigned neighbors of the selected center are added to the cluster and marked as assigned. The algorithm terminates when there is no unassigned vertex left.

This simple algorithm suffers from too many rounds making it unsuitable for very large graphs. Some studies therefore proposed parallel solutions [3,16] that select multiple centers in each round. They also address the newly introduced concurrency problem to avoid that a vertex is assigned to more than one center at a time. We implemented the parallel pivot approach of [3], called CCPivot, since it fits well the Scatter-Gather paradigm. In each round of this algorithm, several vertices are considered as active nodes, i.e., as candidates for becoming a cluster center (or pivot). In the next step, active nodes that are adjacent to each other are removed from the set of active nodes; the remaining vertices become centers. Then adjacent vertices of each center are assigned to that center and form a cluster. If one vertex is adjacent of more than one center at the same time, it will belong to the one with higher priority. As in the other algorithms, the vertex priorities are determined in a preprocessing phase.

Our Scatter-Gather implementation of this algorithm uses three Scatter-Gather phases: one for computing the current maximum degree of the graph, one for selecting active nodes and applying the concurrency-aware rule to select final centers, and one for growing clusters around centers.

5   Evaluation

The goal of our evaluation is to comparatively evaluate the effectiveness and efficiency of the considered clustering approaches and their distributed implementations for different datasets and configurations. We first describe the used datasets from three domains and the considered configurations. We then analyze the relative match and clustering effectiveness of the clustering schemes. Finally we evaluate the runtime performance and scalability of the approaches.

5.1   Datasets and Configuration Setup

For our evaluation we use datasets from three domains for different numbers of duplicate-free sources. Table 3 shows the main characteristics of the datasets in particular the number of clusters and match pairs of the perfect ER result. The smallest dataset DS1 contains geographical real-world entities from four different data sources (DBpedia, Geonames, Freebase, NYTimes) and has already been used in the OAEI competition[2]. For our evaluation we focused on a subset of settlement entities as we had to manually determine the perfect clusters and thus the perfect match pairs.

For the two larger evaluation datasets DS2 and DS3 we applied advanced data generation and corruption tools [9] to be able to evaluate the ER quality and scalability for larger datasets and a controlled degree of corruption. DS2 is based on real records about songs from the MusicBrainz database but uses the DAPO data generator to create duplicates with modified attribute values [9]. The generated dataset consists of five sources and contains duplicates for

---

[2] OAEI 2011 IM: http://oaei.ontologymatching.org/2011/instance/.

Table 3. The specifications of datasets.

Domain	Attributes	#entities	#sources	#perfect match pairs	#clusters
Geographical (DS1)	label, longitude, latitude	3,054	4	4,391	820
Music (DS2)	title, length, artist, album, year, language	20,000	5	16,250	10,000
Persons (DS3)	name, surname, suburb, postcode	5,000,000 10,000,000	5 10	3,331,384 14,995,973	3,500,840 6,625,848

50% of the original records in two to five sources. All duplicates are generated with a high degree of corruption to stress-test the ER and clustering approaches. DS3 is based on real person records from the North-Carolina voter registry and synthetically generated duplicates using the tool GeCo [5]. We consider two configurations with either 5 or 10 sources each having 1 million entities; i.e. we process up to 10 million person records. Each source is duplicate free, but 50% of the entities are replicated in all sources without any corruption. Moreover, 25% of entities are corrupted and replicated in all sources, and the remaining 25% are corrupted but present in only some sources. For the generation of corrupted records we applied a moderate corruption rate of 20%, i.e., most attribute values remained unchanged.

Table 4. Default blocking and match configuration for different datasets.

Dataset	Blocking key	Similarity functions	Match rule
DS1	prefixLength1(label)	sim1: Jarowinkler (name)	$sim1 \geq \theta$ &
		sim2: geographical distance	$sim2 \leq 1358\,km$
DS2	prefixLength1(album)	sim1: 3Gram (title)	$sim1 \geq \theta$
DS3	prefixLength3(surname)	sim1: Jarowinkler (name)	$sim1 \geq 0.9$ &
		sim2: Jarowinkler (surname)	$sim2 \geq 0.9$ &
		sim3: Jarowinkler (suburb)	$sim3 \geq \theta$ &
		sim4: Jarowinkler (postcode)	$sim4 \geq \theta$

To generate the similarity graphs for the different datasets as the input of the clustering schemes we experimented with a large spectrum of blocking and match configurations. Due to space restrictions, we will mostly report results only for the default configurations specified in Table 4 that resulted already in good match quality even without clustering. All configurations apply standard blocking with different blocking keys. The match rules compute different attribute similarities using either string similarity functions (Jarowinkler, 3gram) or geographical distance as well as variable similarity thresholds θ.

5.2   Match Quality of Clustering Approaches

To evaluate the ER quality of our clustering results we use the standard metrics precision, recall and their harmonic mean, F-Measure. These metrics are determined by comparing the computed match pairs (derived from the computed clusters assuming that all entities in a cluster match) with the perfect match results.

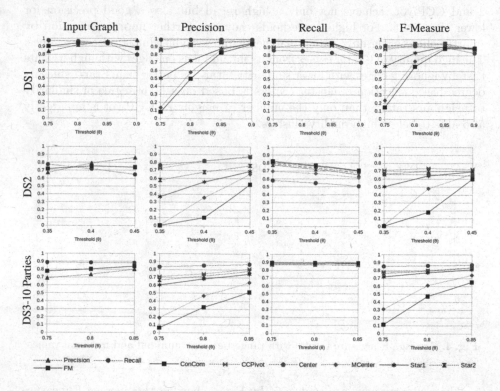

Fig. 3. Match quality of clustering-based ER approaches.

In Fig. 3, we compare the obtained precision, recall and F-measure results for the six clustering schemes, different similarity thresholds θ and our three datasets using the default configurations from Table 4 to determine the initial similarity graphs. On the left, we also show the precision, recall and F-measure values obtained already with the similarity graphs. We observe that for DS1 and DS3 we achieve a relatively high F-measure of more than 0.9 and 0.8 for the considered θ range between 0.75 and 0.9. By contrast, for the noisy data records of DS2 we had to lower the similarity thresholds to values between 0.35 and 0.45 and still could not exceed a maximal F-measure of 0.73 underlining that DS2 represents a more difficult match problem than DS1 or DS3.

Comparing the clustering schemes, we observe that there are substantial differences in their relative match quality. Connected components reaches the lowest F-measure for all datasets and almost all threshold values because it suffers from very poor precision values. Merge Center (MCenter) shows a similar behavior in terms of poor precision and F-measure, indicating that the merging of clusters can often lead to wrong cluster decisions. From the four better clustering schemes, Star-1 has the lowest F-measure (especially for lower values of the similarity threshold values) while the other three are close together and relatively robust against changes in the threshold value. These approaches, Center, Star-2 and CCPivot, achieve not only a high recall but also a good precision for lower thresholds. For higher thresholds they can further improve precision by smartly eliminating only wrong matches while keeping almost all correct ones. The high quality of Center comes from its initial focus on edges with high weights thereby ignoring edges with lower similarity. Star-2 is better than Star-1 since its degree-based selection of cluster centers is based on a high degree of similarity to neighbors rather than only the number of neighbors. CCPivot is apparently also able to select high quality clusters.

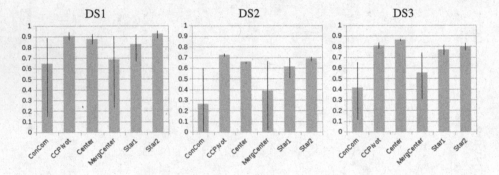

Fig. 4. Average F-measure results with range between minimal and maximal values

These observations are confirmed by Fig. 4 showing the average F-measure results of the clustering schemes over all threshold configurations. The vertical lines also show the F-measure spread between the minimal and maximal value. We again observe the low and highly variable match quality of connected components and MergeCenter. By contrast, the algorithms CCPivot, Center, and Star-2 are more robust and achieve the highest F-measure values.

5.3 Runtimes and Speedup

We determined the runtimes of the clustering algorithms on a cluster with 16 worker nodes. Each worker consists of an E5-2430 6(12) 2.5 GHz CPU, 48 GB RAM, two 4 TB SATA disks and runs openSUSE 13.2. The nodes are connected via 1 Gigabit Ethernet. Our evaluation is based on Hadoop 2.6.0 and Flink 1.1.2. We run Apache Flink standalone with 6 threads and 40 GB memory per worker. In our experiments, we vary the number of workers by setting the parallelism parameter to the respective number of threads (e.g., 4 workers correspond to 24 threads). The runtime of all algorithms is measured for the largest dataset DS3 with 5 and 10 parties applying the configuration from Table 4 with $\theta = 0.80$. The DS3 input datasize is thus doubled for 10 parties compared to 5 parties. We only evaluate the runtimes for the clustering algorithms since the time to determine the similarity graphs is the same for all clustering approaches. Some clustering approaches could not be executed for 1 or 2 workers only due to high memory requirements. We thus evaluate the runtimes for configurations between 4 and 16 workers. Table 5 shows that the runtimes for the two DS3 datasets. The increased dataset size for 10 parties leads to higher runtimes for all algorithms although to different degrees. As expected, the fastest runtimes are achieved by the simple connected components approach. By contrast, CCPivot has the worst runtimes due to large memory requirements and a high message overhead. For the bigger dataset (10 parties) the approach suffered from out-of-memory errors and could only be executed for 16 workers. From the three clustering schemes achieving the best matching quality (Star-2, Center, CCPivot), Star-2 achieves by far the fastest runtimes in all configurations making it a good default strategy for clustering.

Except for connected components, all algorithms can reduce their runtimes by applying more workers, especially for the larger dataset with 10 parties. Figure 5 shows the resulting speedup values. For DS3 with 5 parties, all algorithms except the slow CCPivot achieve an almost linear speedup. For the bigger dataset with 10 parties, speedup values are even better and partly super-linear. The latter, however, is an artifact for the slower algorithms like MCenter that perform poorly for 4 workers because of memory bottlenecks (its runtime for 4 workers is almost 6 times higher for 10 parties than for 5 parties). The substantially increased aggregate memory capacity for 8 and 16 workers thus enabled super-linear runtime improvements but without reaching the absolute runtimes of fast algorithms like Star-2.

Table 5. Runtimes (seconds)

dataset	DS3 - 5 parties			DS3 - 10 parties		
#workers	4	8	16	4	8	16
ConCom	51	57	55	101	79	79
CCPivot	1530	1008	688	–	–	1303
Center	390	208	117	1986	864	423
MCenter	640	349	194	3767	1592	695
Star-1	288	149	85	783	367	197
Star-2	214	124	67	720	317	173

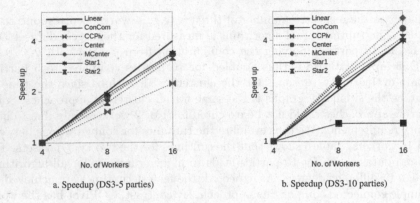

a. Speedup (DS3-5 parties)   b. Speedup (DS3-10 parties)

Fig. 5. Runtimes and speedup

6   Conclusions and Outlook

We presented a new framework called Famer enabling the parallel execution
of ER workflows using the Big Data framework Apache Flink. Famer supports
entity resolution for multiple data sources and groups all matching entities within
clusters. For parallel clustering we currently support six approaches that have
been evaluated for datasets from three domains. The evaluation showed that
three clustering approaches (Center, Star-2 and CCPivot correlation clustering) achieve a similarly high match quality that is clearly superior to a simple
connected components scheme. The parallel implementations of the clustering
approaches mostly achieve good speedups, especially for larger datasets thereby
supporting high scalability. Star-2 achieves lower runtimes than Center and especially CCPivot so that it is a good default approach for clustering-based ER.

In future work, we will further extend and improve Famer, e.g., by post-
processing cluster results to find additional matches or resolve overlapping clusters for Star clustering. We also aim at developing incremental ER strategies that
can incorporate new entities and data sources into already existing clusters.

Acknowledgement. This work was partly funded by the German Federal Ministry of
Education and Research within the project Competence Center for Scalable Data Services and Solutions (ScaDS) Dresden/Leipzig (BMBF 01IS14014B). Also, evaluations
partly performed on the Galaxy-Infrastructure at Leipzig University.

References

1. Aslam, J., Pelekhov, E., Rus, D.: The star clustering algorithm for static and
   dynamic information organization. J. Graph Algorithms Appl. **8**, 95–129 (2004)
2. Bansal, N., Blum, A., Chawla, S.: Correlation clustering. In: Proceedings of the
   Foundations of Computer Science, pp. 238–247. IEEE (2002)
3. Chierichetti, F., Dalvi, N., Kumar, R.: Correlation clustering in MapReduce. In:
   Proceedings of the ACM SIGKDD Conference, pp. 641–650 (2014)

4. Christen, P.: Data Matching: Concepts and Techniques for Record Linkage, Entity Resolution, and Duplicate Detection. Springer, Heidelberg (2012)
5. Christen, P., Vatsalan, D.: Flexible and extensible generation and corruption of personal data. In: Proceedings of CIKM, pp. 1165–1168 (2013)
6. Gionis, A., Mannila, H., Tsaparas, P.: Clustering aggregation. ACM Trans. Knowl. Discov. Data (TKDD) **1**(1), 4 (2007)
7. Hassanzadeh, O., Chiang, F., Lee, H., Miller, R.: Framework for evaluating clustering algorithms in duplicate detection. PVLDB **2**(1), 1282–1293 (2009)
8. Hassanzadeh, O., Miller, R.: Creating probabilistic databases from duplicated data. VLDB J. **18**(5), 1141–1166 (2009)
9. Hildebrandt, K., Panse, F., Wilcke, N., Ritter, N.: Large-scale data pollution with Apache Spark. IEEE Trans. Big Data (2017)
10. Junghanns, M., Petermann, A., Neumann, M., Rahm, E.: Management and analysis of big graph data: current systems and open challenges. In: Zomaya, A.Y., Sakr, S. (eds.) Handbook of Big Data Technologies, pp. 457–505. Springer, Cham (2017). doi:10.1007/978-3-319-49340-4_14
11. Junghanns, M., Petermann, A., Teichmann, N., Gómez, K., Rahm, E.: Analyzing extended property graphs with Apache Flink. In: Proceedings of the ACM SIGMOD Workshop on Network Data Analytics (2016)
12. Kolb, L., Thor, A., Rahm, E.: Dedoop: efficient deduplication with Hadoop. PVLDB **5**(12), 1878–1881 (2012)
13. Köpcke, H., Rahm, E.: Frameworks for entity matching: a comparison. Data Knowl. Eng. **69**(2), 197–210 (2010)
14. Mestre, D., Pires, C., Nascimento, D., de Queriroz, A., Santos, V., Araujo, T.: An efficient Spark-based adaptive windowing for entity matching. J. Syst. Softw. **128**, 1–10 (2017)
15. Nentwig, M., Groß, A., Rahm, E.: Holistic entity clustering for linked data. In: IEEE ICDMW (2016)
16. Pan, X., Papailiopoulos, D., Oymak, S., Recht, B., Ramchandran, K., Jordan, M.: Parallel correlation clustering on big graphs. In: Advances in Neural Information Processing Systems, pp. 82–90 (2015)
17. Rahm, E.: The case for holistic data integration. In: Pokorný, J., Ivanović, M., Thalheim, B., Šaloun, P. (eds.) ADBIS 2016. LNCS, vol. 9809, pp. 11–27. Springer, Cham (2016). doi:10.1007/978-3-319-44039-2_2

Query Optimization, Recovery, and Databases on Modern Hardware

Cost-Function Complexity Matters:
When Does Parallel Dynamic Programming
Pay Off for Join-Order Optimization

Andreas Meister[✉] and Gunter Saake

University of Magdeburg, Magdeburg, Germany
{andreas.meister,gunter.saake}@ovgu.de

Abstract. The execution time of queries can vary by several orders
of magnitude depending on the join order. Hence, an efficient query
execution can be ensured by determining optimal join orders. Dynamic
programming determines optimal join orders efficiently. Unfortunately,
the runtime of dynamic programming depends on the characteristics of
the query, limiting the applicability to simple optimization problems.
To extend the applicability, different parallelization strategies were proposed. Although existing parallelization strategies showed benefits for
complex cost functions, the effects of the cost-function complexity was
not evaluated.

Therefore, in this paper, we compare different sequential and parallel
dynamic programming variants with respect to different query characteristics and cost-function complexities. We show that the parallelization
of a parallel dynamic programming variant is most often only useful for
complex cost functions. For simple cost functions, we show that most
often sequential variants are superior to their parallel counterparts.

1   Introduction

Relational Database Management Systems (DBMSs) provide a high usability by
using declarative query languages, such as the SQL. In declarative query languages, users only need to provide what result is needed, but not how the result
is determined. Hence, DBMSs need to transform queries into efficient query execution plans, defining the way to execute queries. Hereby, not only one but plenty
of equivalent query execution plans exist. Since the execution time of equivalent
query execution plans can vary by several orders of magnitude depending on
the join order, one of the most important challenges for determining an efficient
query execution plan is join-order optimization [5]. For example, for Query 5
of the TPC-H benchmark, the execution time varies from few milliseconds to
several minutes based on the chosen join order [8]. Hence, determining an optimal join order ensures an efficient query execution. Unfortunately, finding an
optimal join order is in general NP-complete [7]. Furthermore, DBMSs apply
time limits for query optimization, because otherwise the optimization might
take longer than executing an inefficient query execution plan. Therefore, two

© Springer International Publishing AG 2017
M. Kirikova et al. (Eds.): ADBIS 2017, LNCS 10509, pp. 297–310, 2017.
DOI: 10.1007/978-3-319-66917-5_20

aspects have to be considered to determine optimal join orders within the time limit of query optimization: efficient execution strategies and resource utilization. Dynamic Programming (DP) guarantees an optimal join order while ensuring an efficient execution strategy for the optimization. In the past, three different sequential variants of DP were proposed DP_{SIZE} [9], DP_{SUB} [11], DP_{CCP} [6]. While DP_{SIZE} and DP_{SUB} suffer from additional overhead based on invalid computations, DP_{CCP} ensures an efficient execution based on an enumeration of valid computations only.

Although the sequential DP variants provide efficient execution strategies, the sequential DP variants share the same drawback: the resource utilization. Current hardware architectures provide their computational power not through a fast Central Processing Unit (CPU) core, but by providing multiple CPU cores. Unfortunately, sequential approaches use only one CPU core and, hence, do not use available resources efficiently given a single query. Sequential approaches can only be used efficiently on multiple CPU cores, when multiple queries are available and each CPU core optimizes a single query independent of the other CPU cores. PDP_{SVA} [2] and $DPE_{GENERIC}$ [3] enable the parallel evaluation of the sequential DP variants, utilizing multiple CPU cores. By using all available CPU cores, the applicability of dynamic programming approaches is extended from queries containing 12 tables to queries containing upto 25 tables [2] with respect to the time limit of query optimization.

Independent of the used DP variant, cost functions are needed to compare different and to select optimal join orders. For each considered join pair, the cost function need to be called. Hence, the complexity and the runtime of cost functions has also a direct impact on the different DP variant. Although in different DBMSs different cost functions are used, to the best of our knowledge, the impact of different complexities of cost functions on the different DP variants was not evaluated yet.

Therefore, in this paper, we will make the following contributions:

- We evaluate sequential and parallel DP variants with respect to different query topologies, different complexities of cost functions, and an increasing number of tables.
- We show that not only the characteristic of queries but also the complexity of cost functions has significant impact on the parallel DP variants. Hereby, existing parallel DP variants mainly provide an advantage only with complex cost functions. For simple cost functions, the sequential DP variants are mostly superior to their parallel counterparts.

The remainder of this paper is structured as follows. In Sect. 2, we will provide background information about DP for join-order optimization in general and specifics about the sequential and parallel DP variants. In Sect. 3, we will evaluate the different sequential and parallel DP variants with respect to different complexity of the cost function, query toplogies, and table number. In the last section, we conclude our work.

2   Dynamic Programming for Join-Order Optimization

DP for join-order optimization is a deterministic exhaustive search approach [10]. Hence, DP guarantees an optimal join order with respect to the cost function. In contrast to the brute-force approach, DP skips unneeded evaluations by applying a bottom-up construction of (intermediate) solutions. First, the optimization problems is split into subproblems. These subproblems are solved in an optimal way. Optimal intermediate solutions are combined to provide solutions for more complex (sub-)problems until an optimal solution for the overall optimization problem can be provided.

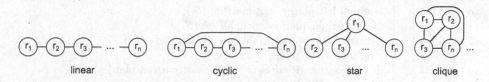

linear   cyclic   star   clique

Fig. 1. Query topologies. From left to right the complexity of join-order optimization increases based on the increasing number of join pairs.

2.1   Complexity

In general, join-order optimization is an NP-complete problem [7]. Hence, the runtime of DP depends on the query characteristics, such as topology and number of contained tables. In the field of join-order optimization, four relevant categories of query topologies exist: linear, cyclic, star, and clique, see Fig. 1.

In linear and cyclic queries, a table is joined with a maximum of two other tables, leading to a limited number of possible join pairs. In contrast, star and clique queries have a higher number of possible join pairs. Hence, star and clique queries are more complex to optimize than linear and cyclic queries. Hereby, star queries are representative for analytical workloads, while clique queries represent an optimization with cross joins.

2.2   Sequential Dynamic Programming Approaches

For the general concept of DP not only one but three different sequential DP variants were proposed in the past: DP_{SIZE} [9], DP_{SUB} [11], and DP_{CCP} [6].

In 1979, Selinger et al. proposed the first DP variant for join-order optimization, DP_{SIZE} [9], see Algorithm 1. For DP_{SIZE}, the different subproblems are related to the number of contained tables. This means for a given query, first, efficient options for accessing each table are evaluated, see Lines 1–2. Afterwards, these given solutions are combined with each other to provide solutions for the next (sub-)problem containing an additional table, see Lines 3–14. For each iteration, first, the number of tables for each join partner needs to be determined, see Lines 4–5. Afterwards, all available solutions containing the number of tables

Algorithm 1. DP_{SIZE} [9]

    Input   : Join query Q with n tables $T = \{T_1, \ldots, T_n\}$
    Output: an optimal bushy join tree

1 **foreach** $T_i \in T$ **do**
2 | optimalPlan(T_i) = T_i ;
3 **for** s = 2 **to** n **do**
4 | **for** $s_l = 1$ **to** $s - 1$ **do**
5 | | $s_r = s - s_l$;
6 | | **foreach** $S_l \subset T : |S_l| = s_l$ **do**
7 | | | **foreach** $S_r \subset T : |S_r| = s_r$ **do**
8 | | | | **if** $S_l \cap S_r \neq \emptyset$ **then continue**;
9 | | | | **if** S_l **not connected to** S_r **then continue**;
10 | | | | optimal-left-plan = optimalPlan(S_l);
11 | | | | optimal-right-plan = optimalPlan(S_r);
12 | | | | current-plan = createJoinTree(optimal-left-plan,
                          optimal-right-plan);
13 | | | | **if** cost(optimalPlan($S_l \cup S_r$)) > cost(current-plan) **then**
14 | | | | | optimalPlan($S_l \cup S_r$) = current-plan ;
15 **return** optimalPlan(T) ;

---

for each join partner are combined, see Lines 6–12. Since each pair of solutions containing a specific number of tables is evaluated, first, each join pair need to be checked whether the join pair is valid, see Lines 8–9, before the join pair can be evaluated, see Line 12. Since multiple equivalent plans exist, equivalent solutions need to be pruned, see Lines 13–14. At the end, an optimal join order is returned, see Line 15.

One problem of DP_{SIZE} is that within each iteration, intermediate solutions are combined naively. A join pair is only selected by the number of contained tables, but not by the actual available tables, see Lines 6–7. Hence, also invalid join pairs need to be evaluated, see Line 9. Based on the evaluation of these invalid join pairs, the efficiency of optimization is reduced. To avoid this inefficiency, Vance et al. proposed DP_{SUB} [11], see Algorithm 2. DP_{SUB} uses integer representations to enumerate all possible subsets in order to construct all valid join pairs. If the i-th bit is set, it means that the i-th table is available, e.g., $1 = \{T_1\}$, $2 = \{T_2\}$, $3 = \{T_1, T_2\}$, $4 = \{T_3\}$. Similar to DP_{SIZE}, in DP_{SUB}, first, different options for accessing each table are evaluated, see Lines 1–2. Afterwards, the solution that should be evaluated next, is determined by the integer representation, see Lines 3–4. Based on the solution, all valid join pairs are determined, see Lines 6–11, and a new solution is constructed, see Line 12. Since equivalent solutions exist, the evaluated solution has to be pruned, see Lines 13–14. After evaluating all valid join pairs, the optimal join order is returned, see Line 15.

Although DP_{SUB} avoids inefficiencies based on invalid join pairs of intermediate solutions, new inefficiencies are introduced based on different query topologies, see Line 7 and Lines 10–11. DP_{SUB} was proposed for optimizing queries,

Algorithm 2. DP_{SUB} [11]
___
 Input : Join query Q with n tables $T = \{T_1, \ldots, T_n\}$
 Output: an optimal bushy join tree
1 **foreach** $T_i \in T$ **do**
2 | optimalPlan$(T_i) = T_i$;
3 **for** k = 2 **to** n **do**
4   **for** $S = 2^{k-1} + 1$ **to** $2^k - 1$ **do**
5   **foreach** $S_l \subsetneq S$ **do**
6   optimal-left-plan = optimalPlan(S_l);
7   **if** optimal-left-plan $= \emptyset$ **then continue**;
8     $S_r = S - S_l$;
9   optimal-right-plan = optimalPlan(S_r);
10   **if** optimal-right-plan $= \emptyset$ **then continue**;
11   **if** optimal-left-plan **not connected to** optimal-right-plan $\neq \emptyset$
     then continue;
12   current-plan = createJoinTree(optimal-left-plan,
     optimal-right-plan);
13   **if** cost(optimalPlan(S)) > cost(current-plan) **then**
14 | optimalPlan(S) = current-plan ;
15 **return** optimalPlan$(2^n - 1)$;
___

considering cross-joins similar to clique queries. When other query topologies, such as linear, cyclic or star query topologies, are considered, not every subset is valid. Hence, depending on the query topology, unneeded join pairs need to be evaluated leading to inefficiencies. To consider the query topology, Moerkotte and Neumann proposed DP_{CCP} [6], see Algorithm 3. Within DP_{CCP}, only required join pairs for (intermediate) solutions are evaluated based on the topology of queries. Hereby, all join pairs are enumerated based on pairs of connected-sub-graphs and complements. For the evaluation, first, equivalent table accesses are evaluated for each table, see Lines 1–2. Afterwards, all connected-sub-graphs are determined, see Line 3. For each determined subgraph, all complements are determined, see Lines 4–5. The evaluation of connected-sub-graphs, as well as, complements are based on a breadth-first node enumeration and a recursive evaluation of neighboring nodes [6]. Afterwards, each pair of subgraph and complement is evaluated, see Lines 6–14. Since each pair is only created once, within DP_{CCP}, both commutative options need to be evaluated and pruned respectively, see Lines 9–14. Afterwards, the optimal join order is returned, see Line 15.

2.3   Parallel Approaches

Since join-order optimization is a NP-complete problem [7], enough computational resources need to be provided in order to determine an optimal join order within the time limit of query optimization. So far, all discussed approaches are sequential algorithms. Hence, these algorithms only use the computational power

---

Algorithm 3. DP_{CCP} [6]

    Input   : Join query Q with n tables $T = \{T_1, \ldots, T_n\}$
    Output: an optimal bushy join tree
1 **foreach** $T_i \in T$ **do**
2 | optimalPlan(T_i) = i ;
3 $csgs$ = enumerateCSG(Q);
4 **foreach** $S_l \in csgs$ **do**
5 | $cmps$ = enumerateCMP(Q, S_l);
6 | **foreach** $S_r \in cmps$ **do**
7 | | optimal-left-plan = optimalPlan(S_l);
8 | | optimal-right-plan = optimalPlan(S_r);
9 | | current-plan = createJoinTree(optimal-left-plan, optimal-right-plan);
10 | | **if** cost(optimalPlan($S_l \cup S_r$)) > cost(current-plan) **then**
11 | | | optimalPlan($S_l \cup S_r$) = current-plan ;
12 | | current-plan = createJoinTree(optimal-right-plan, optimal-left-plan);
13 | | **if** cost(optimalPlan($S_l \cup S_r$)) > cost(current-plan) **then**
14 | | | optimalPlan($S_l \cup S_r$) = current-plan ;
15 **return** optimalPlan(R) ;

---

of a single CPU core. Current systems provide most computational power not by the speed of one single CPU core, but by parallel processors, such as multi-core CPUs, or specialized co-processors, such as GPUs. Hence, to further extend the applicability of DP to more complex optimization problems, parallelization is necessary. Han et al. proposed two different parallel DP variants: PDP_{SVA} [2] and $DPE_{GENERIC}$ [3].

Basically, PDP_{SVA} follows the execution strategy of DP_{SIZE}, see Algorithm 4. Therefore, at the beginning, an efficient access method for each table is determined, see Lines 1–2. Although this step can easily be parallelized, the little number of available access methods and tables do not provide much benefit for a parallel evaluation, and, hence, is neglected. Afterwards the parallel optimization of the join order is performed, see Lines 3–11. Because PDP_{SVA} is similar to DP_{SIZE}, new solutions containing a specific number of tables are created in each iteration, see Line 3. One thread enumerates and assigns join pairs to all available threads using Search Space Description Vectors (SSDVs) [2], see Line 4. After the assignment is finished, join pairs are submitted to threads and each thread evaluates all assigned join pairs completely independent of other threads, see Lines 5–7. Since equivalent plans can be created by different threads, a final pruning step needs to be performed, combining the intermediate solutions of each thread, see Line 8. Since PDP_{SVA} uses the execution strategy of DP_{SIZE}, PDP_{SVA} also need to evaluate invalid join pairs, see Lines 8–9 of Algorithm 1. To skip not only a single invalid join pair but all following, PDP_{SVA} uses Skip Vector Arrays (SVAs). Hence, for new intermediate solutions, SVAs are created in parallel, see Lines 9–11. After all join pairs are evaluated, the optimal join order is returned, see Line 12.

Algorithm 4. PDP_{SVA} [2]

    Input : Join query Q with n tables $T = \{T_1, \ldots, T_n\}$
    Output: an optimal bushy join tree
1 **foreach** $T_i \in T$ **do**
2 | optimalPlan(T_i) = T_i ;
3 **for** $s = 2$ **to** n **do**
4 | SSDVs = AllocateSearchSpace(S,m);
5 | **for** $i = 1$ **to** MAX_THREAD_ID **do**
6 | | threadPool.SubmitJob(MutiplePlanJoin(SSDVs[i],S));
7 | threadPool.sync();
8 | MergeAndPrunePlanPartitions(S);
9 | **for** $i = 1$ **to** MAX_THREAD_ID **do**
10 | | threadPool.SubmitJob(BuildSkipVectorArray(i));
11 | threadPool.sync();
12 **return** optimalPlan(R) ;

Although PDP_{SVA} enables a parallel execution of DP based on the execution strategy of DP_{SIZE}, the other more efficient sequential DP variants cannot be parallelized given the execution strategy of PDP_{SVA}. Hence, Han et al. introduced $DPE_{GENERIC}$ to provide a general applicable parallelization strategy, parallelizing all existing sequential DP variants [3], see Algorithm 5. The idea of $DPE_{GENERIC}$ is to use the producer-consumer model instead of explicitly assigning join pairs to specific threads. The producer needs to create a partial order following the sequential DP variants to determine dependent join pairs, see Line 3, whereas consumers evaluate independent join pairs in parallel. After creating the partial order, the producer parses a specific amount of join pairs and stores them into a concurrent buffer for consumers, see Line 4. Hereby, in order to reduce synchronization conflicts a Double Buffer is used, one buffer for consumers and one for the producer. Furthermore, in order to avoid conflicting memory accesses, join pairs are grouped into equivalence classes and memory accesses of solution and inputs are determined [3]. By this, no concurrent evaluation of the data structure storing intermediate solutions is needed during the parallel evaluation and the synchronization overhead is reduced. Join pairs are evaluated in parallel until no more join pairs are available, see Line 5. Before the parallel evaluation starts, the buffers must be switched to make the parsed join pairs available for all consumers, see Line 6. After switching the buffers, the parsed join pairs are evaluated by all consumers, see Lines 7–8. While the consumers are evaluating the parsed join pairs, the producer parses the join pairs for the next iteration, see Line 9. To utilize all available threads, the producer also starts to consume and evaluate join pairs of the current iteration, after all join pairs for the next iteration are determined, see Line 10. At the end of each iteration, consumers are synchronized, see Line 11. In contrast to PDP_{SVA}, the consumers directly perform the pruning of equivalent solutions. Hence, the pro-

---

Algorithm 5. $DPE_{GENERIC}$ [3]

    Input   : Join query Q with n tables $T = \{T_1, \ldots, T_n\}$
    Output: an optimal bushy join tree
1 EnumerationBuffer B_c, B_p;
2 Hash-Table Memo;
3 partial_order = buildPartialOrder(R);
4 e = parseCalculations(partial_order, Memo, B_p, MAX_ENUM_CNT);
5 **while** e $\neq NO\_MORE\_PAIR$ **do**
6   switchBuffers()   // $B_c = B_p$ and $B_p = B_c$;
7   **for** $i = 0$ to $MAX\_THREAD\_ID - 1$ **do**
8   threadPool.SubmitJob(GenerateQEPs(B_c,Memo));
9   e = parseCalculations(partial_order, Memo, B_p, MAX_ENUM_CNT);
10   GenerateQEPs(B_c,Memo));
11   threadPool.sync();
12 **return** Memo(R) ;

---

ducer does not need to perform a final pruning step. When all iterations are finished, the optimal join order is returned, see Line 12.

2.4 Cost Estimation

Independent of the used variant of DP for join-order optimization, a cost function is needed to determine optimal solutions. In the related work, a variety of different cost functions were proposed. Sometimes simple cost functions are used, which only consider operator cardinality [11], whereas sometimes more complex cost functions are used, considering resource consumption, such as disk access and CPUs utilization [1]. Depending on the system requirements different aspects can be considered, such as execution time, energy consumption, main memory utilization, transfer costs, disk or page accesses, cache utilization, CPU utilization, or a combination of multiple aspects. Depending on the aspects themselves and the number of considered aspects, the complexity and, hence, the runtime of cost functions increases. The goal of using complex cost functions is more accurate and robust cost estimations. Unfortunately, this goal is often not accomplished [4]. Hence, complex cost functions are not necessarily superior.

3 Impact of Cost-Function Complexity

The considered aspects and, hence, the complexity of the cost function determines the runtime of the cost function. Since the cost function is called several times during the join-order optimization, differences in the runtime of cost functions, will also affect the runtime of the different DP variants. In this section, we will evaluate the effects of different complexities of cost functions on the different variants of DP.

3.1 Evaluation Setup

In our evaluation, we will consider the runtime of each sequential and parallel DP variant, DP_{SIZE}, DP_{SUB}, DP_{CCP}, PDP_{SVA}, and $DPE_{GENERIC}$, with respect to the different query topologies, different number of tables, and different complexities of the cost function. Therefore, we randomly create queries based on the specified topology and number of tables. Other parameters of the query, such as size of the tables or selectivity of join operators, were assigned randomly. For our evaluation, we used a different maximal number of tables for different topologies. Since clique queries are complex to optimize, we only used maximal 15 tables to achieve reasonable execution times. For the other topologies, we used maximal 20 tables. For each combination of DP variant, topology, and number of tables, and complexity of the cost function, we ran the optimization process 30 times and aggregated the measures using the average.

As cost function, we used a simple cost function, estimating the cost only based on operator cardinality [6]. In order to achieve different runtimes of the cost function, we simulated complex cost functions by adding additional overhead to the used simple cost function. The use of additional overhead instead of a different cost function, ensures that in both setups the same calculations are performed and, hence, the optimization is not altered. Hereby, the runtimes of the different DP variants are comparable to the reported runtimes of previous publications for both types of cost functions, simple [6] and complex [2,3].

As evaluation system, we used a system with the operating system CentOS 7, Intel Xeon CPU E5-2630v3 (8 cores each with 2,4 GHz) and 1 TB main memory.

Furthermore. since the used CPU provides 8 cores, we evaluated the parallel DP variants PDP_{SVA} and $DPE_{GENERIC}$ with a varying number of threads (1–8 threads). For the PDP_{SVA}, we used the allocation scheme round-robin inner and SVAs as it provided the best results in the initial evaluation [2]. For similar reasons, we used the grouping strategy SLQS combined with the enumeration of DP_{CCP} for $DPE_{GENERIC}$ [3].

3.2 Results

Linear and Cyclic Queries. In Fig. 2, we show the evaluation results for linear queries using complex (left) and simple (right) cost functions.

For few tables (complex: 12, simple: 8), the sequential DP variants, DP_{SIZE}, DP_{SUB}, and DP_{CCP}, show a similar performance. Only for a larger number of tables, the execution time of DP_{SUB} increases significantly, based on the enumeration of all possible not only the needed join pairs, see Lines 3–4 of Algorithm 2.

The parallel DP variants PDP_{SVA} and $DPE_{GENERIC}$ have an initialization overhead and provide no significant benefit. Only in case of a large number tables and a complex cost function PDP_{SVA} is superior to the sequential DP variants. In this case, PDP_{SVA} reduces the execution time by 52% for 20 tables. Although a speedup is achieved, PDP_{SVA} uses 8 threads in contrast to 1 by the best sequential variant DP_{CCP}. Hence, the parallelism of PDP_{SVA} is still limited.

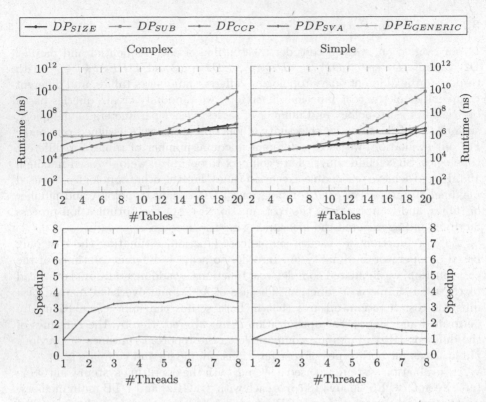

Fig. 2. Runtime and speedup of parallel dynamic programming variants for linear queries. A simple cost function leads to a reduced parallelism compared to a complex cost function.

Hereby, the main problem is that within linear queries, only few possible join pairs need to be evaluated, but each thread needs to be initialized. Complex cost functions can partially hide this initialization overhead, but simple cost functions show that the initialization overhead of an increased number of threads can reduce the efficiency of the optimization.

The limitation regarding the parallelism becomes even more obvious considering $DPE_{GENERIC}$. In the $DPE_{GENERIC}$ approach another problem arises besides the few number of join pairs: Resource contention. $DPE_{GENERIC}$ uses only a single buffer to pass join pairs to the consumers, see Line 8 of Algorithm 2. Hereby, not only a single join pair but a group of join pairs for an equivalent solution is stored in one entry of the buffer. Based on the few number of join pairs, less entries are available within the groups. Therefore, consumers finish the evaluation of an equivalence class faster and need to access the consumer buffer more frequently. The access to the consumer buffer needs to be serialized to ensure the correctness of the optimization. Hence, consumers need to wait to pull new equivalence classes from the buffer. The overhead of synchronizing the buffer reduces the efficiency of the evaluation and even leads to a slowdown for

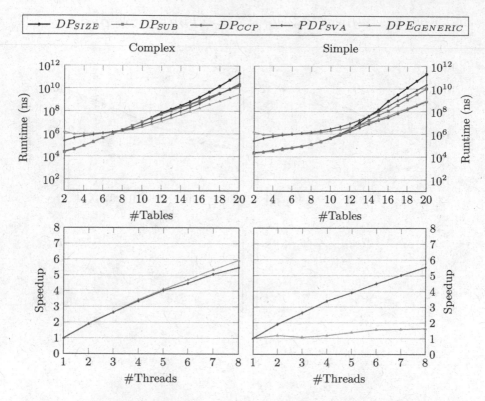

Fig. 3. Runtime and speedup of parallel dynamic programming variants for star queries. A simple cost function leads to a reduced parallelism compared to a complex cost function only for $DPE_{GENERIC}$.

simple cost functions. Similar to PDP_{SVA}, a complex cost function can partially hide the overhead of synchronization.

All DP variants show a similar performance for cyclic queries. Hence, we will not report the evaluation results for cyclic queries.

Star Queries. In Fig. 3, we show the evaluation results for star queries using complex (left) and simple (right) cost functions. Considering star queries, the behavior of all DP variants is more homogeneous, although the runtimes of approaches still differ. For a larger number of tables (complex: >12, simple: >8), DP_{SIZE} performs worse compared to the other sequential DP variants. The reason is the additional overhead introduced by the evaluation of combining all solutions containing a specific number of tables, see Lines 6–7 of Algorithm 1. Hence, not only valid join pairs need to be evaluated, but also invalid join pairs. DP_{SUB} and DP_{CCP} provide similar performance. Based on the enumeration overhead of DP_{CCP}, see Lines 3–6 of Algorithm 3, DP_{SUB} performs better for queries containing less queries. Using simple cost functions, the overhead of

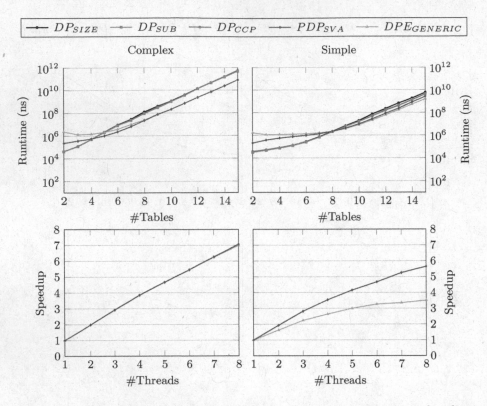

Fig. 4. Runtime and speedup of parallel dynamic programming variants for clique queries. A simple cost function leads to a reduced parallelism compared to a complex cost function.

evaluating invalid join pairs, see Lines 7–8 of Algorithm 2, introduces a significant overhead. Therefore, DP_{CCP} is superior to DP_{SUB} with more than 9 tables. Considering complex cost functions, the impact of this overhead is reduced, because the ratio between overhead and runtime of the cost function is less. Hence, considering complex cost functions, DP_{CCP} is only superior for queries with more than 18 tables.

The parallel variant PDP_{SVA} shows a good parallel evaluation of join pairs. Unfortunately, PDP_{SVA} suffers from the same drawback as DP_{SIZE}, the evaluation of invalid join pairs. Although the parallel evaluation can compensate the additional overhead upto 17 queries. Afterwards the number of invalid join pairs dominate, making the approach slower compared to the sequential approach DP_{CCP}. Since only valid join pairs need to be evaluated considering $DPE_{GENERIC}$, $DPE_{GENERIC}$ can ensure an efficient optimization with complex cost functions, leading to a speedup of upto 5.7 for 20 tables.

Considering simple cost functions, the execution of the parallel DP variants PDP_{SVA} and $DPE_{GENERIC}$ differ compared to complex cost functions. Based on the overhead of initialization and parallelism, and the evaluation of invalid join pairs, PDP_{SVA} provide no benefit to a sequential evaluation. Similar to linear queries, $DPE_{GENERIC}$ suffers from the synchronization overhead of the buffer, limiting the parallelism of this approach considering simple cost functions. Therefore, the sequential approach DP_{CCP} is faster than the parallel variant $DPE_{GENERIC}$ with 8 threads.

Clique Queries. In Fig. 4, we show the evaluation results for clique queries using complex (left) and simple (right) cost functions. In contrast to the other topologies, the main bottleneck for clique queries is the evaluation of valid join pairs. The additional overhead of enumerating valid join pairs (DP_{CCP}), enumerating all possible join pairs (DP_{SUB}), or evaluating invalid join pairs (DP_{SIZE}) is negligible. Hence, all sequential DP variants perform similar.

The parallel DP variants PDP_{SVA} and $DPE_{GENERIC}$ show a similar behavior. In contrast to the other topologies, in clique queries, enough computations are available for a parallel evaluation. Hence, for both complex and simple cost functions the parallel DP variants provide a benefit compared to the sequential DP variants (Speedup of upto 6 for 20 tables). Nevertheless, considering simple cost functions, both parallel DP variants PDP_{SVA} and $DPE_{GENERIC}$ show a reduced parallelism, based on the already discussed issues, overhead of parallelism and resource contention. Hereby, the reduction of parallelism of $DPE_{GENERIC}$ is greater compared to PDP_{SVA}. Nevertheless, since $DPE_{GENERIC}$ only needs to evaluate valid join pairs, $DPE_{GENERIC}$ provides a better performance compared to PDP_{SVA}.

3.3 Discussion

In our evaluation, we showed that existing parallelization strategies of DP highly depends on the runtime of the cost functions. For complex cost functions, in most of the cases, parallelism provides an advantage compared to the sequential evaluation. In contrast to previous evaluations, we could even observe a small advantage for simple optimization problems, such as linear or cyclic queries. However, this changes with simple cost functions. Using simple cost functions, in most of the cases, existing parallel DP variants cannot provide any benefits for many optimization problems. Only for clique queries with a larger number of tables, the parallel optimization provide an advantage.

Although the presented evaluation results are comparable to the results of previous evaluations with respect to the used cost function, we show that the applicability of the parallelization highly depends on the used cost function. Hence, for join-order optimization, there is neither an optimal sequential nor optimal parallel variant of DP. To select a suitable variant several factors need to be considered: complexity of cost functions, query topologies, and number of tables contained in queries. Depending on these factors different DP variants provide advantages and disadvantages. The parallel DP variants ensure an efficient join-order optimization for queries with a larger number of tables using complex cost functions or complex topologies, such as clique queries. For the optimization of queries with few tables or simple cost functions, sequential DP variants are superior to their parallel counterparts.

4   Conclusion

For DP for join-order optimization several sequential and parallel variants exist. We showed that the benefit of the parallel DP variants not only depends on the characteristics of queries, such as query topology and number of tables, but also highly depends on the complexity of the used cost function. For simple optimization problems, such as few tables, simple cost function, or simple query topologies, existing parallel DP variants provide no benefit compared to suitable sequential DP variants. Only for complex optimization problems, such as a larger number of tables, complex cost functions, or complex query topologies, existing parallel DP variants provide an advantage.

Acknowledgments. This work was partially funded by the DFG (grant no.: SA 465/50-1).

References

1. Fong, Z.: The design and implementation of the POSTGRES query optimizer. Technical report, University of California, Berkeley, August 1986
2. Han, W.S., Kwak, W., Lee, J., Lohman, G.M., Markl, V.: Parallelizing query optimization. PVLDB **1**(1), 188–200 (2008)
3. Han, W.S., Lee, J.: Dependency-aware reordering for parallelizing query optimization in multi-core CPUs. In: SIGMOD, pp. 45–58. ACM (2009)
4. Leis, V., Gubichev, A., Mirchev, A., Boncz, P., Kemper, A., Neumann, T.: How good are query optimizers, really? PVLDB **9**(3), 204–215 (2015)
5. Moerkotte, G., Fender, P., Eich, M.: On the correct and complete enumeration of the core search space. In: SIGMOD, pp. 493–504. ACM (2013)
6. Moerkotte, G., Neumann, T.: Analysis of two existing and one new dynamic programming algorithm for the generation of optimal bushy join trees without cross products. In: VLDB, pp. 930–941. VLDB Endowment (2006)
7. Moerkotte, G., Scheufele, W.: Constructing optimal bushy processing trees for join queries is NP-hard. Technical report Informatik-11/1996 (1996)
8. Neumann, T.: Engineering high-performance database engines. PVLDB **7**(13), 1734–1741 (2014)
9. Selinger, P.G., Astrahan, M.M., Chamberlin, D.D., Lorie, R.A., Price, T.G.: Access path selection in a relational database management system. In: SIGMOD, pp. 23–34. ACM (1979)
10. Steinbrunn, M., Moerkotte, G., Kemper, A.: Heuristic and randomized optimization for the join ordering problem. VLDB J. **6**(3), 191–208 (1997)
11. Vance, B., Maier, D.: Rapid bushy join-order optimization with cartesian products. In: SIGMOD, pp. 35–46. ACM (1996)

Instant Restore After a Media Failure

Caetano Sauer[1(✉)], Goetz Graefe[2], and Theo Härder[1]

[1] TU Kaiserslautern, Kaiserslautern, Germany
{csauer,haerder}@cs.uni-kl.de
[2] Google, Madison, WI, USA
goetzg@google.com

Abstract. Media failures usually leave database systems unavailable for several hours until recovery is complete, especially in applications with large devices and high transaction volume. Previous work introduced a technique called single-pass restore, which increases restore bandwidth and thus substantially decreases time to repair. Instant restore goes further as it permits read/write access to any data on a device undergoing restore—even data not yet restored—by restoring individual data segments on demand. Thus, the restore process is guided primarily by the needs of applications, and the observed mean time to repair is effectively reduced from several hours to a few seconds.

This paper presents an implementation and evaluation of instant restore. The technique is incrementally implemented on a system starting with the traditional ARIES design for logging and recovery. Experiments show that the transaction latency perceived after a media failure can be cut down to less than a second. The net effect is that a few "nines" of availability are added to the system using simple and low-overhead software techniques.

1 Introduction

Advancements in hardware technology have significantly improved the performance of database systems over the last decade, allowing for throughput in the order of thousands of transactions per second and data volumes in the order of petabytes. Availability, on the other hand, has not seen drastic improvements, and the research goal postulated by Jim Gray in his ACM Turing Award Lecture of a system "unavailable for less than one second per hundred years" [12] remains an open challenge. Improvements in reliable hardware and data center technology have contributed significantly to the availability goal, but proper software techniques are required to not only avoid failures but also repair failed systems as quickly as possible. This is especially relevant given that a significant share of failures is caused by human errors and unpredictable defects in software and firmware, which are immune to hardware improvements [11]. In the context of database logging and recovery, the state of the art has unfortunately not changed much since the early 90's, and no significant advancements were achieved in the software front towards the availability goal.

© Springer International Publishing AG 2017
M. Kirikova et al. (Eds.): ADBIS 2017, LNCS 10509, pp. 311–325, 2017.
DOI: 10.1007/978-3-319-66917-5_21

Instant restore is a technique for media recovery that drastically reduces mean time to repair by means of simple software techniques. It works by extending the write-ahead logging mechanism of ARIES [19] and, as such, can be incrementally implemented on the vast majority of existing database systems. The key idea is to introduce a different organization of the log archive to enable efficient on-demand, incremental recovery of individual data pages. This allows transactions to access recovered data from a failed device orders of magnitude faster than state-of-the-art techniques, all of which require complete restoration of the entire device before access to the application's working set is allowed.

The problem of inefficient media recovery in state-of-the-art techniques, including ARIES and its optimizations, can be attributed to two major deficiencies. First, the media recovery process has a very inefficient random access pattern, which in practice

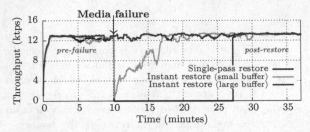

Fig. 1. Effect of instant restore (Color figure online)

encourages excessive redundancy and frequent incremental backups—solutions that only alleviate the problem instead of eliminating it. The second deficiency is that the recovery process is not incremental and requires full recovery before any data can be accessed—on-demand schedules are not possible and there is no prioritization scheme to make most needed data available earlier. Previous work addressed the first problem with a technique called single-pass restore [24], while this paper focuses on the second one.

The effect of instant restore is illustrated in Fig. 1, where transaction throughput is plotted over time and a media failure occurs after 10 min. In single-pass restore, as in ARIES, transaction processing halts until the device is fully restored (the red line in the chart), while instant restore continues processing transactions, using them to guide the restore process (blue and green lines). In a scenario where the application working set fits in the buffer pool (blue line), there is actually no visible effect on transaction throughput.

In the remainder of this paper, Sect. 2 describes related work, both previous work leading to the current design as well as competing approaches. Then, Sect. 3 describes the instant restore technique. Finally, Sect. 4 presents an empirical evaluation, while Sect. 5 concludes this paper.

A high-level description of instant restore was previously published in a book chapter [6] among other instant recovery techniques. The additional contribution here is a more detailed discussion of the design and implementation aspects as well as an empirical evaluation of the technique with an open-source prototype.

2   Related Work

2.1   Failure Classes and Assumptions

Database literature traditionally considers three classes of database failures [14], which are summarized in Table 1 (along with single-page failures, a fourth class to be discussed in Sect. 2.5). In the scope of this paper, it is important to distinguish between system and media failures, which are conceptually quite different in their causes, effects, and recovery measures. System failures are usually caused by a software fault or power loss, and what is lost—hence what must be recovered—is the state of the server process in main memory; this typically entails recovering page images in the buffer pool (i.e., "repeating history" [19]) as well as lists of active transactions and their acquired locks, so that they can be properly aborted. The process of recovery from system failures is called *restart*.

Table 1. Failure classes, their causes, and effects

Failure class	Loss	Typical cause	Response
Transaction	Single-transaction progress	Deadlock	Rollback
System	Server process (in-memory state)	Software fault, power loss	Restart
Media	Stored data	Hardware fault	Restore
Single page	Local integrity	Partial writes, wear-out	Repair

Instant restart [6] is an orthogonal technique that provides on-demand, incremental data access following a system failure. While the goals are similar, the design and implementation of instant restore require quite different techniques.

In a media failure, which is the focus here, a persistent storage device fails but the system might continue running, serving transactions that only touch data in the buffer pool or on other healthy devices. If a system and media failures happen simultaneously, or perhaps one as a cause of the other, their recovery processes are executed independently, and, by recovering pages in the buffer pool, the processes coordinate transparently.

The present work makes the same assumptions as most prior research on database recovery. The log and its archive copy reside on "stable storage", i.e., they are assumed to never fail. We consider failures on the database device only, i.e., the permanent storage location of data pages. Recovery from such failures requires a backup copy (possibly days or weeks old) of the lost device and all log records since the backup was taken; these may reside either in the active transaction log or in the log archive. The process of recovery from media failures is called *restore*. The following sections briefly describe previous restore methods.

2.2   ARIES Restore

Techniques to recover databases from media failures were initially presented in the seminal work of Gray [10] and later incorporated into the ARIES family of recovery algorithms [19]. In ARIES, restore after a media failure first loads a backup image and then applies a redo log scan, similar to the redo scan of restart after a system failure. Figure 2 illustrates the process, which we now briefly describe. After loading full and incremental backups into the replacement device, a sequential scan is performed on the log archive and each update is replayed on its corresponding page in the buffer pool. A global *minLSN* value (called "media recovery redo point" by Mohan et al. [19]) is maintained on backup devices to determine the begin point of the log scan.

Because log records are ordered strictly by LSN, pages are read into the buffer pool in random order, as illustrated in the restoration of pages A and B in Fig. 2. Furthermore, as the buffer pool fills up, they are also written in random order into the replacement device, except perhaps for some minor degree of clustering. As the log scan progresses, evicted pages might be read again multiple times, also randomly. This mechanism is quite inefficient, especially for magnetic drives with high access latencies. Thus, it is no surprise that multiple hours of downtime are required in systems with high-capacity drives and high transaction rates [24].

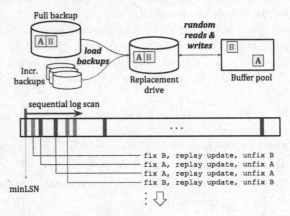

Fig. 2. Random access pattern of ARIES restore

Another fundamental limitation of the ARIES restore algorithm is that it is not incremental, i.e., pages cannot be restored to their most up-to-date version one-by-one and made available to running transactions incrementally. As shown in the example of Fig. 2, the last update to page A may be at the very end of the log; thus, page A remains out-of-date until almost the end of the log scan. Some optimizations may alleviate this situation (e.g., reusing checkpoint information), but there is no general mechanism for incremental restoration. Furthermore, even if pages could somehow be released incrementally when their last update is replayed, the hottest pages of the application working set are most likely to be released only at the very end of the log scan, and probably not even then, because they might contain updates of uncommitted transactions and thus require subsequent undo. This leads to yet another limitation of this approach: even if pages could be restored incrementally, there is no effective way to provide on-demand restoration, i.e., to restore most important pages first.

Despite a variety of optimizations proposed to the basic ARIES algorithm [19–21], none of them solves these problems in a general and effective manner. In summary, all proposed techniques that enable earlier access to recovered data items suffer from the same problem: early access is only provided for data for which early access is not really needed—hot data in the application working set is not prioritized and most accesses must wait for complete recovery.

Finally, industrial database systems that implement ARIES recovery suffer from the same problems. IBM's DB2 speeds up log replay by sorting log records after restoring the backup and before applying the log records to the replacement database [13]. While a sorted log enables a more efficient access pattern, incremental and on-demand restoration is not provided. Furthermore, the delay imposed by the offline sort may be as high as the total downtime incurred by the traditional method. As another example, Oracle attempts to eliminate the overhead of reading incremental backups by incrementally maintaining a full backup image [22]. While this makes recovery slightly more efficient, it does not address the deficiencies discussed earlier.

2.3 Replication

Given the extremely high cost of media recovery in existing systems, replication solutions such as disk mirroring or RAID [2,3] are usually employed in practice to increase mean time to failure. However, it is important to emphasize that, from the database system's perspective, a failed disk in a redundant array does not constitute a media failure as long as it can be repaired automatically. Restore techniques aim to improve mean time to repair whenever a failure occurs that cannot be masked by lower levels of the system. Therefore, replication techniques can be seen largely as orthogonal to media restore techniques as implemented in database recovery mechanisms.

Nevertheless, a substantial reduction in mean time to repair, especially if done solely with simple software techniques, opens many opportunities to manage the trade-off between operational costs and availability. One option can be to maintain a highly-available infrastructure (with whatever costs it already requires) while availability is increased by deploying software with more efficient recovery. Alternatively, replication costs can be reduced (e.g., downgrading RAID-10 into RAID-5) while maintaining the same availability. Such level of flexibility, with solutions tackling both mean time to failure and mean time to repair, are essential in the pursuit of Gray's availability goal [12].

2.4 In-Memory Databases

Early work on in-memory databases focused mainly on restart after a system failure, employing traditional backup and log-replay techniques for media recovery [4,16]. The work of Levi and Silberschatz [17] was among the first to consider the challenge of incremental restart after a system failure. While an extension of their work for media recovery is conceivable, it would not address the efficiency problem discussed in Sect. 1. Thus, it would, in the best case and with a more complex algorithm, perform no better than the algorithm discussed later in Sect. 2.5.

Recent proposals for recovery on both volatile and non-volatile in-memory systems usually ignore the problem of media failures, employing the unspecific term "recovery" to describe system restart only [1,18,23]. Therefore, recovery from media failures in modern systems either relies on the traditional techniques or is simply not supported, employing replication as the only means to maintain service upon storage hardware faults. As discussed above, while relying on replication is a valid solution to increase mean time to failure, a highly available system must also provide efficient repair facilities. In this aspect, traditional database system designs—using ARIES physiological logging and buffer management—provide more reliable behavior. Therefore, we believe that improving traditional techniques for more efficient recovery with low overhead on memory-optimized workloads is an important open research challenge.

2.5   Single-Page Repair

Single-page failures are considered a fourth class of database failures [8], along with the other classes summarized in Table 1. It covers failures restricted to a small set of individual pages of a storage device and applies online localized recovery to that individual page instead of invoking media recovery on the whole device. The single-page repair algorithm, illustrated in Fig. 3 (with backup and replacement devices omitted for simplification), has two basic requirements: first, the LSN of the most recent update of each page is known (i.e., the current PageLSN value) without having to access the page; second, starting from the most recent log record, the complete history of updates to a page can be retrieved. The former requirement can be provided with a page recovery index— a data structure mapping page identifiers to their most recent PageLSN value. Alternatively, the current PageLSN can be stored together with the parent-to-child node pointer in a B-tree data structure [9]. The latter requirement is provided by per-page log record chains, which are straight-forward to maintain using the PageLSN fields in the buffer pool.

In principle, single-page repair could be used to recover from a media failure, by simply repairing each page of the failed device individually. One advantage of this technique is that it yields incremental and on-demand restore, addressing the second deficiency of traditional

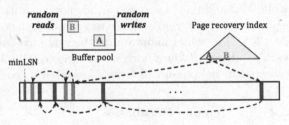

Fig. 3. Single-page repair

media recovery algorithms mentioned in Sect. 1. To illustrate how this would work in practice, consider the example of Fig. 3. If the first page to be accessed after the failure is A, it would be the first to be restored. Using information from the page recovery index (which can be maintained in main memory or fetched directly from backups), the last red log record on the right side of the diagram would be fetched first. Then, following the per-page chain, all red log records until *minLSN* would be retrieved and replayed in the backup image of page A, thus yielding its most recent version to running transactions.

While the benefit of on-demand and incremental restore is a major advantage over traditional ARIES recovery, this algorithm still suffers from the first deficiency discussed in Sect. 1—namely the inefficient access pattern. The authors of the original publication even foresee the application to media failures [8], arguing that while a page is the unit of recovery, multiple pages can be repaired in bulk in a coordinated fashion. However, the access pattern with larger restoration granules would approach that of traditional ARIES restore—i.e., random access during log replay. Thus, while the technique introduces a useful degree of flexibility, it does not provide a unified solution for the two deficiencies discussed.

2.6 Single-Pass Restore

Our previous work introduced a technique called single-pass restore, which aims to perform media recovery in a single sequential pass over both backup and log archive devices [24]. Eliminating random access effectively addresses the first deficiency discussed in Sect. 1. This is achieved by partially sorting the log on page identifiers, using a stable sort to maintain LSN order within log records of the same page. The access pattern is essentially the same as that of a sort-merge join: external sort with run generation and merge followed by another merge between the two inputs—log and backup in the media recovery case.

The idea itself is as old as the first recovery algorithms (see Sect. 5.8.5.1 of Gray's paper [10]) and is even employed in DB2's "fast log apply" [13]. However, the key advantage of single-pass restore is that the two phases of the sorting process—run generation and merge—are performed independently: runs are generated during the log archiving process

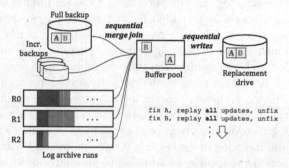

Fig. 4. Single-pass restore

(i.e., moving log records from the latency-optimized transaction log device into high-capacity, bandwidth-optimized secondary storage) with negligible overhead; the merge phase, on the other hand, happens both asynchronously as a maintenance service and also during media recovery, in order to obtain a single sorted log stream for recovery. Importantly, merging runs of the log archive and applying the log records to backed-up pages can be done in a sequential pass, similar to a merge join. The process is illustrated in Fig. 4. We refer to the original publication for further details [24].

Having addressed the access pattern deficiency of media recovery algorithms, single-pass restore still leaves open the problem of incremental and on-demand restoration. Nevertheless, given its superiority over traditional ARIES restore (see [6,24] for an in-depth discussion), it is a promising approach to use as starting point in addressing the two deficiencies in a unified way. Therefore, as mentioned in Sect. 1, single-pass restore is taken as the baseline for the present work.

3   Instant Restore

The main goal of instant restore is to preserve the efficiency of single-pass restore while allowing more fine-granular restoration units (i.e., smaller than the whole device) that can be recovered incrementally and on demand. We propose a generalized approach based on segments, which consist of contiguous sets of data pages. If a segment is chosen to be as large as a whole device, our algorithm behaves exactly like single-pass restore; on the other extreme, if a segment is chosen to be a single page, the algorithm behaves like single-page repair.

This section starts by introducing the log data structure employed to provide efficient access to log records belonging to a given segment or page; after that, we present the restore algorithm based on this data structure.

3.1   Indexed Log Archive

In order to restore a given segment incrementally, instant restore requires efficient access to log records pertaining to pages in that segment. In single-page repair, such access is provided for individual pages, using the per-page chain among log records [8]. As already discussed, this is not efficient for restoration units much larger than a single page. Therefore, we build upon the partially sorted log archive organization introduced in single-pass restore [24].

In instant restore, the partially sorted log archive is extended with an index. The log archiving process sorts log records in an in-memory workspace and saves them into runs on persistent storage. These runs must then be indexed, so that log records of a given page or segment identifier can be fetched directly. Sorting and indexing of log records is done online and without any interference on transaction processing, in addition to standard archiving tasks such as compression.

In an index lookup for instant restore, the set of runs to consider would be restricted by the given *minLSN* (see Sect. 2.2) of the backup image, since runs older than that LSN are not needed. Furthermore, Bloom filters can be appended to each run to restrict this set even further. The result of the lookup in each indexed run is then fed into a merge process that delivers a single stream of log records sorted primarily by page identifier and secondarily by LSN. This stream is then used by the restore algorithm to replay updates on backup segments.

Multiple choices exist for the physical data structure of the indexed log archive. Ideally, the B-tree component of the indexing subsystem can be reused, but there is an important caveat in terms of providing atomicity and durability to this structure. A typical index relies on write-ahead logging, but that is not an option for the indexed log archive because it would introduce a kind of self-reference loop—updates to the log data structure itself would have to be logged and used later on for recovery. This self-reference loop could be dealt with by introducing special logging and recovery modes (e.g., a separate "meta"-log for the indexed log archive), but the resulting algorithm would be too cumbersome. In our prototype, we chose a simpler solution: each partition of the log archive is maintained in its own read-only file; temporary shadow files are then used for merges and appends. In this scheme, atomicity is provided by the file rename operation, which is atomic in standard filesystems [5].

3.2 Restore Algorithm

When a media failure is detected, a restore manager component is initialized and all page read and write requests from the buffer pool are intercepted by this component. The diagram in Fig. 5 illustrates the interaction of the restore manager with the buffer pool and all persistent devices involved in the restore process: failed and replacement devices, log archive, and backup. For reasons discussed in previous work [24], incremental backups are made obsolete by the partially sorted log archive; thus, the algorithm performs just as well with full backups only. Nevertheless, incremental backups can be easily incorporated, and the description below considers a single full backup without loss of generality.

In the following discussion, the numbers in parentheses refer to the numbered steps in Fig. 5. The restore manager keeps track of which segments were already restored using a segment recovery bitmap, which is initialized with zeros. When a page access occurs, the restore manager first looks up its segment in the bitmap (1). If set to one, it indicates that the segment

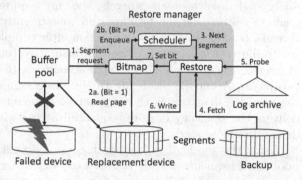

Fig. 5. Instant restore flow chart

was already restored and can be accessed directly on the replacement device (2a). If set to zero, a segment restore request is placed into a restore scheduler (2b), which coordinates the restoration of individual segments (3).

To restore a given segment, an older version is first fetched from the backup directly (4). This is in contrast to ARIES restore, which first loads entire backups into the replacement device and then reads pages from there [19]. This has the implication that backups must reside on random-access devices (i.e., not on tape) and allow direct access to individual segments, which might require an index if backup images are compressed. These requirements, which are also present in single-page repair [8], seem quite reasonable given the very low cost per byte of current high-capacity hard disks. For moderately-sized databases, it is even advisable to maintain log archive and backups on flash storage.

While the backed-up image of a segment is loaded, the indexed log archive data structure is probed for the log records pertaining to that segment (5); the results of each probe are merged to form a single sorted log stream.

Then, log replay is performed to bring the segment to its most recent state, after which it can be written back into a replacement device (6).

Finally, once a segment is restored, the bitmap is updated (7) and all pending read and write requests can proceed. Typically, a requested page will remain in the buffer pool after its containing segment is restored, so that no additional I/O access is required on the replacement device.

All read and write operations described above—log archive index probe, segment fetch, and segment write after restoration—happen asynchronously with minimal coordination. The read operations are essentially merged index scans—a very common pattern in query processing. The write of a restored segment is also

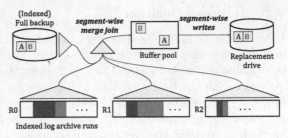

Fig. 6. Instant restore

easily made asynchronous, whereby the only requirement is that marking a segment as restored on the bitmap, and consequently enabling access by waiting threads, be done by a callback function after completion of the write.

To illustrate the access pattern of instant restore, similarly to the diagrams in Sect. 2, Fig. 6 shows an example scenario with three log archive runs and two pages, A and B, belonging to the same segment. The main difference to the previous diagrams is the segment-wise, incremental access pattern, which delivers the efficiency of pure sequential access with the responsiveness of on-demand random reads.

Using this mechanism, user transactions accessing data either in the buffer pool or on segments already restored can execute without any additional delay, whereby the media failure goes completely unnoticed. Access to segments not yet restored are used to guide the restore process, triggering the restoration of individual segments on demand. As such, the time to repair observed by transactions accessing data not yet restored is multiple orders of magnitude lower than the time to repair the whole device. Furthermore, time to repair observed by an individual transaction is independent of the total capacity of the failed device. This is in contrast to previous methods, which require longer downtime for larger devices.

4   Experiments

Our experimental evaluation covers three main measures of interest during recovery from a media failure: restore latency, restore bandwidth, and transaction throughput. Before presenting the empirical analysis, a brief summary of our experimental environment is provided.

4.1   Environment

We implemented instant restore in a fork of the Shore-MT storage manager [15] called *Zero*. The code is available as open source[1]. The workload consists of the TPC-C benchmark as implemented in Shore-MT, but adapted to use the Foster B-tree [7] data structure for both table and index data.

All experiments were performed on dual six-core CPUs with HyperThreading. The system has 100 GB of high-speed RAM and several Samsung 840 Pro 250 GB SSDs. The operating system is Ubuntu Linux 14.04 with Kernel 3.13.0-68 and all code is compiled with gcc 4.8 and -O3 optimization.

The experiments all use the same workload, with media failure and recovery set up as follows. Initial database size is 100 GB, with full backup and log archive of the same size—i.e., recovery starts from a full backup of 100 GB and must replay roughly the same amount of log records. Log archive runs are a little over 1.5 GB in size, resulting in 64 inputs in the restore merge logic. All persistent data is stored on SSDs and 24 worker threads are used at all times.

4.2   Restore Latency and Bandwidth

Our first experiment evaluates restore latency by analyzing the total latency of individual transactions before and after a media failure. The hypothesis under test is that average transaction latency immediately following a media failure is in the order of a few seconds or less, after which is gradually decreases to the pre-failure latency. Furthermore, with larger memory, i.e., where a larger portion of the working set fits in the buffer pool, average latency should remain at the pre-failure level throughout the recovery process.

The results are shown in Fig. 7a. After ten minutes of normal processing, during which the average latency is 1–2 ms, a media failure occurs. The immediate effect is that average transaction latency spikes up (to about 100 ms in the buffer pool size of 30 GB) but then decreases linearly until pre-failure latency is reestablished. For the largest buffer pool size of 45 GB, there is a small perturbation in the observed latency, but the average value seems to remain between 1 and 2 ms. From this, we can conclude that for any buffer pool size above 45 GB, a media failure goes completely unnoticed.

These results successfully confirm our hypothesis: average latency of a transaction accessing failed media is reduced from several minutes to 100 ms, which corresponds to three orders of magnitude or three additional 9's of availability. Note that the average restore latency is independent of total device capacity, and thus of total recovery time. Therefore, the availability improvement could be in the order of four or five orders of magnitude in certain cases. This would be expected, for instance, for very large databases (in the order of terabytes) stored on relatively low-latency devices. In these cases, the gap between a full sequential read and a single random read—hence, between mean time to repair with single-pass restore and with instant restore—is very pronounced.

---

[1] http://github.com/caetanosauer/zero.

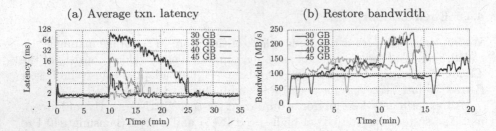

Fig. 7. Transaction latency and restore bandwidth observed with instant restore

Next, we evaluate restore bandwidth for the same experiment. The hypothesis here is that, in general, restore bandwidth gradually increases throughout the recovery process until it reaches the bandwidth of single-pass restore. From these two general behaviors, two special cases are, again, the small and large buffer pools. In the former, bandwidth may not reach single-pass speeds due to prioritization of low latency for the many incoming requests (recall that each buffer pool miss incurs a read on the replacement device, which, in turn, incurs a restore request). In the latter case, restore bandwidth should be as large as single-pass restore.

Figure 7b shows the results of this experiment for four buffer pool sizes. For the smallest buffer pool of 30 GB, restore bandwidth remains roughly constant in the first 15 min. This indicates that during this initial period, most segments are restored individually in response to an on-demand request resulting from a buffer pool miss. As the buffer size increases, the rate of on-demand requests decreases as restore progresses, resulting in more opportunities for multiple segments being restored at once. In all cases, restore bandwidth gradually increases throughout the recovery process, reaching the maximum speed of 240 MB/s towards the end in the larger buffer pool sizes.

4.3   Transaction Throughput

The next experiments evaluate how media failure and recovery impact transaction throughput with instant restore. We take the same experiment performed in the previous section and look at transaction throughput for each buffer pool size individually. As instant restore progresses, transactions continue to access data in the buffer pool, triggering restore requests for each page miss. Therefore, we expect that the larger the buffer pool is (i.e., more of the working set fits into main memory), the less impact a media failure has on transaction throughput. This effect was already presented in the diagram of Fig. 1—the present section analyzes that in more detail.

Figure 8 presents the results. In the four plots shown, transaction throughput is measured with the red line on the left y-axis. At minute 10, a media failure occurs, after which a green straight line shows the pre-failure average throughput. The number of page reads per second is shown with the blue line on the right y-axis.

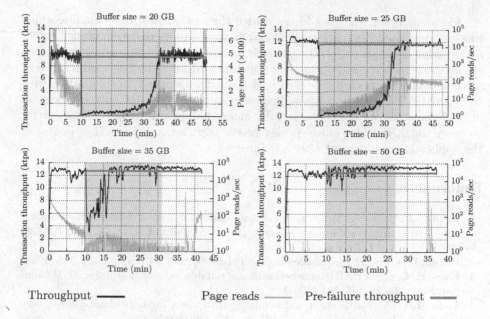

Fig. 8. Impact of instant restore on txn. Throughput at varying buffer pool sizes.

Moreover, total recovery time, which also varies depending on the buffer pool size, is also shown as the shaded interval on the x-axis.

The goal of instant restore in this experiment is to re-establish the pre-failure transaction throughput (i.e., the dotted green line) as soon as possible. Similar to the evaluation on previous experiments, our hypothesis is that this occurs sooner the larger the buffer pool is. The results show that for a small buffer pool of 20 GB, transaction throughput drops substantially, and it only regains the pre-failure level at the very end of the recovery process. As the buffer size is increased to 25 and then 35 GB, pre-failure throughput is re-established at around minute 7, i.e., 1/3 of the total recovery time. Lastly, for the largest buffer pool of 50 GB, the media failure does not produce any noticeable slowdown, as predicted in our hypothesis.

5   Conclusions

Instant restore improves perceived mean time to repair and thus database availability in the presence of media failures. We identified two main deficiencies with traditional recovery techniques, such as the ARIES design [19]: (i) media recovery is very inefficient due to its random access pattern on database pages, which means that time to repair is unacceptably long; and (ii) data on a failed device cannot be accessed before recovery is completed. The first deficiency was addressed with single-pass restore [24], which introduces a partial sort order on the log archive, eliminating the random access pattern of log replay.

The second deficiency is addressed with the instant restore technique, which was first described in earlier work [6] and discussed in more detail, implemented, and evaluated in this paper. By generalizing single-pass restore and other recovery methods such as single-page repair, instant restore is the first media recovery method to effectively eliminate the two deficiencies discussed. In comparison with traditional ARIES media restore, instant restore delivers not only the benefits of single-pass restore (i.e., substantially higher bandwidth and therefore shorter recovery time), but also much quicker access (e.g., seconds instead of hours) to the application working set after a failure.

References

1. Arulraj, J., Pavlo, A., Dulloor, S.: Let's talk about storage & recovery methods for non-volatile memory database systems. In: Proceedings of SIGMOD, pp. 707–722 (2015)
2. Bitton, D., Gray, J.: Disk shadowing. In: Proceedings of VLDB, pp. 331–338 (1988)
3. Chen, P.M., et al.: RAID: high-performance, reliable secondary storage. ACM Comput. Surv. **26**(2), 145–185 (1994)
4. Eich, M.H.: A classification and comparison of main memory database recovery techniques. In: Proceedings of ICDE, pp. 332–339 (1987)
5. GLIBC: The GNU C Library Reference Manual (2014), http://www.gnu.org/software/libc/manual/html_node/Renaming-Files.html. Accessed 06 Oct 2014
6. Graefe, G., Guy, W., Sauer, C.: Instant Recovery with Write-Ahead Logging: Page Repair, System Restart, Media Restore, and System Failover, 2nd edn. Synthesis Lectures on Data Management. Morgan & Claypool Publishers (2016)
7. Graefe, G., Kimura, H., Kuno, H.A.: Foster B-trees. ACM Trans. Database Syst. **37**(3), 17 (2012)
8. Graefe, G., Kuno, H.A.: Definition, detection, and recovery of single-page failures, a fourth class of database failures. PVLDB **5**(7), 646–655 (2012)
9. Graefe, G., Kuno, H.A., Seeger, B.: Self-diagnosing and self-healing indexes. In: Proceedings of DBTest, p. 8 (2012)
10. Gray, J.N.: Notes on data base operating systems. In: Bayer, R., Graham, R.M., Seegmüller, G. (eds.) Operating Systems. LNCS, vol. 60, pp. 393–481. Springer, Heidelberg (1978). doi:10.1007/3-540-08755-9_9
11. Gray, J.: Why do computers stop and what can be done about it? In: Symposium on Reliability in Distributed Software and Database Systems, pp. 3–12 (1986)
12. Gray, J.: What next?: a dozen information-technology research goals. J. ACM **50**(1), 41–57 (2003)
13. Haderle, D.J., Majithia, T.: Fast log apply, US Patent 6,289,355, 11 September 2001
14. Härder, T., Reuter, A.: Principles of transaction-oriented database recovery. ACM Comput. Surv. **15**(4), 287–317 (1983)
15. Johnson, R., Pandis, I., Hardavellas, N., Ailamaki, A., Falsafi, B.: Shore-MT: a scalable storage manager for the multicore era. In: Proceedings of EDBT, pp. 24–35 (2009)
16. Lehman, T.J., Carey, M.J.: A recovery algorithm for a high-performance memory-resident database system. In: Proceedings of SIGMOD, pp. 104–117 (1987)
17. Levy, E., Silberschatz, A.: Incremental recovery in main memory database systems. IEEE Trans. Knowl. Data Eng. **4**(6), 529–540 (1992)

18. Malviya, N., Weisberg, A., Madden, S., Stonebraker, M.: Rethinking main memory OLTP recovery. In: Proceedings of ICDE, pp. 604–615 (2014)
19. Mohan, C., Haderle, D., Lindsay, B., Pirahesh, H., Schwarz, P.: ARIES: a transaction recovery method supporting fine-granularity locking and partial rollbacks using write-ahead logging. ACM Trans. Database Syst. **17**(1), 94–162 (1992)
20. Mohan, C., Narang, I.: An efficient and flexible method for archiving a data base. SIGMOD Rec. **22**(2), 139–146 (1993)
21. Mohan, C., Treiber, K., Obermarck, R.: Algorithms for the management of remote backup data bases for disaster recovery. In: Proceedings of ICDE, pp. 511–518 (1993)
22. Oracle Corporation: RMAN Incremental Backups, Oracle Database Documentation 10g, Sect. 4.4 (2015)
23. Oukid, I., et al.: SOFORT: a hybrid SCM-DRAM storage engine for fast data recovery. In: Proceedings of DaMoN, pp. 8:1–8:7 (2014)
24. Sauer, C., Graefe, G., Härder, T.: Single-pass restore after a media failure. In: Proceedings of BTW. LNI, vol. 241, pp. 217–236 (2015)

Rethinking DRAM Caching for LSMs in an NVRAM Environment

Lucas Lersch[1,2]([⊠]), Ismail Oukid[1,2], Ivan Schreter[2], and Wolfgang Lehner[1]

[1] TU Dresden, Dresden, Germany
{lucas.lersch,i.oukid}@sap.com, wolfgang.lehner@tu-dresden.de
[2] SAP SE, Walldorf, Germany
ivan.schreter@sap.com

Abstract. The rise of NVRAM technologies promises to change the way we think about system architectures. In order to fully exploit its advantages, it is required to develop systems specially tailored for NVRAM devices. Not only this imposes great challenges, but also developing full system architectures from scratch is undesirable in many scenarios due to prohibitive development costs. Instead, we analyze in this paper the behavior of an existing log-structured persistent key-value store, namely LevelDB, when run on top of an emulated NVRAM device. We investigate initial opportunities for improvement when adapting a system tailored for HDD/SSDs to run on top of an NVRAM environment. Furthermore, we analyze the behavior of the DRAM caching components of LevelDB and whether more suitable caching policies are required.

Keywords: Log-structured merge-tree · Persistent memory · Caching · Storage

1 Introduction

For some years already, NVRAM technologies have been announced as the next evolution step for persistent storage. These devices are expected to offer latencies much closer to those of DRAM, while providing higher storage capacity. Furthermore, even if accessible as a block device through the file system layer, these non-volatile memories are byte-addressable and can be directly accessed by the processor through its caches. While more convenient, the persistence aspect makes it non-trivial to develop systems that directly access NVRAM the same way as DRAM while leveraging its non-volatility; data consistency, persistent memory leaks, and partial writes are some of the challenges to be considered. The focus of this work is to analyze caching trade-offs involved in a hybrid DRAM-NVRAM environment. We take a log-structured merge-tree system (LSM) [14] as a case of study and investigate its behavior running on NVRAM. Not only LSMs are in the core of many modern systems, but their architecture handle reads and writes separately. This separation is particularly interesting, as it enables an analysis of the impact of caching in read and write operations in isolation.

© Springer International Publishing AG 2017
M. Kirikova et al. (Eds.): ADBIS 2017, LNCS 10509, pp. 326–340, 2017.
DOI: 10.1007/978-3-319-66917-5_22

In this work, we put LevelDB as an example of an LSM system on top of an emulated NVRAM device. LevelDB is designed to make efficient use of HDDs/SSDs. While a block device driver can be easily wrapped around NVRAM to make it look like an HDD/SSD, this does not exploit its full potential. Therefore, the first contribution of this work is the design of a persistent memory environment for LevelDB, named *pmemenv*, to enable a more efficient management of persistent storage at a cache-line granularity. We consider that this approach offers a good compromise, as we are adapting an existing system instead of proposing a completely new architectural design or relaying all storage management to general purpose file systems originally designed for block devices.

While many file systems support Direct Access (DAX) [3, 19] to bypass the page cache when accessing NVRAM, most database systems implement their own cache at the application level. Since DRAM will still have a lower latency than NVRAM, the second contribution of this work is to investigate if better performance can be achieved by using DRAM for caching in LevelDB.

The remaining of the paper is organized as follows: Sect. 2 presents the required background. Section 3 covers related works. Section 4 describes architectural details of LSMs and LevelDB. Section 5 introduces the implementation of *pmemenv*. Section 6 discusses caching policies better suited for NVRAM. Section 7 presents an experimental analysis. Finally, Sect. 8 concludes the paper.

2 Background

The Storage Networking Industry Association (SNIA) [18] defines that NVRAM devices should be managed by an NVRAM-aware file system which supports direct access. In order to acquire direct access via load/store semantics, the user application has to create a file and memory-map it to its virtual memory space, creating a *persistent memory pool*. The direct access provided by the file system guarantees that operations are done in the persistent memory pool without any sort of DRAM caching.

However, when dealing with load/store operations, one must consider any instruction re-ordering that might be introduced by the compiler or the processor. The non-volatility of NVRAM makes instruction re-ordering critical, as, in case of a system failure, it might result in problems such as loss of data consistency, partial writes, and persistent memory leaks. To avoid that, applications have to be carefully designed to enforce a proper durability order of store instructions. This can be currently achieved with the help of hardware instructions such as SFENCE, CLFLUSH, and *non-temporal stores*. SFENCE guarantees that all preceding store-to-memory instructions have been executed. CLFLUSH evicts a cache line and writes its content to memory. Non-temporal stores ensure that data is written directly to memory, bypassing the processor cache. Intel has also announced the CLFLUSHOPT and CLWB instructions to enhance performance over CLFLUSH: CLFLUSHOPT is ordered with a smaller set of memory traffic, allowing multiple cache lines to be flushed in parallel within a single logical processor's instruction stream, and CLWB writes back a cache line to memory, without invalidating it, making future reads and writes much faster.

L. Lersch et al.

Furthermore, a persistent memory pool might be memory mapped to a different virtual memory space at restart time. Therefore, an application accessing NVRAM through this interface must support persistent pointers [15] to keep track of its persistent allocated memory regions. Finally, in order to avoid partial writes, the application should consider that, while the unit of transfer between the processor and NVRAM is a cache line, durable atomicity of writes is only guaranteed at a smaller granularity (8 Bytes on Intel x86 architectures).

In order to guarantee a certain level of consistency, database systems require careful control of when data is written to persistent storage. In an NVRAM scenario, even if this level of control can be achieved with the aid of the aforementioned hardware instructions, these introduce additional overhead and complexity. In this context, even if NVRAM provides fast random access, log-structured systems are still interesting for the following reasons: First, in contrast to update-in-place systems, maintaining consistency and atomicity of a log-structured write is simpler, as only the tail of the log is updated to reflect the operation, thus avoiding partial writes. Second, it is possible to employ non-temporal stores to bypass the CPU cache and write directly to NVRAM, avoiding operations such as CLFLUSH. Third, persistent log-structured key-value stores, such as LevelDB and RocksDB, aim at reducing the amount of data written to persistent storage (write amplification). This is even more appealing considering that NVRAM supports a limited number of writes.

We take the LSM as an example of a log-structured system to analyze its behavior on NVRAM. The LSM was initially proposed to improve write performance of update-in-place data structures (e.g., B+Tree) in HDDs. The write-optimized nature of LSMs makes them appealing to cloud systems that experience high write and data injection rates. Systems that must ingest an event log and query the ingested data with acceptable response time are common examples. The popularity of LSMs increased following the trend of Google's Bigtable [9]. Other systems that implement a similar LSM-based architecture at the storage layer are: HBase [2], Cassandra [1], Riak [6], LevelDB [4], and RocksDB [7]. We use LevelDB since it is a smaller, simpler, and easier-to-modify system.

3 Related Work

A significant amount of research has been conducted in the past years on NVRAM technologies. It is still unclear which role such devices will to play in the storage hierarchy [8]. Nevertheless, an assumption considered by the vast majority of research is that DRAM will keep co-existing with NVRAM in hybrid environments. In the following, we highlight approaches closely related to this paper.

Pelley et al. [16] analyze the behavior of Shore-MT, an update-in-place storage manager developed for HDDs, when running on an NVRAM device. The authors investigate the buffering components and the overhead of re-ordering writes in different recovery algorithms. Dulloor et al. [10] presented PMFS, a lightweight file system to manage files at the user-space level, avoiding going through the block layer and the page cache of the operating system. Also, the ext4 file system was extended with Direct Access (DAX) capabilities to bypass the operating system cache to better support NVRAM devices. Xu et al. [20] present a log-structured file system adapted to make efficient use of hybrid environments and exploit the fast random access of NVRAM.

More recently, Li et al. [13] investigated adapting RocksDB to NVRAM. They aim at improving the recovery time of RocksDB by replacing the votatile MemTable by a persistent one, thus avoiding any logging for durability. Furthermore, they also analyze the caching behavior when having NVRAM in between DRAM and SSDs in the storage hierarchy. RocksDB originated as a fork of LevelDB and provides currently many improvements over LevelDB. Nevertheless, while RocksDB shows better numbers performance-wise, both systems implement a log-structured merge-tree (LSM) at their core.

4   LSM and LevelDB Architecture

Figure 1 illustrates the general architecture of an LSM. Updates in an LSM are made in a separate in-memory data structure (C_0) which is made durable by logging. When C_0 reaches a certain threshold size, it is migrated to a lower level persistent component C_1, which can be, for instance, a tree structure. This can be generalized to a hierarchy of multiple persistent components $C_1..C_n$, in such a way that the size of components grows exponentially in relation to the preceding component in the hierarchy. All components store records in a sorted order.

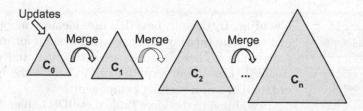

Fig. 1. General architecture of a log-structured merge-tree (LSM).

A lookup operation has to consider the multiple components. A rolling-merge process between components usually runs in the background. This is required to reclaim space and keep a predictable performance by alleviating the read penalty introduced by multiple components. Indeed, components are sorted by recency and a lookup has to inspect from the most recent to the oldest component until the key is found. The improved write performance is achieved by two different characteristics: converting random writes into sequential ones and reducing the amount of data written to persistent storage (i.e., decreasing write-amplification).

4.1 LevelDB

LevelDB is an open-source, embeddable, persistent key-value store originally developed by Google. It treats keys and values as arbitrary byte arrays and stores data sorted by key. The user can interact with the system through a basic interface including *Put(key,value)*, *Get(key)*, and *Delete(key)*. In the following, we detail some relevant implementation details required to understand the remaining of this paper. The architecture of LevelDB is summarized in Fig. 3. LevelDB uses a skip-list [17] as its in-memory data structure, called *MemTable*. The persistent components are organized as levels L_0 to L_n. Each level is composed of a certain number of *Sorted String Tables* (SST).

An SST has a fixed size (around 2 MB) and consists of four types of blocks. A *data block*, usually 4 KB, contains keys and values in sorted order (possibly compressed). Additionally, an SST has a single *index block* used to locate the data block of a given key. Optionally, there can be *meta blocks* that store information such as bloom filters, in which case there will also be a *meta index block* to locate them. The layout of an SST is represented in Fig. 2. Whenever the MemTable reaches a threshold size (4 MB by default), it is compressed and converted into an SST of L_0. With the exception of L_0, SSTs of the same level do not have overlapping key ranges, meaning that at most one SST per level has to be read. When the number of SSTs in L_n reaches its threshold, they are merged with the SSTs of L_{n+1} that have overlapping key ranges to generate new SSTs. LevelDB has 7 levels by default, with L_0 containing a maximum of 4 SSTs, L_1 10 SSTs, and each level after contains a maximum of factor 10 the amount of SSTs in the level above. Finally, a system catalog keeps track of information on all levels, e.g., current SSTs and their key ranges.

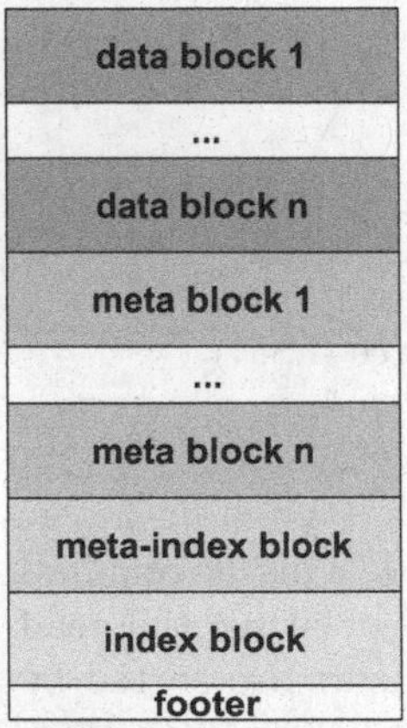

Fig. 2. Sorted String Table layout.

Caching. By default, LevelDB uses memory-mapped files and the page cache of the operating system for improved performance. These features are orthogonal and we have disabled them in our experiments to isolate the behavior of LevelDB's own caching component.

In addition to the MemTable, LevelDB implements two DRAM read-only caches: *table cache* and *block cache*. The table cache is used to hold entries containing metadata about SSTs and index blocks (possibly meta blocks) of SSTs recently accessed. The block cache holds exclusively the data blocks from SSTs. In order to improve concurrency, both caches are composed of 32 shards (default), and each shard implements least recently used (LRU) as the default block replacement policy. A read operation must first check if the given key is present in the MemTable. If not, then it will locate the candidate SST in the next level based on the SST key ranges contained in the catalog. Once the relevant SST is found, the table

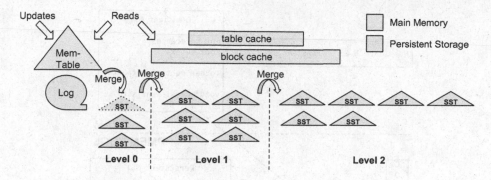

Fig. 3. LevelDB architecture represented with the first three levels for simplicity.

cache and block cache are searched for the corresponding index block and data block, respectively.

5   *Pmemenv*: Persistent Memory Environment

We have implemented a lightweight persistent memory environment, denoted *pmemenv*, using Intel's NVML library [5]. *Pmemenv* is tailored specifically for accessing and managing LevelDB's components in NVRAM directly in the user-space. This contrasts with the usual interface, where the application has to go through the file system layer to access the persistent storage. *Pmemenv* has two main advantages: First, it enables zero-copy reads, meaning that data can be read directly from NVRAM without loading it to DRAM. Second, it enables read and write operations to NVRAM at a cache-line granularity.

LevelDB manages SSTs through a virtual interface, enabling users to customize the required behavior. NVML enables the user to create a persistent memory pool and manage objects in it through a persistent allocator interface, hiding from the user the complexity required to properly enforce the order of write operations. The library uses a lightweight logging scheme to guarantee the fail-safe atomicity of these persistent allocations. A persistent pointer for each allocated block is stored in a collection located in a fixed memory region of the pool. The user is able to iterate over this collection and retrieve every allocated object in the persistent memory pool, thus preventing persistent memory leaks. As a side note, NVML also offers transactional support to enable more complex atomic memory operations than allocation/deallocation of blocks of memory. These memory transactions, however, are not explicitly used in this work.

Through NVML, *pmemenv* implements the following atomic operations on SSTs required by LevelDB: create/allocate, write bytes (append-only), read bytes, delete/deallocate, rename. Figure 4 illustrates the general architecture of *pmemenv*. It comprises two main parts: an in-memory hash table and a persistent memory pool. Every SST is composed of a transient part and a persistent part. The hash table is used to map the unique identifier of an SST to its transient part. The transient part of an SST includes non-critical metadata that is

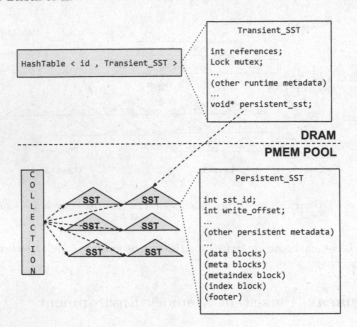

Fig. 4. *pmemenv* architecture.

only required during normal execution, such as reference counters, mutexes, and status flags. The transient part also contains a pointer indicating the location of the persistent part in the persistent memory pool. The persistent part contains critical data required by basic operations and to recover the hash table after a system failure. In our implementation, the persistent part of an SST is composed of the unique identifier, the current append offset, and the remaining data and index blocks shown in Fig. 2. The NVML library already stores the size of allocated persistent memory blocks, therefore, even if this is critical data, we do not store it and rely on the library to provide this information. In case of a system failure, the hash table can be rebuilt by iterating over the collection of pointers pointing to the allocated SSTs and retrieving the SST identifiers. The metadata in the transient SST part is set to default values and the pointer is set to the corresponding address in the persistent memory pool.

The separation of SSTs into transient and persistent parts allows metadata to be moved between them, enabling the system to possibly slide a *persistence bar* to choose which parts to make persistent [15]. In one extreme scenario, all data, metadata, and data structures are allocated in the persistent memory pool. This would reduce the recovery time to a minimum at a possible performance cost and additional complexity. Nevertheless, this is out of the scope of this paper.

Finally, since all writes to SSTs are log-structured (i.e., append-only), they can use non-temporal stores to bypass the processor cache. Writes of an arbitrary number of bytes to an SST are protected from partial writes by updating the current write offset of the SST (8 Bytes), which is guaranteed to be an atomic operation. The MemTable log and other auxiliary files, such as the catalog of SSTs, are also managed by *pmemenv* the same way as SSTs.

6   2Q Cache Policy for NVRAM

LevelDB implements the LRU as its default replacement policy, meaning that whenever the cache is full and a miss occurs, the least recently accessed block is evicted to make space for the requested one. However, when the SSTs are in NVRAM, the processor is able to directly read these blocks without bringing them to DRAM. On one hand, DRAM has a lower latency and enables faster access to data blocks. On the other hand, not only a cache policy introduces additional complexity, but there is also an overhead for transferring data from NVRAM to DRAM when a miss occurs. This overhead might not be worthwhile when compared to the cost of simply accessing the data directly from NVRAM.

Ideally, we would like the cache replacement policy to keep track of accesses not only to cached blocks, but also to un-cached ones. This would enable the policy to make a better decision whether it is advantageous to transfer a given block to DRAM, avoiding a hotter block to be evicted following a miss to a colder one. As an example, this behavior would avoid that a table scan trashes the cache by evicting all of its contents.

The 2Q replacement policy [12] considers similar goals. While the original 2Q was proposed in the context of main memory and hard disks, we have adapted the concept to NVRAM by enabling zero-copy reads from the persistent storage. Similar to the original, our 2Q policy has two components: AM and A1. AM holds cached blocks and is managed by some replacement policy (LRU in our case). A1 does not hold any blocks, but only keeps track of accesses to un-cached blocks (blocks read directly from NVRAM). Since only references to blocks are kept, the space consumption of A1 is minimal. The size of A1 is tunable and references are kept in a FIFO queue.

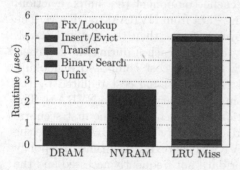

Fig. 5. Average runtime of binary search over 4 KB of integers.

Algorithm 1 shows the pseudo-code for the two main functions of the replacement policy. The function *FIX* is called to request access to a block. If the block is already cached, it is directly returned (line 3). Otherwise, if it is not going to cause another block to be evicted or if there was another reference to it in the recent past (line 5), the block is transferred to DRAM (line 6). If the block was accessed recently, it means that it is probably hot and is a good candidate to be transferred to DRAM. If both conditions are false, the block is read directly from NVRAM and a reference to it is added to A1 (line 9–10). A hash table is used for efficient containment test of both AM and A1. The *VICTIM* function is called when the cache is full and we need to pick a block for eviction. The block picked is the least recently used one (line 16), but a reference to it is additionally added to A1 (line 17). Since A1 is a FIFO with a limited size, it discards its oldest reference when a new one is inserted.

Algorithm 1. 2Q policy pseudo-code

```
 1: function FIX(blk_id)
 2:     if AM.contains(blk_id) then
 3:         return AM.get(blk_id)
 4:     else
 5:         if !AM.full() OR A1.contains(blk_id) then
 6:             AM.load(blk_id)
 7:             return AM.get(blk_id)
 8:         else
 9:             A1.add(blk_id)
10:             return NVRAM.get(blk_id)
11:         end if
12:     end if
13: end function
14:
15: function VICTIM( )
16:     blk_id ← AM.remove_lru()
17:     A1.add(blk_id)
18: end function
```

Figure 5 shows the runtime of a binary search over 4 KB of integers in DRAM, NVRAM (latency set to 4x that of DRAM), and simulating a miss of LRU in a DRAM cache. In the breakdown of the cost of a LRU miss it is possible to see the constant overhead introduced by the cache component (fix, unfix, eviction, etc.), as well as the huge cost of transferring data from NVRAM to DRAM. The binary search represented in the breakdown is faster than the one in DRAM, because the data was already cached by the CPU caches during the transfer. The cache miss in the LRU policy has a constant cost comprised of lookup, eviction, and move of a block to DRAM. The additional cost required by the policy exceeds by far the cost of simply doing the binary search in NVRAM. The 2Q policy introduces two different scenarios for a cache miss: the first scenario simply adds a reference to the FIFO and reads directly from NVRAM, while the second scenario is similar to LRU. In an NVRAM context, a well-tuned 2Q policy should prefer to pay the lower miss cost for data not frequently accessed and the higher miss cost for data expected to be frequently accessed in the near future. While most proposed replacement strategies (LRU, LFU, CLOCK, etc.) focus on improving the hit ratio, the idea of being able to choose between two different miss costs adds a new dimension. Assuming that the hit ratio is determined by the replacement strategy and the cache size, a smaller 2Q cache is likely to have more misses than a larger LRU cache. However, if most of the 2Q misses pay the lower cost, similar or even better performance than LRU can be achieved with a lower memory consumption. This introduces the non-intuitive idea that a higher hit ratio does not necessarily translate to better performance, as the costs for misses might differ.

7   Evaluation

We use the Intel NVM Emulation platform that emulates an NVRAM device by accessing a dedicated area of DRAM with higher, tunable latency. The higher access latency to DRAM is achieved thanks to a special BIOS. A full description of this system can be found in [11]. The system is equipped with two Intel Xeon E5 processors. Each one has 8 cores, running at 2.6GHz, and featuring 32 KB L1 data and 32 KB L1 instruction cache as well as 256 KB L2 cache. The 8 cores of one processor share a 20 MB last-level cache. The system has 64 GB of DRAM and 192 GB of emulated NVRAM. The emulated NVRAM device is mounted with the ext4 file system with DAX support. In the experiments, we set the latency of NVRAM to 360 ns, approximately 4x the latency of DRAM (90 ns). Considering HDD/SSD is out of the scope of this work, since all experiments are based on the assumption that the storage device can be directly accessed through the CPU caches. The system runs Linux with kernel version 4.4.21.

We use NVML Release Version 1.2 and LevelDB Release Version 1.19. All the source code was compiled using GCC 4.8.5. We disabled the memory mappings of SSTs and operating system caching in LevelDB. Compression and filtering of SSTs (bloom filters) are also not used. The MemTable is set to its default size of 4 MB. All requests to LevelDB are made from a single thread.

7.1   Write Performance

We first analyze runtime and latency of two approaches for writing to NVRAM: through the ext4 file system with Direct Access (DAX) support (as a drop-in replacement for HDDs/SSDs) and through *pmemenv*. As mentioned in Sect. 5, the file system manages these operations at a block granularity (usually 4 KB), while *pmemenv* allows finer control over written data. This implies that, in a scenario where durability of single operations must be guaranteed, *pmemenv* is able to write only the changed cache-lines. However, most systems implement some sort of group commit to hide write latencies. LevelDB enables this by batching many *Put* operations in a *WriteBatch* that is accumulated in DRAM and is later made durable as a single *Put*. We consider scenarios with different *WriteBach* sizes. Additionally, we investigate if batching writes in DRAM still offers benefits for *pmemenv*. We run a write-only workload of YCSB with 100M key-value records, where each key is 16 bytes and each value is 112 Bytes, giving a total of 128 Bytes per record. The results are depicted in Fig. 6.

First, for a group size of one, the ext4+DAX configuration has to persist data at the granularity of pages via *fsync*, which incurs a high cost if single operations are to be made durable. *Pmemenv* is able to avoid the kernel path and to persist data at a much smaller granularity, which reduces the overhead related to write operations. Increasing the group size drastically improves the performance of ext4+DAX (16 times faster when increasing group size from 1 to 100), as the cost of *fsync* is amortized across many insertions. For *pmemenv*, an improvement of approximately 50% in runtime is observed when increasing the group size from 1 to 10. For larger group sizes the difference is not significant.

Grouping insert operations in batches introduces a trade-off between throughput (runtime) and latency, as the first requests to arrive in a group are delayed. Figure 6b shows the average latency of single insert requests for different group sizes. The standard deviation can be observed in Table 1. For smaller group sizes (1 to 100) in ext4+DAX, the increased latency is justified by the large gains in runtime, making the batching of operations an obvious choice for most applications. However, for *pmemenv*, this trade-off is not so clear and the decision of sacrificing latency for better runtime might become a matter of Service Level Agreements, like response time required by applications. Finally, not only *pmemenv* presents lower runtime and lower average latency than ext4+DAX, but also a lower standard deviation for group sizes 1 to 100, which translates to a more predictable performance over time.

7.2 Read Performance

We also analyze if better performance can be achieved by dedicating a portion of DRAM for caching hot data. Since writes in LevelDB are made in a

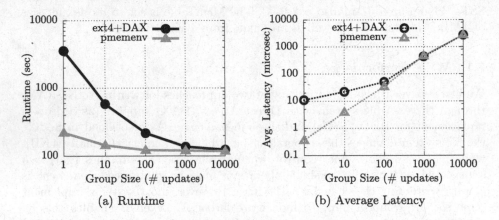

(a) Runtime   (b) Average Latency

Fig. 6. Insertion of 100 million key-value records with varying *WriteBatch* size.

Table 1. Latency standard deviation of Fig. 6b in microseconds.

File system	1	10	100	1000	10000
ext4 + DAX	23	56	161	586	4750
pmemenv	12	39	125	517	4749

separate data structure, the remaining caching components benefit mainly read operations. Therefore, we have considered read-only YCSB workloads to better outline the performance impact. Each workload issues 50 million lookups over 10 million key-value pairs. Before each workload is executed, the caches are warmed up by executing read requests until they become completely full. The warmup time is not considered. We analyze two different scenarios: uniform and skewed distribution (80% of requests to 20% of the records) of key requests. It is worth noting that each *Get* request for a key translates into two block requests, one for the index block and one for the data block. Hence, even a uniform distribution of keys presents a skewed distribution of block accesses. Figure 7 illustrates the number of accesses of each block sorted from the most to the last accessed.

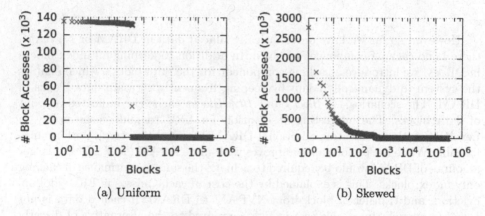

(a) Uniform   (b) Skewed

Fig. 7. Distribution of accesses to blocks.

Figure 8 presents the runtime of both read-only workloads for different system configurations. The X axis represents which portions of SSTs are cached in DRAM. We start with a *NoCache* approach, where the caching components were completely removed and all SSTs are read directly from NVRAM through *pmemenv*. Later, we gradually increase the DRAM consumption by statically placing portions of every SSTs in DRAM. The *Footers* scenario has the footers of all SSTs in DRAM. The footer of an SST contains pointers to the index blocks, as well as checksums and additional status flags. Footers are frequently accessed (it is where each read in an SST starts) and, since they are relatively small (around 64 Bytes), keeping all of them in DRAM improves performance at a minimal cost of memory consumption. Next, *IndexBlocks* considers holding the index blocks of all SSTs in DRAM. For our workloads, every index block is approximately 18 KB and there are around 500 SSTs, giving a total of about 10 MB additional DRAM consumption (less than 1% of the total size). The observed performance gains are significant and justify the additional memory consumption. At this point, we can conclude that a careful placement of frequently accessed data in DRAM is beneficial despite the low latency of NVRAM.

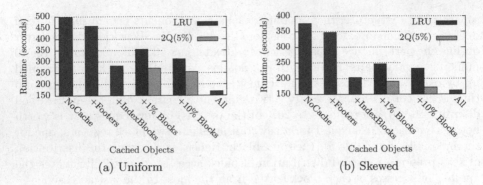

Fig. 8. Runtime of read-only workload.

However, so far we have only statically placed data in DRAM or NVRAM, there is no caching component involved. In addition to keeping all index blocks in DRAM we introduce a caching component for the data blocks, which enables the system to dynamically adapt by keeping frequently accessed data blocks in DRAM. The scenarios *1% Blocks* and *10% Blocks* cache the indicated amount of data blocks. The interesting observation for LRU (default cache policy in LevelDB) is that dedicating additional DRAM harms the system's performance initially. While the performance improves with more DRAM (*10% Blocks*), larger amounts of DRAM would be required to achieve the same performance of caching only index blocks. This is explained by the cost of cache misses in LRU: lookup, eviction, and transfer of block from NVRAM to DRAM. If cached data is not accessed enough times, this cost is high compared to the alternative of directly accessing data in NVRAM and avoiding the overhead caching. As discussed in Sect. 6, the cost of transferring data to DRAM is only worthwhile if the policy can predict that this block will be accessed frequently in the near future.

To alleviate the high cost of misses, we have implemented 2Q to enable a more lightweight policy. We have set the A1 size to 5% of the AM size. Our initial goal with 2Q is to avoid the observed behavior where the system gets slower when more DRAM is dedicated for caching. In contrast to LRU, there is always some performance improvement with larger caches for data blocks. This comes from the fact that 2Q avoids evicting a cached block and moving a new block to DRAM when a miss occurs. Finally, the *All* scenario represents the runtime with the whole dataset cached in DRAM. It is possible to see in Fig. 8b that the 2Q cache with enough DRAM to hold 10% of the data blocks can achieve similar performance of holding all blocks in DRAM.

7.3 Mixed Workloads

Based on the observations from the previous experiments, we analyze the overall behavior of the system in workloads containing both updates and lookups. Two mixed workloads with skewed access are considered: 25% and 50% of updates. We also run the experiments with varying NVRAM latencies to show

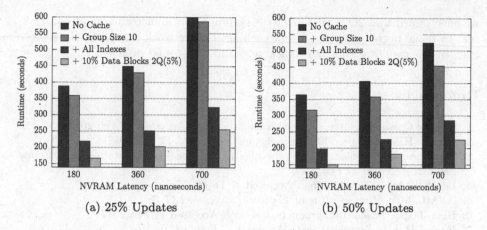

(a) 25% Updates   (b) 50% Updates

Fig. 9. Runtime of skewed mixed workload.

that the behavior is the same regardless of the slowdown/speedup incurred by higher/lower latencies.

Similar to previous results, we start with a *NoCache* approach and gradually dedicate more DRAM for caching purposes. The *Group Size 10* has enough DRAM for holding 10 update operations and persist them as a single *WriteBatch*. Later, in addition to that, we reserve enough DRAM to hold *All Indexes*. Finally, we hold up to 10% of the data blocks in a 2Q cache. Figure 9 presents the gradual performance gains achieved in each of these steps. The biggest improvement happens when keeping all index blocks in DRAM. Not only index blocks are frequently accessed, but their additional DRAM consumption is minimal, making it realistic to hold all of them in DRAM and avoiding any replacement policy overhead. While a cache for data blocks with 2Q replacement policy offers some benefits in terms of performance, it is up to the user to decide if the cost of additional DRAM justify these gains. Nevertheless, we consider that enabling the system to manage hot and cold data is important and better caching policies can probably achieve this behavior with even better performance.

8   Conclusion

We have adapted LevelDB to run on top of an emulated NVRAM device. We considered a hybrid DRAM-NVRAM environment and discussed relevant implementation details. We implemented *pmemenv*, a data accessor to enable fine-grained management of SSTs in NVRAM. Furthermore, we analyzed the LevelDB caching components for improving write and read performance. Even with its lower latency, simply dedicating a portion of DRAM to cache data from NVRAM is not necessarily beneficial. On one hand, we have observed that poor caching policies might even harm performance. On the other hand, a careful placement of data offers significant benefits. A dynamic management of hot and cold data can be achieved through lightweight caching policies. In this context, we have shown that 2Q never harms the system performance and enables the system to make better decisions about caching. Future work can probably achieve even better results with other lightweight and well-tuned policies.

References

1. Apache Cassandra. http://cassandra.apache.org/. Accessed 17 Feb 2017
2. Apache HBase. https://hbase.apache.org/. Accessed 17 Feb 2017
3. Direct Access for files. https://www.kernel.org/doc/Documentation/filesystems/dax.txt. Accessed 17 Feb 2017
4. LevelDB. http://leveldb.org/. Accessed 17 Feb 2017
5. NVML. http://pmem.io/nvml/libpmem/. Accessed 17 Feb 2017
6. Riak. http://basho.com/products/riak-kv/. Accessed 17 Feb 2017
7. RocksDB. http://rocksdb.org/. Accessed 17 Feb 2017
8. Bonnet, P.: What's up with the storage hierarchy? In: 8th Biennial Conference on Innovative Data Systems Research, CIDR 2017 (Online Proceedings) (2017)
9. Chang, F., Dean, J., Ghemawat, S., Hsieh, W.C., Wallach, D.A., Burrows, M., Chandra, T., Fikes, A., Gruber, R.E.: Bigtable: a distributed storage system for structured data. In: Proceedings of the 7th USENIX Symposium on Operating Systems Design and Implementation (2006)
10. Dulloor, S., Kumar, S., Keshavamurthy, A., Lantz, P., Reddy, D., Sankaran, R., Jackson, J.: System software for persistent memory. In: Eurosys Conference (2014)
11. Dulloor, S.R.: Systems and Applications for Persistent Memory. Ph.D. Thesis (2015). https://smartech.gatech.edu/bitstream/handle/1853/54396/DULLOOR-DISSERTATION-2015.pdf
12. Johnson, T., Shasha, D.E.: 2Q: a low overhead high performance buffer management replacement algorithm. In: PVLDB (1994)
13. Li, J., Pavlo, A., Dong, S.: NVMRocks: RocksDB on non-volatile memory systems. http://istc-bigdata.org/index.php/nvmrocks-rocksdb-on-non-volatile-memory-systems. Accessed 17 Feb 2017
14. O'Neil, P.E., Cheng, E., Gawlick, D., O'Neil, E.J.: The log-structured merge-tree (LSM-Tree). Acta Inf. **33**, 351–385 (1996)
15. Oukid, I., Booss, D., Lehner, W., Bumbulis, P., Willhalm, T.: SOFORT: a hybrid SCM-DRAM storage engine for fast data recovery. In: International Workshop on Data Management on New Hardware (2014)
16. Pelley, S., Wenisch, T.F., Gold, B.T., Bridge, B.: Storage management in the NVRAM era. In: PVLDB (2013)
17. Pugh, W.: Concurrent maintenance of skip lists. Univ. of Maryland Institute for Advanced Computer Studies Report No. UMIACS-TR-90-80 (1990)
18. SNIA: NVM Programming Model V1.1 (2015). http://www.snia.org/sites/default/files/NVMProgrammingModel_v1.1.pdf
19. Wilcox, M.: Add support for NV-DIMMs to ext4. https://lwn.net/Articles/613384/. Accessed 17 Feb 2017
20. Xu, J., Swanson, S.: NOVA a log-structured file system for hybrid volatile/non-volatile main memories. In: Proceedings of the 14th USENIX Conference on File and Storage Technologies (FAST) (2016)

Semantic Data Processing

SPARQL Query Containment
with ShEx Constraints

Abdullah Abbas$^{(\boxtimes)}$, Pierre Genevès, Cécile Roisin, and Nabil Layaïda

University Grenoble Alpes, CNRS,
Institute of Engineering University Grenoble Alpes, Inria, LIG,
38000 Grenoble, France
{abdullah.abbas,cecile.roisin,nabil.layaida}@inria.fr,
pierre.geneves@cnrs.fr

Abstract. ShEx (Shape Expressions) is a language for expressing constraints on RDF graphs. We consider the problem of SPARQL query containment in the presence of ShEx constraints. We first propose a sound and complete procedure for the problem of containment with ShEx, considering several SPARQL fragments. Particularly our procedure considers OPTIONAL query patterns, that turns out to be an important fragment to be studied with schemas. We then show the complexity bounds of our problem with respect to the fragments considered. To the best of our knowledge, this is the first work addressing SPARQL query containment in the presence of ShEx constraints.

1   Introduction

ShEx (or Shape Expressions) is intended to be an RDF constraint language [19]. It can be used to validate documents and communicate expected graph patterns. Static analysis and query optimisation can make a considerable benefit from the presence of schemas when used to infer satisfiability/unsatisfiabliity of queries and relations between queries (such as containment and equivalence) by utilising the additional information provided by the schemas.

In this work we investigate the SPARQL query containment with ShEx constraints. Given two SPARQL queries, and a set of ShEx constraints, our purpose is to statically analyse such queries, namely determining the containment relation between them before being actually executed on the data.

For the fragments of SPARQL including OPTIONAL patterns, the containment of queries is normally investigated with the notion of subsumption [1]. A solution mapping is a mapping from a set of variables to a set of values, thus designating an answer for a query. A solution mapping σ_1 is subsumed by another solution mapping σ_2 written as $\sigma_1 \sqsubseteq \sigma_2$ if all the variables of σ_1 are also in σ_2 and have the same mapping values. Given a set of mappings Ω_1 (resembling a SPARQL query solution), it is subsumed by another set of mappings Ω_2 written as $\Omega_1 \sqsubseteq \Omega_2$ if for every $\sigma_1 \in \Omega_1$ there exists $\sigma_2 \in \Omega_2$ such that $\sigma_1 \sqsubseteq \sigma_2$.

The consideration of ShEx constraints in query containment is important, because such constraints may affect the results of containment checking. Consider the following two SPARQL query graph patterns:

M. Kirikova et al. (Eds.): ADBIS 2017, LNCS 10509, pp. 343–356, 2017.
DOI: 10.1007/978-3-319-66917-5_23

Q_1: {?x :producer :p1 . ?x :feature "feature1"}
    OPT {?x :feature "feature2". ?x :expiryDate ?d}
Q_2: {?x :producer ?y . ?x :feature "feature1"}

Without constraints, no containment relation holds between these two queries. However, consider the following ShEx constraints defined for a $\langle Product \rangle$ node type:

```
<Product> {
:name xsd:string ,
:expiryDate xsd:date ? ,
:producer @<Company> + ,
:feature xsd:string }
```

The previous ShEx shape definition means that a node of type "Product" should have a name of type string, optionally have an expiry date, have at least one producer which belongs to a another ShEx shape $\langle Company \rangle$, and have exactly one feature of type string. Given that these ShEx constraints apply to the data, we can deduce that a containment relation $Q_1 \sqsubseteq Q_2$ holds between the two queries. This is due to the constraint that a "feature" predicate is allowed to occur only once, and thus in query Q_1 the right hand side of the optional pattern will never return results. In such case, we can deduce that the containment relation $Q_1 \sqsubseteq Q_2$ holds between the two queries.

There are several kinds of ShEx constraint violations that may lead to a new conclusion about the containment of two queries. These include (1) cardinality constraint violations, (2) basic data type constraint violations (like xsd:string, xsd:data ...), and (3) ShEx type definition violations (like @$\langle Company \rangle$ type).

Data on the web are getting larger, and distribution of data is getting more applicable. Different data sources are often being managed by different authorities. The need of schemas becomes increasingly necessary in order to manage the big amounts of data. While different sources in the same domain may share the same vocabulary, their constraints on data may vary. While these slight differences in data shapes may become a hassle for users to track individually, the use of OPT patterns in SPARQL provides a way to ask for constraints that are not necessarily applicable, and that is why the study of the optional fragment is particularly interesting.

In this work we define a sound and complete procedure for containment of SPARQL query fragments in the presence of a ShEx schema, based on the usage of ShEx validators and query containment solvers that don't consider any schema constraints. We also study the complexity of the problem. The results vary from NP-c to Π_2^P-c according to the fragment considered. We also provide a higher NEXP complexity bound for further fragment extensions (FILTERS, MINUS, and property path patterns).

Paper Outline. In Sect. 2 we comment on the related works. In Sect. 3 we introduce some preliminaries necessary for understanding the rest of the paper. In Sect. 4 we define a query transformation function which is necessary for the definition of the containment procedure. In Sect. 5 a sound and complete containment procedure is given for different SPARQL fragments. In Sect. 6 we derive complexity bounds of our problem. Finally, we conclude in Sect. 7.

2   Related Works

In [8], the authors proposed a schema language for edge-labeled data graphs (like RDFs), and then studied the satisfiability of 3 different classes of query languages (RPQs, NREs, and CRPQs) when such constraints are considered, but this study did not include containment. In [10,11] the authors studied static analysis aspects of XPath using mu-calculus and Monadic Second-order Logic respectively, then the authors provided in [12] a tool related to their studies. XPath is a query language on tree structures while in our work SPARQL is a query language on graphs, yet these works inspire our work for SPARQL fragment extensions using logical formulas.

The work in [5] studied containment of PSPARQL, an extension of SPARQL 1.0 with paths and path constraints. In [14], the authors explored the complexity of containment and evaluation problems for fragments of SPARQL 1.1 property paths. The study in [18] provides complexity analysis for several fragments of SPARQL. Additionally, in [16] the containment of well-designed OPT queries is investigated. None of these works consider schemas in the study of containment.

The works in [4,6,7] study the containment problem with ontology languages and entailment regimes (SHI, RDFS, OWL...). Ontology languages put constraints on data, like schemas, but also allows for entailment of implicit data relations. The works on containment with ontology languages focus on entailment regimes employed in these languages, but not on the fragments of SPARQL with optional patterns which we want to consider.

For the works on ShEx, in [2,20,23] the expressiveness and validation complexity of ShEx was studied. The work in [9] proposes an implementation of shape expressions to RDF graphs.

With the deep static analysis of SPARQL queries - including containment - on one side, and the emergence of schema languages for RDF (like ShEx) on the other side, we find a value in investigating these targets in a common framework.

3   Definitions

3.1   SPARQL

SPARQL is an RDF query language and a W3C Recommendation, where RDF is a directed, labeled graph data format for representing information in the web [15,21]. SPARQL contains capabilities for querying required and optional graph patterns along with their conjunctions and disjunctions [22].

A SPARQL graph pattern is defined inductively from triple patterns. Given disjoint infinite sets of IRIs - Internationalised Resource Identifiers - (I), blank nodes (B), literals (L), and variables (V), we define a triple pattern as an instance of $(I \cup B \cup V) \times (I \cup V) \times (I \cup B \cup L \cup V)$ denoted by $IBV \times IV \times IBLV$. A SPARQL graph pattern q is defined inductively from triple patterns as follows:

$q ::= t \mid q \; AND \; q' \mid q \; UNION \; q' \mid q \; OPT \; q' \mid q \; MINUS \; q' \mid q \; FILTER \; C$

where t is a triple pattern and C is a condition or a conjunction and/or disjunction of conditions on variables.

In order to reference different SPARQL fragments later, we define them as follows:

- BGP: This is the conjunctive fragment of SPARQL, i.e. the fragment that only allows using the AND operator between triples.
- AND-OPT: The fragment of SPARQL allowing the AND and OPT operators only. We particularly consider the well-designed patterns within this fragment (defined just after this list).
- AND-OPT-(UNION): The AND-OPT fragment extended with UNION on the top level only (external).
- AND-OPT-(UNION)-FILTER: The AND-OPT-(UNION) fragment extended with the FILTER operator. For this fragment we only consider filters that are decidable for query satisfiability [24], which is a necessary requirement for query containment, namely $FILTER(bound, =, \neq_c)$ and $FILTER(bound, \neq, \neq_c)$. Where $bound(?x)$ means that the variable $?x$ should be bound to a value in the query results. $=/\neq$ are the equality/inequality relations between variables. $\neq_c$ is the inequality of variable with respect to a constant belonging to $I \cup L$.
- AND-OPT-(UNION)-PP: The AND-OPT-(UNION) fragment extended with property path patterns from the SPARQL 1.1 syntax. These are regular expressions allowed in the predicate position.
- AND-OPT-(UNION)-MINUS: The AND-OPT-(UNION) fragment extended with the MINUS operator which puts constraints on situations that must not occur in the results.

Well-Designed OPT Patterns. Well-designed OPT patterns define a class of OPTIONAL patterns that have several desired properties [17], such as evaluation performance advantages.

A query q is well-designed if for every subpattern $q' = (q_1 \; OPT \; q_2)$ of q and every variable x occurring in q, it holds that: if x occurs inside q_2 and outside q', then x also occurs inside q_1.

It is also shown in [17] that any well-designed graph pattern can be equivalently rewritten in the normal form:

$(\ldots(t_1 \; AND \; \ldots \; AND \; t_k) \; OPT \; O_1) \; OPT \; O_2) \ldots) \; OPT \; O_n)$ *where each t_i is a triple pattern, and each O_j has the same form (also in normal form).*

These normal forms can be represented as pattern trees as described in [16]. For example, a query of the form $((P_1 OPT \; (P_{11} \; OPT \; P_{111} \; OPT \; P_{112}))$ $OPT \; P_{12}) \; OPT \; P_{13}$, where each P_i is a BGP, can be represented as a pattern tree as shown in Fig. 1.

In our work, we use pattern tree representations of the queries in order to study their containment with ShEx.

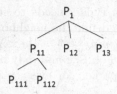

Fig. 1. Pattern tree example

3.2 ShEx

ShEx (or Shape Expressions) is intended to be an RDF constraint language. Logical operators in Shape Expressions such as grouping, conjunction, disjunction and cardinality constraints, are defined to make as closely as possible to their counterparts in regular expressions and grammar languages like BNF [20]. Shape Expressions correlate an ordered pattern of pairs of predicate and object classes (called NameClass and ValueClass) and logical operators against an unordered set of edges in a graph. For example, $\langle Shape1 \rangle$ is a definition of a shape in ShEx, where a ShEx document contains definitions of several shapes.

```
<Shape1> {
ex:name xsd:string ,
ex:phone xsd:string }
```

In the previous example, ex:name and ex:phone are NameClasses and xsd:string is a ValueClass. This definition means that for a node belonging to this shape there must strictly exist the predicates ex:name and ex:phone, each once. The objects corresponding to these predicates must be of type xsd:string.

Abstract Syntax of ShEx. Given a finite set of edge labels Σ and a finite set of types Γ, we define a shape expression e over $\Sigma \times \Gamma$ as follows: $e ::= \epsilon \mid \Sigma \times \Gamma \mid e^* \mid (e\text{"|"}e) \mid (e\text{"||"}e)$, where "|" is a disjunction, "||" is an unordered concatenation, and "*" is an unordered Kleene star. This definition also allows us to further define as macros $e^?$ (optional), e^+ (positive closure), and $(\Sigma \times \Gamma)^{([m;n])}$ (interval from m to n), which are all parts of the ShEx syntax. In the sequel we write $(a,t) \in \Sigma \times \Gamma$ simply as $a :: t$.

A shape expression schema (ShEx), or simply schema, is a tuple $S = (\Sigma, \Gamma, \delta)$, where Σ is a finite set of edge labels, Γ is a finite set of types, and δ is a type definition function that maps elements of Γ to shape expressions e over $\Sigma \times \Gamma$. If the δ is not defined for some type $t \in \Gamma$, the default definition is $\delta(t) = \epsilon$.

Semantics of ShEx. [20] Semantically, an RDF graph is valid against a ShEx schema if it is possible to assign types to the nodes of the graph in a manner that satisfies the type definitions of the schema.

We assume a fixed graph $G = (V, E)$ which resembles an RDF graph, and a fixed schema $S = (\Sigma, \Gamma, \delta)$. A typing of G w.r.t. S is a function $\lambda : V \to 2^{\Gamma}$ that associates with every node of G a set of types.

Next, the conditions that a typing needs to satisfy are identified. Given a typing λ and a node $n \in V$ we define the neighborhood-typing of n w.r.t. λ as bag over $\Sigma \times 2^\Gamma$ as $neighborTyping_G^\lambda(n) = \{|a :: \lambda(m) \mid (n, a, m) \in E|\}$.

Now, λ is a valid typing of S on G if and only if every node satisfies the type definitions of its associated type i.e., for every $n \in V$, $neighborTyping_G^\lambda(n) \in \delta(t)$, for all $t \in \lambda(n)$.

4   Query Transformation

Query transformation is a process in which we rewrite a query, where the resulting query is equivalent to the original query given that the ShEx constraints hold on the data sets. Two queries are considered to be equivalent if they always give the same execution results.

The resulting query transformations defined in this section has several utilisations, namely for optimisation purposes, especially that they are equivalent to and smaller than the original queries. We use them in this work particularly for defining containment in Sect. 5.

Before defining the transformation procedures, we give some preliminary definitions.

Definition 1. *Given a set of triple patterns P, $\mathcal{RDF}(P)$ is a function that yields a set of RDF triples by replacing each variable in P by a fresh IRI. The replacement is unique for each variable name.*

According to the previous definition, there always exists a homomorphism from the triples graph of P to the triples graph of $P' = \mathcal{RDF}(P)$. In fact, P' is an RDF data set that can be validated against a ShEx schema.

Definition 2. *Given two sets of RDF triples D_1 and D_2 and a ShEx schema S, we say that D_2 is a complement of D_1 w.r.t. S, if:*

1. $D_1 \subseteq D_2$
2. D_2 is valid w.r.t. S

Definition 3. *Given a ShEx schema S, the minimals discarding ShEx schema of S is given by the function $\mathcal{MIN}_0(S)$, and is defined by replacing all minimal cardinality constraints of S by zeros. (i.e. all cardinality constraints $[m, n]$, $+$ and 1 respectively, are replaced with $[0, n]$, $*$ and ? (optional) respectively).*

4.1   BGP Transformation

Query transformation of a BGP query is based on the RDF document validation. RDF validation against ShEx is defined with its NP-complete complexity in [23].

Definition 4 (Query Transformation). *For a BGP SPARQL query Q and a ShEx schema S, the query transformation function $\mathcal{T}_S$ is defined as follows:*

$$\mathcal{T}_S(Q) = \begin{cases} Q, & \text{if } \mathcal{RDF}(Q) \text{ is valid w.r.t. } \mathcal{MIN}_0(S) \\ empty\ query, & otherwise \end{cases}$$

The validation against $\mathcal{MIN}_0(S)$ is due to the fact that the query triples do not catch the complete data structure. Indeed, queries by nature are just partial representations of the constraints on the data that should be extracted.

4.2 AND-OPT Transformation

We extend the BGP transformation to a more interesting SPARQL fragment for our problem, the AND-OPT fragment. The results in this case will be a modified AND-OPT query that is equivalent to the original query, by applying two steps: (1) Eliminating non-valid OPT patterns, and (2) replacing some OPT operators with AND operators.

For the step (1), if we find out that some OPTIONAL pattern will never return results due to the ShEx constraints, the new query that results from this transformation is by omitting this OPTIONAL pattern.

Consider the following SPARQL query:

Q: `{:p1 :producer ?y}` OPT `{:p1 :review ?z}`

and the following ShEx schema (a minimals discarding ShEx schema):

```
<product> {
:name xsd:string ? ,
:expiryDate xsd:date ? ,
:producer @<company> * ,
:feature xsd:string ? }
```

We consider two RDF triple sets for validation against the ShEx schema, `{:p1 :producer :y}` which is valid, and `{:p1 :producer :y. :p1 :review :z}` - the optional pattern with its parent - which is not valid. As a result of this validation step, we rewrite the query by removing the optional pattern which corresponds to the RDF triple set which is not valid, and thus we get:

Q': `{:p1 :producer ?y}`

For step (2), we check if it is possible to replace some OPT operators with the AND operator. Considering the pattern tree representation of a query this operation can be described by uniting two directly connected nodes into one node, one of which is a child node, and the other is a parent node.

To show this by example, consider the following query:

Q: `{?x :name ?n}` OPT `{?x :phone ?p}`

and the following ShEx shape:

```
<Person> {
:name xsd:string ,
:phone xsd:string }
```

According to this ShEx shape definition, we know that :name and :phone will always occur together. Thus the right hand side of the OPT pattern will always occur with the left hand side of it. We therefore deduce that the previous query Q is equivalent to another query Q' without an OPT pattern.

Q: `{?x :name ?n. ?x :phone ?p}`

Two nodes in a query pattern tree must be merged into one node (the parent node), if and only if the triples of the child node will necessarily return results whenever the parent node returns results. We apply this check on every pair of parent-child nodes in the query pattern tree in order to get the final transformation of the query.

The transformations described in the latter examples for the AND-OPT SPARQL fragment are given formally in Definition 6.

Definition 5. *Given a pattern tree* $\mathcal{P}$*, and a node n of P, we define* $\mathcal{R}_\mathcal{P}(n)$ *to be the union of the set of triples of n and the set of triple of all its parent nodes up to the root node.*

Definition 6 (Query Transformation). *For an AND-OPT SPARQL query Q, its pattern tree representation* $\mathcal{P}$*, and a ShEx schema S, the query transformation function* $\mathcal{T}_S$ *is defined by the following steps:*

1. *For each node n of* $\mathcal{P}$*, if* $\mathcal{R}_\mathcal{P}(n)$ *is not valid w.r.t.* $\mathcal{MIN}_0(S)$*, then eliminate n and all its descendants from* $\mathcal{P}$*. Let* $\mathcal{P}'$ *be the new pattern tree after the validation of all the nodes of* $\mathcal{P}$*.*
2. *For each pair of nodes n_1 and n_2 of* $\mathcal{P}'$*, such that n_1 is the parent of n_2, if it is necessary for every complement of* $\mathcal{RDF}(n_1)$ *to include the RDF triples of* $\mathcal{RDF}(n_2)$ *according to S, then merge n_1 and n_2 into one node. Let P'' be the new pattern tree obtained.*

We define $\mathcal{T}_S(\mathcal{P}) = \mathcal{P}''$*.*

4.3   AND-OPT-(UNION) Transformation

For the AND-OPT SPARQL fragment extended with UNION at the top level, the same procedure can be applied on each UNION pattern separately.

5   Query Containment with ShEx

In this section we show how SPARQL query containment with ShEx can be done by benefiting from the transformations of Sect. 4.

We first apply the transformation procedure on the two queries to be checked for containment based on a given ShEx schema. The resulting transformations are then checked for containment without considering the ShEx document using query containment solvers as the one proposed in [18]. If the containment of the query transformations hold, then the containment of the original queries with the consideration of ShEx holds.

We briefly describe the containment procedure - without ShEx, displayed in the following lemma taken from [16]. The lemma provides the necessary and sufficient conditions for the deciding containment of a well-designed OPT SPARQL queries. The conditions are formulated in terms of pattern trees.

Lemma 1 *Consider two well-designed pattern trees* $\mathcal{T}_1$ *and* $\mathcal{T}_2$ *with roots r_1 and r_2, respectively. Then* $\mathcal{T}_1 \sqsubseteq \mathcal{T}_2$ *if and only if for every subtree* $\mathcal{T}_1'$ *of* $\mathcal{T}_1$ *rooted at r_1, there exists a subtree* $\mathcal{T}_2'$ *of* $\mathcal{T}_2$ *rooted at r_2 such that:*

1. $vars(T_1') \subseteq vars(T_2')$, and
2. there exists a homomorphism from the triples in T_2' to the triples in T_1' that is the identity over $vars(T_1')$.

We notice that a pattern tree containment relation $T_1 \sqsubseteq T_2$ also yields the query containment corresponding to these pattern trees (lets say $q_1 \sqsubseteq q_2$) [16].

Definition 7. *Given two queries q_1 and q_2, we define the relation $q_1 \sqsubseteq_S q_2$ to mean that $q_1 \sqsubseteq q_2$ holds in the presence of a ShEx schema S.*

Theorem 1. *Given two queries q_1 and q_2, and their corresponding transformations q_1' and q_2' according to a ShEx schema S, the containment relation $q_1 \sqsubseteq_S q_2$ holds if and only if $q_1' \sqsubseteq q_2'$ holds.*

Proof. The soundness of our procedure is evident from the fact that the transformations are equivalent to the original queries in the presence of the ShEx constraints. We use the empty schema as a transformation, or we eliminate parts of the queries only when we are sure that these parts will not return results according to the given schema.

For the completeness of the procedure, we prove it according to the corresponding fragment for each case.

1. For the BGP SPARQL fragment, assume we have two BGP queries q_1 and q_2 and their corresponding transformations q_1' and q_2' according to a ShEx schema S. For the completeness of the procedure, our purpose now is to prove that if $q_1' \not\sqsubseteq q_2'$, then $q_1 \not\sqsubseteq_S q_2$. For the case where q_1' is an empty query, $q_1' \sqsubseteq q_2'$ always holds, since the empty query is contained in every other query, and therefore the assumption condition can never happen. For the case where only q_2' is an empty query, that means that also q_2 will never return results due to a violation to the ShEx rules. No query can be contained in a query that does not return results except the empty query, and since we know that q_1 may return results due to the absence of any ShEx violation, then $q_1 \not\sqsubseteq_S q_2$ always holds. The final case is when both q_1' and q_2' are kept exactly the same as q_1 and q_2. In the latter case, if $q_1' \not\sqsubseteq q_2'$, then there exists no homomorphism from q_2' to q_1'. Given that the triples of q_1' don't violate the ShEx schema rules, then there exists a data set D which is a complement of $\mathcal{RDF}(q_1')$ w.r.t. S. BGP query solving is based on homomorphism from the set of query triple patterns to the set of RDF triples ([22]). A solution for q_1 necessarily exists in the proposed data set since the homomorphism exists by our proposal. Now we assume that the same solution also holds for q_2 and conclude a contradiction. If the same solution holds for q_2, then there exists a homomorphism from its triples patterns to D. Since all variables of q_1 are replaced with fresh IRIs, then a homomorphism is also necessary to hold from the triple patterns of q_2' to the triple patterns of q_1', and thus we conclude a contradiction because this homomorphism is a sufficient condition for deriving that the containment $q_1' \sqsubseteq q_2'$ holds (condition from [18]).

2. For the AND-OPT SPARQL fragment, assume we have two AND-OPT queries q_1 and q_2 and their corresponding transformations q_1' and q_2' according to a ShEx schema S. We show that for a transformation q' of any query q as proposed in Sect. 4, the containment $q' \sqsubseteq q$ always holds. This follows from the fact that our transformation includes only elimination of optional parts of the query and transformation of other optional conditions into necessary conditions (transformation of OPT operators into AND operators). Both of the transformations make the query more restrictive in the meaning that it eliminates some solutions but never adds solutions to the original query. Assume $q_1' \not\sqsubseteq q_2'$, our purpose is to show that $q_1 \not\sqsubseteq_S q_2$. Since $q_1' \not\sqsubseteq q_2'$, then for some subtree T_1' of q_1', there doesn't exist a subtree of q_2' with the homomorphism condition of Lemma 1. On the other hand, there exists a data set D which is a complement of $\mathcal{RDF}(T_1')$ w.r.t. S. A solution for q_1' necessarily exists in D. If this solution is also a solution for q_2', and thus for q_2, then a homomorphism must hold from T_2' of q_2' to the D, and thus there exists a homomorphism from T_2' to T_1', that necessarily doesn't hold due to the fact that $q_1' \not\sqsubseteq q_2'$, and therefore a contradiction is derived.

3. For the AND-OPT-(UNION) SPARQL fragment, the same proof holds as for the AND-OPT fragment, except that instead of proposing complement data set that has a solution for q_1, leading to a contradiction when assuming it to have a solution for q_2, we alternatively propose multiple data sets, each corresponding to a top level UNION part of the query, and deriving a contradiction for each of the proposed data sets.   □

6   Complexity

In this section we study the complexity of SPARQL query containment with ShEx with respect to different SPARQL fragments.

We show that the complexity varies from NP-complete to Π_2^P-complete for the SPARQL fragments BGP, AND-OPT, and AND-OPT-(UNION). We also extend these fragments to include filter, property path patterns, and the MINUS operator whose containment problem is in the NEXP time complexity class, yet this is not shown to be an upper bound.

6.1   SPARQL AND-(OPT)-(UNION) Fragments

Theorem 2. *Containment with ShEx for the SPARQL BGP fragment is NP-complete.*

Proof. The complexity of containment of the SPARQL BGP fragment is NP-complete [3]. In the presence of ShEx constraints, a sufficient procedure to check containment is to first validate the BGP of each of the considered queries against the ShEx document. RDF validation against ShEx is NP-complete. An invalid query will return no results, and thus is contained in any other query. Otherwise, the normal containment procedure (without ShEx) is applied. Then the BGP fragment containment with ShEx is also in NP.

To show the NP-hardness of the problem, we argue that containment with ShEx is at least as hard as containment without ShEx which is shown to be NP-complete for the considered fragment. A reduction from the containment problem to the containment with ShEx problem can be easily shown by assuming an empty schema.   □

Theorem 3. *Containment with ShEx for the well-designed OPT SPARQL fragment is NP-complete.*

Proof. In [18], the authors studied the problem of containment of well-designed OPT SPARQL queries. The authors provide a procedure for solving the problem, and show the complexity of the problem to be NP-complete for this fragment.

The procedure we follow for deciding query containment of this SPARQL fragment with ShEx is based on both the query transformation described previously in this work, and the query containment procedures of [18]. Given two SPARQL queries in the well designed OPT fragment, their containment with ShEx can be decided by the two following steps:

1. Transform both queries. The results of these transformations are two new queries equivalent to the original queries respectively.
2. The two new resulting queries from the first step are used as an input of a general SPARQL containment solver (like the solver described in [18] for this fragment).

Validation of an RDF document against a ShEx document is NP-complete [23]. Step (1) of the procedure is a series of ShEx validations each of which is in NP. The number of validation considered is polynomial since for a given query pattern tree, the validation occurs on all possible branches, rather than subtrees. While the number of subtrees is exponential in a pattern tree, the number of branches is polynomial. As each branch is validated independently, so the result of each branch validation doesn't affect the ones of other branches. Step (2), which is the query containment problem, is NP-complete for the well-designed OPT fragment. Thus the complexity of containment with ShEx is in NP for this fragment.

To show the NP-hardness of the problem, we argue that containment with ShEx is at least as hard as containment without ShEx which is shown to be NP-complete for the considered fragment. A reduction from the containment problem to the containment with ShEx problem can be easily shown by assuming an empty schema.   □

Theorem 4. *Containment with ShEx for the well-designed OPT SPARQL fragment extended with top level UNION is Π_2^P-complete.*

Proof. In [18], the authors also studied the problem of containment of the AND-OPT-(UNION) fragment. The authors provide a procedure for solving the problem, and show the complexity of the problem to be Π_2^P-complete.

The procedure we follow to for deciding query containment of this fragment with ShEx is similar to the one followed for the AND-OPT fragment, except that in step (2) we use the solver designed for the corresponding fragment. The usage of such solver will rise the complexity to Π_2^P.

To show the Π_2^P-hardness of the problem, we argue that containment with ShEx is at least as hard as containment without ShEx which is shown to be Π_2^P-complete for the considered fragment. A reduction from the containment problem to the containment with ShEx problem can be easily shown by assuming an empty schema.   □

6.2   SPARQL AND-OPT-(UNION)-FILTER/PP/MINUS Fragment

For extending the query containment problem with ShEx to the SPARQL fragments including filters, property path patterns, and the MINUS operator, we use an imitation of our procedures with first-order logic (FOL). The basic idea is based on generating an FOL formula corresponding to our procedure and checking its validity with existing FOL theorem provers.

We use a decidable fragment of FOL with only 2 variables, known as FOL^2, whose satisfiability (and thus validity) is NEXP-complete [13]. An advantage of this method is that it allows to benefit from the highly optimised implementations of theorem provers.

A drawback of the problem solving with FOL is that the Kleene closure can be expressed only on atomic ShEx rules, but not on compound rules, which restricts the ShEx fragment allowed. We notice that we can avoid the Kleene closure limitation by adopting the normal procedure and using FOL particularly for the fragment extensions, and giving FOL validity feedback for the original procedure which supports Kleene closure everywhere.

Table 1 summarises our complexity results for the containment problem studied for the different fragments.

Table 1. Containment complexity

SPARQL fragment	No ShEx	ShEx-All
BGP	[Chandra, 1977] NP-c	NP-c
AND-OPT	[Pichler, 2014] NP-c	NP-c
AND-OPT-(UNION)	[Pichler, 2014] Π_2^P-c	Π_2^P-c
AND-OPT-(UNION)-Minus	[FOL^2] NEXP	[FOL^2] NEXP
AND-OPT-(UNION)-FILTER	[FOL^2] NEXP	[FOL^2] NEXP
AND-OPT-(UNION)-PP	[FOL^2] NEXP	[FOL^2] NEXP
AND-OPT-(UNION)-FILTER-PP-MINUS	[FOL^2] NEXP	[FOL^2] NEXP

We currently have two working implementation prototypes, one directly based on the procedures described in this paper utilising existing ShEx validators and containment solvers, and another implementation based on the FOL imitation of the problem, with the drawback and advantages of each as mentioned previously.

7   Conclusion

In this paper we studied the problem of SPARQL query containment with ShEx constraints, and the OPT patterns were shown to be particularly interesting to this study, due to its flexibility with constraints and the absence of similar studies in the literature. We showed how transformation of queries can be done based on customised validation procedures. Then we proposed a procedure for the problem of containment with ShEx, and the complexity related to the AND-OPT SPARQL fragment was shown to be NP-complete, and that of the AND-OPT SPARQL fragment extended with external UNION to be Π_2^P-complete. We finally mentioned that other fragment extensions can be adopted with FOL.

As a perspective for future work, we manage to adopt modified versions of the same techniques provided in this work for query constructs other than the SELECT. Other SPARQL constructs that can be adopted with further discussions include INSERT, DELETE, and CONSTRUCT.

References

1. Arenas, M., Pérez, J.: Querying semantic web data with SPARQL. In: Proceedings of the Thirtieth ACM SIGMOD-SIGACT-SIGART Symposium on Principles of Database Systems, PODS 2011, pp. 305–316. ACM, New York (2011)
2. Boneva, I., Gayo, J.E.L., Hym, S., Prud'hommeaux, E.G., Solbrig, H.R., Staworko, S.: Validating RDF with shape expressions. CoRR, abs/1404.1270 (2014)
3. Chandra, A.K., Merlin, P.M.: Optimal implementation of conjunctive queries in relational data bases. In: Proceedings of the Ninth Annual ACM Symposium on Theory of Computing, STOC 1977, pp. 77–90. ACM, New York (1977)
4. Chekol, M.W.: On the containment of SPARQL queries under entailment regimes. In: Proceedings of the Thirtieth AAAI Conference on Artificial Intelligence, AAAI 2016, pp. 936–942. AAAI Press (2016)
5. Chekol, M.W., Euzenat, J., Genevès, P., Layaïda, N.: PSPARQL query containment. In: DBPL (2011)
6. Chekol, M.W., Euzenat, J., Genevès, P., Layaïda, N.: SPARQL query containment under RDFS entailment regime. In: Gramlich, B., Miller, D., Sattler, U. (eds.) IJCAR 2012. LNCS, vol. 7364, pp. 134–148. Springer, Heidelberg (2012). doi:10.1007/978-3-642-31365-3_13
7. Chekol, M.W., Euzenat, J., Genevès, P., Layaïda, N.: SPARQL query containment under SHI axioms. In: Proceedings of the Twenty-Sixth AAAI Conference on Artificial Intelligence, AAAI 2012, pp. 10–16. AAAI Press (2012)
8. Colazzo, D., Sartiani, C.: Typing regular path query languages for data graphs. In: Proceedings of the 15th Symposium on Database Programming Languages, DBPL 2015, pp. 69–78. ACM, New York (2015)
9. Gayo, J.E.L., Prud'hommeaux, E., Boneva, I., Staworko, S., Solbrig, H.R., Hym, S.: Towards an RDF validation language based on regular expression derivatives, pp. 197–204 (2015)
10. Genevès, P., Layaïda, N.: A system for the static analysis of XPath. ACM Trans. Inf. Syst. **24**(4), 475–502 (2006)

11. Genevès, P., Layaïda, N.: Deciding XPath containment with MSO. Data Knowl. Eng. **63**(1), 108–136 (2007). Data Warehouse and Knowledge Discovery (DAWAK ?05) 7th International Congress on Data Warehouse and Knowledge Discovery (DAWAK?05)
12. Genevès, P., Layaïda, N.: XML reasoning made practical. In: 2010 IEEE 26th International Conference on Data Engineering (ICDE 2010), pp. 1169–1172, March 2010
13. Grädel, E., Kolaitis, P.G., Vardi, M.Y.: On the decision problem for two-variable first-order logic. Bull. Symbolic Logic **3**(1), 53–69 (1997)
14. Kostylev, E.V., Reutter, J.L., Romero, M., Vrgoč, D.: SPARQL with property paths. In: Arenas, M., Corcho, O., Simperl, E., Strohmaier, M., d'Aquin, M., Srinivas, K., Groth, P., Dumontier, M., Heflin, J., Thirunarayan, K., Staab, S. (eds.) ISWC 2015. LNCS, vol. 9366, pp. 3–18. Springer, Cham (2015). doi:10.1007/978-3-319-25007-6_1
15. Lanthaler, M., Cyganiak, R., Wood, D.: RDF 1.1 concepts and abstract syntax. W3C recommendation, W3C, February 2014. http://www.w3.org/TR/2014/REC-rdf11-concepts-20140225/
16. Letelier, A., Pérez, J., Pichler, R., Skritek, S.: Static analysis and optimization of semantic web queries, pp. 89–100 (2012)
17. Pérez, J., Arenas, M., Gutierrez, C.: Semantics and complexity of SPARQL. ACM Trans. Database Syst. **34**(3), 16:1–16:45 (2009)
18. Pichler, R., Skritek, S.: Containment and equivalence of well-designed SPARQL. In: Proceedings of the 33rd ACM SIGMOD-SIGACT-SIGART Symposium on Principles of Database Systems, PODS 2014, pp. 39–50. ACM, New York (2014)
19. Prud'hommeaux, E.: Shape expressions (ShEx) primer. Technical report, W3C and MIT, February 2017. http://shexspec.github.io/primer/
20. Prud'hommeaux, E., Gayo, J.E.L., Solbrig, H.: Shape expressions: an RDF validation and transformation language. In: Proceedings of the 10th International Conference on Semantic Systems, SEM 2014, pp. 32–40. ACM, New York (2014)
21. Schreiber, G., Raimond, Y.: RDF 1.1 primer. W3C note, W3C, June 2014. http://www.w3.org/TR/2014/NOTE-rdf11-primer-20140624/
22. Seaborne, A., Harris, S.: SPARQL 1.1 query language. W3C recommendation, W3C, March 2013. http://www.w3.org/TR/2013/REC-sparql11-query-20130321/
23. Staworko, S., Boneva, I., Gayo, J.E.L., Hym, S., Prud'hommeaux, E.G., Solbrig, H.: Complexity and expressiveness of ShEx for RDF. In: Arenas, M., Ugarte, M. (eds.) 18th International Conference on Database Theory (ICDT 2015). Leibniz International Proceedings in Informatics (LIPIcs), vol. 31, pp. 195–211. Dagstuhl, Germany (2015). Schloss Dagstuhl-Leibniz-Zentrum fuer Informatik
24. Zhang, X., Van Den Bussche, J., Picalausa, F.: On the satisfiability problem for sparql patterns. J. Artif. Int. Res. **56**(1), 403–428 (2016)

Updating RDF/S Databases Under Constraints

Mirian Halfeld-Ferrari[1]([✉]) and Dominique Laurent[2]([✉]) ·

[1] Université d'Orléans, INSA CVL - LIFO EA, Orléans, France
[2] ETIS - Université Paris Seine - CNRS, Cergy-Pontoise, France
`mirian@univ-orleans.fr`, `dominique.laurent@u-cergy.fr`

Abstract. We address the issue of database updating in the presence of constraints, using the first order logic formalism. Based on a chasing technique, we propose a deterministic update strategy. While generalizing key-foreign key constraints, our approach satisfies the consistency and minimal change requirements, with a polynomial time complexity. Our custom version of the chase allows improvement in updating non-null RDF/S databases with constraints.

1 Introduction

The Resource Description Framework (RDF) is a flexible graph-based data model now largely adopted for data publishing and sharing [21]. The increasing number of distributed RDF datasets, together with their dynamic nature, bring the need of providing reliable information to users worrying about data quality and validity. The maintenance of data consistency is an essential problem in databases and constraints *still* are an imperative quality label. As the commonly used key-foreign key constraints are not enough in most of the RDF applications, we are impelled to consider tuple-generating (TGD) and denial dependencies as constraints. However, the use of these richer constraints is restricted by our goal of dealing with *deterministic updates*.

Focusing on situations where the evaluation of update impacts is required *before* changes on data instances, our main goals include:

(1) A *deterministic* update strategy ensuring database *consistency* of a *non-null* instance and satisfying the *minimal change* requirement.
(2) An update side effect computation *without* accessing the database instance.
(3) Dealing with a set of insertions and deletions as an *update transaction*.

Our approach is illustrated by a running example (borrowed from [8]) where an RDF/S database is defined by a schema, a set of constraints $\mathcal{C}$ and an instance D satisfying the schema and the constraints. We use a first-order logic formalism, recalling that an RDF triple $\langle aPb \rangle$ is seen as the fact $P(a, b)$.

Example 1. Figure 1 shows our instance D together with the set $\mathcal{C}$ of constraints. Predicates Paper and Journal deal, respectively, with paper and journal titles.

© Springer International Publishing AG 2017
M. Kirikova et al. (Eds.): ADBIS 2017, LNCS 10509, pp. 357–371, 2017.
DOI: 10.1007/978-3-319-66917-5_24

Publn informs which journal publishes a paper and Cites indicates paper's citations. Constraints impose papers to be published in journals (c_1 and c_2) and to cite at least one other paper (c_3, c_4, c_7 and c_8). Each paper must be published (c_5), but a paper cannot be published in different journals (c_6).

Noting that the instance D of Fig. 1 satisfies the constraints in $\mathcal{C}$, we consider the following three *simple* updates:

(1) Deletion of Publn(*coolP*, *Cool_J*). The removal of only this fact from D implies the violation of c_5. In [8], consistency is ensured by inserting a fact such as Publn(*coolP*, J), where the journal name J is arbitrarily chosen. In our approach, we avoid arbitrary or non deterministic choices. Thus, the cascading deletes of Paper(*coolP*), Cites(*coolP*, *rdfP*), Cites(*dbP*, *coolP*), Paper(*dbP*), Cites(*rdfP*, *dbP*), Paper(*rdfP*), Publn(*dbP*, *KAIS*) are required in order to satisfy, respectively, c_5, c_3, c_4, c_7, c_4, c_7 and c_1.

(2) Insertion of Publn(*coolP*, *VLDBJ*). Because of c_6, the deletion of all facts of the form Publn(*coolP*, J) where $J \neq VLDBJ$ is necessary to preserve consistency. This insertion thus requires the deletion of Publn(*coolP*, *Cool_J*), generating the same updates as above.

(3) Insertion of Paper(*newP*). Due to c_5 and c_7 we should also insert Publn(*newP*, J) and Cites(*newP*, P) where J (respectively P) stands for an arbitrary journal (respectively an arbitrary paper, other than *newP* due to c_8). As this is clearly a case of non determinism, the insertion is rejected.

Consider now the following *global* update (*i.e.*, several updates in a unique transaction): delete Publn(*coolP*, *Cool_J*) and insert Publn(*coolP*, *VLDBJ*). In this case, the deleted fact is simply replaced by the inserted one, as expected. □

FACTS (D): Paper(*dbP*), Paper(*rdfP*), Paper(*coolP*), Journal(*KAIS*), Journal(*VLDBJ*), Journal(*Cool_J*) Publn(*dbP*, *KAIS*), Publn(*rdfP*, *KAIS*), Cites(*dbP*, *coolP*), Cites(*rdfP*, *dbP*), Publn(*coolP*, *Cool_J*), Cites(*coolP*, *rdfP*)

CONSTRAINTS ($\mathcal{C}$):
$c_1 : (\forall x, y)(\text{Publn}(x, y) \Rightarrow \text{Paper}(x));$
$c_2 : (\forall x, y)(\text{Publn}(x, y) \Rightarrow \text{Journal}(y));$ $c_3 : (\forall x, y)(\text{Cites}(x, y) \Rightarrow \text{Paper}(x))$
$c_4 : (\forall x, y)(\text{Cites}(x, y) \Rightarrow \text{Paper}(y));$ $c_5 : (\forall x)(\text{Paper}(x) \Rightarrow (\exists y)(\text{Publn}(x, y)))$
$c_6 : (\forall x, y, z)((y \neq z) \wedge \text{Publn}(x, y) \wedge \text{Publn}(x, z) \Rightarrow \bot)$
$c_7 : (\forall x)(\text{Paper}(x) \Rightarrow (\exists y)(\text{Cites}(x, y)))$ $c_8 : (\forall x, y)((x = y) \wedge \text{Cites}(x, y) \Rightarrow \bot)$

Fig. 1. RDF/S database: facts composing the instance D and constraints in $\mathcal{C}$

From the above example, we notice that we deal with constraints as in a traditional database viewpoint, and not as so-called ontological constraints [9], seen as inference rules. Then, as c_2 states that all papers must be published in a journal, a database containing the only fact Paper(*rdfP*) is *not* consistent, whereas it is consistent when constraints are seen as inference rules.

Paper organization: After giving background definitions in Sect. 2, Sect. 3 introduces our update method. Section 4 focusses on implementation issues and applications to which our approach is well adapted. Related work in Sect. 5 and concluding remarks in Sect. 6 close the paper. The proofs of the propositions are omitted due to lack of space; they can be found in [12].

2 Background

Alphabet. Let $\mathbf{A}$ be an alphabet consisting of the following pairwise disjoint sets: $\mathbf{A}_C$, a countably infinite set of constant; $\mathbf{A}_N$, a countably infinite set of labelled nulls; VAR an infinite set of variables ranging over $\mathbf{A}_C \cup \mathbf{A}_N$ (we use $\mathbf{X}$ as an abbreviation to denote the set $\{X_1, \ldots, X_k\}$ where $k > 0$); PRED, a *finite* set of predicates, each associated with its arity.

A term is a constant, a null or a variable. An *atomic formula* (or *atom*) has one of the forms: (*i*) $P(t_1, ..., t_n)$, where P is an n-ary predicate and $t_1, ..., t_n$ are terms; (*ii*) $\top$ (meaning true) or $\bot$ (meaning false); (*iii*) $(t_1 \; op \; t_2)$, where t_1 and t_2 are terms and op is a comparison operator $(=, <, >, \leq, \geq)$. A *literal* is an atom of the form $P(t_1, ..., t_n)$. An *instantiated literal* is an atom of the form $P(u)$ where $u \in (\mathbf{A}_C \cup \mathbf{A}_N)^n$, and a *fact* is an atom of form $P(u)$ where $u \in (\mathbf{A}_C)^n$.

Substitution. A *substitution* from the set of symbols E_1 to the set of symbols E_2 is a function $h : E_1 \rightarrow E_2$. A *homomorphism* from the set of atoms A_1 to the set of atoms A_2, both over the same predicate P, is a substitution h from the terms of A_1 to the terms of A_2 such that: (*i*) if $t \in \mathbf{A}_C$, then $h(t) = t$, and (*ii*) if $P(t_1, ..., t_n) \in A_1$, then $P(h(t_1), ..., h(t_n)) \in A_2$. A substitution that associates the variable x with the constant a and the variable y with the null N, is denoted as the set $\{x/a, y/N\}$. Moreover, if h is a homomorphism, $P(h(t_1), ..., h(t_n))$ is simply denoted by $h(P(t_1, ..., t_n))$. The notion of homomorphism naturally extends to conjunctions of atoms.

The following terminology is also used: (*i*) An *endomorphism* h on a finite set of atoms A_1 is a homomorphism such that $h(A_1) \subseteq A_1$; (*ii*) A *valuation* is a homomorphism whose image of each symbol is a constant in $\mathbf{A}_C$.

Database Instance and Constraints. A database instance over alphabet $\mathbf{A}$ is a pair $\Delta = (D, \mathcal{C})$ where D is a set of *facts* and $\mathcal{C}$ is a set of constraints such that D satisfies $\mathcal{C}$. Constraints and constraint satisfaction are defined below.

- A *positive constraint* is of the form: $(\forall \mathbf{X}, \mathbf{Y})(L_1(\mathbf{X}, \mathbf{Y}) \Rightarrow (\exists \mathbf{Z})(L_2(\mathbf{X}, \mathbf{Z})))$ where $L_1(\mathbf{X}, \mathbf{Y})$ and $L_2(\mathbf{X}, \mathbf{Z})$ are atoms.
- A *negative constraint* is either of the form $(\forall \mathbf{X})((comp(\mathbf{X}') \wedge L(\mathbf{X})) \Rightarrow \bot)$ or $(\forall \mathbf{X})((comp(\mathbf{X}') \wedge L_1(\mathbf{X}_1) \wedge L_2(\mathbf{X}_2)) \Rightarrow \bot)$, where $\mathbf{X}_1 \cap \mathbf{X}_2 \neq \emptyset$ and $comp(\mathbf{X}')$ is a (possibly empty) comparison formula with variables $\mathbf{X}'$ that all occur in $\mathbf{X}$, and where $L(\mathbf{X})$, $L_1(\mathbf{X}_1)$ and $L_2(\mathbf{X}_2)$ are atoms.

The left and right hand-sides of a constraint c are respectively denoted by $body(c)$ and $head(c)$. When no confusion is possible, quantifiers are omitted.

A set I of instantiated atoms satisfies a constraint c, denoted by $I \models c$, if for every homomorphism h from the variables in $body(c)$ into constants or nulls in I, the following holds:

- If c is positive: if $h(body(c))$ is in I then there is an extension h' of h such that $h'(head(c))$ is in I.
- If c is negative: if $h(comp(\mathbf{X}'))$ is true in I then depending on c, either $h(L(\mathbf{X}))$ is not in I or one of the two atoms $h(L_1(\mathbf{X}_1))$ or $h(L_2(\mathbf{X}_2))$ is not in I.

Given a set of constraints $\mathcal{C}$, I satisfies $\mathcal{C}$, denoted by $I \models \mathcal{C}$, if for every c in $\mathcal{C}$, $I \models c$ holds.

Positive constraints are *linear* LAV (local-as-a-view) TGD (Tuple Generating Dependency) – *i.e.*, each body and head has a unique atom [2]. Negative constraints are denial dependencies with one or two atoms in their bodies. Restrictions on TGD and denials are exploited to ensure *update determinism*, but, our constraints *still* generalize key-foreign key constraints.

Computing side effects of updates when only positive constraints are considered relies on the notion of *trigger*. Let $c : L_1(\mathbf{X}, \mathbf{Y}) \Rightarrow L_2(\mathbf{X}, \mathbf{Z})$ be a positive constraint and I a set of instantiated atoms. A triple (c, h_1, h_2) is *a trigger in I* if there exists $L_1(\alpha, \beta)$ in I such that $h_1(L_1(\mathbf{X}, \mathbf{Y})) = h_2(L_1(\alpha, \beta))$ where h_1 is a homomorphism from VAR to $\mathbf{A}_C \cup \mathbf{A}_N$ and h_2 is an endomorphism on $\mathbf{A}_C \cup \mathbf{A}_N$. When there is a trigger in I for a positive constraint c, we say that c is *activated* to produce a new instantiated atom. In this case, when existential variables $\mathbf{Z}$ are present in $head(c)$, if (c, h_1, h_2) is a trigger in I then we define an extension h'_1 of h_1 ($h'_1 \supseteq h_1$) such that, for every Z_i in $\mathbf{Z}$, $h'_1(Z_i) = N_i$, where N_i is a fresh labelled null in $\mathbf{A}_N$ not introduced before.

Example 2. If $I = \{A(a, N_1)\}$, and $h_1^1 = \{x/a\}$ and $h_2^1 = \{N_1/a\}$, then the trigger (c_1, h_1^1, h_2^1) activates $c_1 : A(x, x) \Rightarrow B(x)$ to produce $B(a)$. Clearly, $h_1^1(A(x, x)) = h_2^1(A(a, N_1)) = A(a, a)$.

Similarly, if $I = \{E(b)\}$; $h_1^2 = \{x/b\}$ and $h_2^2 = \emptyset$ then the trigger (c_2, h_1^2, h_2^2) activates $c_2 : E(x) \Rightarrow D(x, y)$ to produce $D(b, N_3)$ by means of the extension $(h_1^2)'$ of h_1^2 such that $(h_1^2)' = \{x/b, y/N_3\}$, with $N_3 \in \mathbf{A}_N$.   □

3 Updates

Updating $\Delta = (D, \mathcal{C})$ with respect to *IReq* and *DReq* (insertion and deletion requests, respectively) means either finding that the requested updates are not possible or building $\Delta' = (D', \mathcal{C})$ according to the following policy (P). We remark that a negative constraint cannot be violated by a deletion.

- (P1): If $c : L_1(\mathbf{X}, \mathbf{Y}) \Rightarrow L_2(\mathbf{X}, \mathbf{Z})$ is not satisfied when inserting $\varphi = L_1(\alpha, \beta)$ then, if $\mathbf{Z}$ is empty, insert the corresponding fact $L_2(\alpha)$, otherwise reject the update if no instance of $L_2(\alpha, \mathbf{Z})$ can be found in D or in the insertion side effects.
- (P2): If $c : L_1(\mathbf{X}, \mathbf{Y}) \Rightarrow L_2(\mathbf{X}, \mathbf{Z})$ is not satisfied when deleting $\varphi = L_2(\alpha, \gamma)$ then delete the corresponding instances of $L_1(\alpha, \mathbf{Y})$.

- $(P3)$: If $c : comp(\mathbf{X'}) \wedge L(\mathbf{X}) \Rightarrow \bot$ is not satisfied when inserting $\varphi = L(\alpha, \beta)$ then reject the update.
- $(P4)$: If $c : comp(\mathbf{X'}) \wedge L_1(\mathbf{X}_1) \wedge L_2(\mathbf{X}_2) \Rightarrow \bot$ is not satisfied when inserting $\varphi = L_1(\alpha, \beta)$ then delete the corresponding instances of $L_2(\mathbf{X}_2)$.

We provide algorithms implementing policy (P) in a such a way that updates are deterministic and satisfy the minimal change requirement, in the sense that cancelling one side effect of the given updates violates constraints.

Before considering the case of insertions and deletions under positive or negative constraints, we present separately insertions and deletions under *positive* constraints as a two step processing: (1) build a tableau using the constrains and the updates, and (2) perform the update using the tableau and the database.

3.1   Insertions Under Positive Constraints

According to policy $(P1)$, inserting new facts may require to insert other facts as side effects. When a side effect is an atom with a null, its insertion is accepted only if this atom unifies with a fact being inserted or already in the database. As we are updating a *consistent* non-null database, the existence of this unification ensures consistency maintenance (because the side effects are already in the database or being inserted). In this way, we adapt the chase to our problem.

Algorithm 1. $BuildInsTab(I, \mathcal{C}, \text{T-INS})$

Input: A set of positive constraints $\mathcal{C}$ and a set of facts I.
Output: The tableau T-INS $= \text{C-INS}(I) \cup \text{N-INS}(I)$.
1:   C-INS$(I) := I$; N-INS$(I) := \emptyset$
2:   *continue* := **true**
3:   **while** *continue* **do**
4:        *continue* := **false**
5:        **for all** c having a trigger (c, h_1, h_2) in C-INS(I) **do**
6:             For $c : L_1(\mathbf{X}, \mathbf{Y}) \Rightarrow L_2(\mathbf{X}, \mathbf{Z})$, let h'_1 be an extension of h_1 such that for each $Z_i \in \mathbf{Z}$, $h'_1(Z_i) = N_i$ where N_i is a fresh labelled null in $\mathbf{A}_N$
7:             **if** $h'_1(L_2(\mathbf{X}, \mathbf{Z}))$ contains no null **then**
8:                  **if** $h'_1(L_2(\mathbf{X}, \mathbf{Z})) \notin$ C-INS(I) **then**
9:                       C-INS$(I) :=$ C-INS$(I) \cup \{h'_1(L_2(\mathbf{X}, \mathbf{Z}))\}$
10:                      *continue* := **true**
11:             **else**
12:                  N-INS$(I) :=$ N-INS$(I) \cup \{h'_1(L_2(\mathbf{X}, \mathbf{Z}))\}$
13: **return** T-INS $=$ C-INS$(I) \cup$ N-INS(I)

Side effects are computed by Algorithm 1 as the first step of our approach to insertions. A tableau T-INS is built as the union of two sub-tableaux: C-INS(I) whose rows contain constants only (and no labelled null), and N-INS(I) in which each row contains at least one labelled null. The tableaux C-INS(I) and N-INS(I)

are initialized on line 1 and then, facts in C-INS(I) are used to trigger constraints in $\mathcal{C}$ (line 5). While such triggers exist and facts are produced (line 7), the loop line 3 goes on. This loop stops when only atoms containing nulls are generated (line 12). The two tableaux are returned (line 13) as the tableau T-INS.

The tableau obtained by Algorithm 1 is used by Algorithm 2, where the insertions are performed only when the nulls in N-INS($IReq$) can be instantiated into facts either in the database instance or in the set C-INS($IReq$). It should be clear that insertions are performed deterministically according to policy ($P1$). Moreover, the following proposition holds.

Proposition 1. *For every finite set of positive constraints $\mathcal{C}$ and every finite set of facts IReq, Algorithm 1 applied to $\mathcal{C}$ and IReq always terminates.*

Algorithm 2 computes $\Delta' = (D',\mathcal{C})$ from $\Delta = (D,\mathcal{C})$ (where $D \models \mathcal{C}$) and a set of facts IReq. Either $\Delta' = \Delta$ or the following properties hold: (1) D' contains IReq and $D' \models \mathcal{C}$ and (2) for every φ in C-INS(IReq) \ IReq, $D' \setminus \{\varphi\} \not\models \mathcal{C}$. □

The first part of Proposition 1 implies that our approach applies even if the graph of constraints is non weakly acyclic (see [6,18]). We emphasize that Algorithm 1 does *not* require to access the database instance and that statements (1) and (2) in Proposition 1 respectively show that, when insertions are performed, consistency is maintained and minimal change is satisfied.

Algorithm 2. $PerfIns(\Delta, IReq)$

Input: $\Delta = (D,\mathcal{C})$ where $\mathcal{C}$ is a set of positive constraints and $IReq$ a set of facts.
Output: $\Delta' = (D',\mathcal{C})$ resulting from the insertion of $IReq$ in Δ.
1: $BuildInsTab(IReq,\mathcal{C},\text{T-INS}))$
2: **if** there is a valuation v of the nulls in N-INS($IReq$) such that
    $v(\text{N-INS}(IReq)) \subseteq (D \cup \text{C-INS}(IReq))$ **then**
3:     $D' := D \cup \text{C-INS}(IReq)$
4: **else**
5:     $D' := D$ // The insertion is *not* possible
6: **return** $\Delta' = (D',\mathcal{C})$

Example 3. Let Δ_P be the database instance $(D,\mathcal{C}_P)$ where D is the set shown in Fig. 1 and where $\mathcal{C}_P$ contains the positive constraints c_1-c_5 and c_7. For $IReq = \{\text{Paper}(newP)\}$, when running Algorithm 1, triggers involving c_5 and c_7, produce the atoms Publn($newP$, N_1) and Cites($newP$, N_2). Since no other fact is generated, the algorithm stops, returning T-INS $= \{\text{Paper}(newP)\} \cup \{\text{Publn}(newP, N_1), \text{Cites}(newP, N_2)\}$. As no instantiation of atoms in N-INS($IReq$) is in $D \cup$ C-INS($IReq$), Algorithm 2 returns Δ_P. □

3.2 Deletions Under Positive Constraints

Given $\Delta = (D,\mathcal{C})$ and a set of facts $DReq$ to be deleted, in a first step (Algorithm 3) a tableau T-DEL is built, with no access to the database and in a second step (Algorithm 4) the updated database $\Delta' = (D',\mathcal{C})$ is output.

To do so, every positive constraint $c : L_1(\mathbf{X}, \mathbf{Y}) \Rightarrow L_2(\mathbf{X}, \mathbf{Z})$ is associated with its 'backward' or inverse form $\bar{c} : L_2(\mathbf{X}, \mathbf{Z}) \Rightarrow L_1(\mathbf{X}, \mathbf{Y})$, which is also a positive constraint. The intuition of our method is outlined below.

$\mathcal{C}$	D		T-DEL		
			Del	Query	Ref
$A(x, y) \Rightarrow$	$A(b, a)\ \ B(b, a)$	1	$B(b, b)$	–	–
$B(x, z)$	$A(c, a)\ \ B(b, b)$	2	$B(c, b)$	–	–
	$A(c, c)\ \ B(c, b)$	3	$A(b, N_1)$	$B(b, N_3)$	1
		4	$A(c, N_2)$	$B(c, N_4)$	2

Fig. 2. Database and tableau of Examples 4 and 5

Example 4. Consider the database instance $\Delta = (D, \mathcal{C})$ shown in the left part of Fig. 2. Clearly, $D \setminus \{B(b, b)\}$ satisfies $\mathcal{C}$ whereas $D \setminus \{B(c, b)\}$ does *not*. Therefore, when deleting $B(b, b)$ no further update is necessary, whereas when deleting $B(c, b)$, facts $A(c, a)$ and $A(c, c)$ have to be deleted in order to maintain constraint satisfaction according to policy $(P2)$. To take into account these remarks, we build a chasing tableau T-DEL as shown in the right part of Fig. 2.

(1) The facts to be deleted are first put in the first column of the first two lines.
(2) Applying the constraint backwards generates line 3 with T-DEL$[3, \mathsf{Del}] = A(b, N_1)$ for $B(b, b)$, and line 4 for $B(c, b)$ with T-DEL$[4, \mathsf{Del}] = A(c, N_2)$.
(3) The constraint, now applied forwards, produces T-DEL$[3, \mathsf{Query}] = B(b, N_3)$ and $TAB[4, \mathsf{Query}] = B(c, N_4)$.
(4) Since line 3 has been obtained from T-DEL$[1, \mathsf{Del}]$, 1 is stored in T-DEL$[3, \mathsf{Ref}]$; for similarly reasons 2 is stored in T-DEL$[4, \mathsf{Ref}]$.

Column 2 of T-DEL is a query. The cardinality of its answer allows to decide whether further deletions are needed: for $B(b, N_3)$, the answer is $\{B(b, a), B(b, b)\}$ and no other deletion is needed, whereas for $B(c, N_4)$, the answer is $\{B(c, b)\}$, implying that all instances of $A(c, N_3)$ must be deleted.   □

Notice that the deletion of an atom having a null value implies the deletion of all facts unifying with this atom. Thus, deletions of isomorphic atoms have the same effects. Algorithm 3 takes advantage of this remark and does not generate isomorphic atoms. Once again, instead of working with a general-purpose chase algorithm we offer a version adapted to our problem.

Let us now explain Algorithm 3 in details. After initializing T-DEL (line 3), the loop on line 5 processes the rows just added. For $startline \leq i \leq (endline - 1)$, for each trigger with respect to T-DEL$[i, \mathsf{Del}]$ involving an inverse constraint $\bar{c} : L_2(\mathbf{X}, \mathbf{Z}) \Rightarrow L_1(\mathbf{X}, \mathbf{Y})$, lines 7–17 are performed:

- On line 8, for each trigger $(\bar{c}, h_1, h_2)$ such that $h_1(L_2(\mathbf{X}, \mathbf{Z})) = h_2(\text{T-DEL}[i, \mathsf{Del}])$ an extension of h_1 is defined so as to generate the new atom $h_1'(L_1(\mathbf{X}, \mathbf{Y}))$.

Algorithm 3. $BuildDelTab(\mathcal{C}, I)$

Input: A set of positive constraints $\mathcal{C}$ and a set of facts I.
Output: The tableau T-DEL.
1: $i := 0$;
2: **for all** $\varphi \in I$ **do**
3:     $i := i + 1$;   T-DEL$[i, \mathsf{Del}] := \varphi$
4: $N := i + 1$; $endline := N$; $startline := 1$;
5: **while** $startline < endline$ **do**
6:    **for** $i := startline$ **to** $(endline - 1)$ **do**
7:     **for all** trigger $(\overline{c}, h_1, h_2)$ with respect to T-DEL$[i, \mathsf{Del}]$ **do**
8:      Let h_1' be an extension of h_1 to the variables of $\mathbf{Y}$ built from $(\overline{c}, h_1, h_2)$
9:      Denoting by $h_1'[\mathbf{XY}]$ the restriction of h_1' to $\mathbf{X} \cup \mathbf{Y}$, let $h_1''[\mathbf{XY}]$ be an
        extension of $h_1'[\mathbf{XY}]$, obtained by applying c to $h_1'[\mathbf{XY}](L_1(\mathbf{X}, \mathbf{Y}))$
10:     $IsoRow := Iso(h_1'[\mathbf{XY}](L_1(\mathbf{X}, \mathbf{Y})), h_1''[\mathbf{XY}](L_2(\mathbf{X}, \mathbf{Z})))$
11:     **if** $IsoRow = \emptyset$ **then**
12:      T-DEL$[N, \mathsf{Del}] := h_1'[\mathbf{XY}](L_1(\mathbf{X}, \mathbf{Y}))$
13:      T-DEL$[N, \mathsf{Query}] := h_1''[\mathbf{XY}](L_2(\mathbf{X}, \mathbf{Z}))$
14:      T-DEL$[N, \mathsf{Ref}] := i$
15:      $N := N + 1$
16:     **else**
17:      Add i in T-DEL$[\rho, \mathsf{Ref}]$, where ρ is such that $IsoRow = \{\rho\}$
18:    $startline := endline$; $endline := N$
19: **return** T-DEL

- Denoting by $h_1'[\mathbf{XY}]$ the restriction of h_1' to the variables in $\mathbf{X}$ or in $\mathbf{Y}$, we have $h_1'[\mathbf{XY}](L_1(\mathbf{X}, \mathbf{Y})) = h_1'(L_1(\mathbf{X}, \mathbf{Y}))$. The trigger $(c, h_1'[\mathbf{XY}], \emptyset)$ (line 9) generates $h_1''[\mathbf{XY}](L_2(\mathbf{X}, \mathbf{Z}))$ where $h_1''[\mathbf{XY}]$ is an extension of $h_1'[\mathbf{XY}]$.
- The new row N is expected to be $\langle h_1'[\mathbf{XY}](L_1(\mathbf{X}, \mathbf{Y})), h_1''[\mathbf{XY}](L_2(\mathbf{X}, \mathbf{Z})), i \rangle$. Before inserting this row, we check (line 11) if the tableau contains a row whose atoms in columns Del and Query are respectively equal to $h_1'[\mathbf{XY}](L_1(\mathbf{X}, \mathbf{Y}))$ and $h_1''[\mathbf{XY}](L_2(\mathbf{X}, \mathbf{Z}))$, up to null renaming.
- Calling such rows *isomorphic rows*, this task is achieved by the function $IsoRow$ which returns either the empty set or the row number where isomorphic atoms have been found. If the empty set is returned, the triple is inserted as the row N of T-DEL (cf. lines 12–14); otherwise the row number i whose atom T-DEL$[i, \mathsf{Del}]$ has triggered $\overline{c}$ is added in the column Ref of row ρ (line 17).

As for insertions, the tableau T-DEL, which is built without accessing the database, is used to perform the deletion according to Algorithm 4. In this algorithm, as well as in Algorithm 5, access to D is performed through a function *eval* taking as input an atom φ and returning the set of all instantiations in D of φ. Algorithms 4 and 5 are explained in the following example.

Example 5. Consider the database instance and deletions shown is Fig. 2 and discussed in Example 4. The loop on line 3 of Algorithm 4 is run for each fact in

---

Algorithm 4. $PerfDel(\Delta, DReq)$

Input: $\Delta = (D, \mathcal{C})$ where $\mathcal{C}$ is a set of positive constraints; $DReq$ is a set of facts
Output: $\Delta' = (D', \mathcal{C})$ resulting from the deletion of $DReq$ from Δ
1: T-DEL $:= BuildDelTab(\mathcal{C}, DReq)$
2: $DSet := \emptyset$
3: **for all** $l \in DReq$ **do**
4:   $i := FindRow(\text{T-DEL}, l)$
5:   $AnsQ_0 := eval(\text{T-DEL}[i, \mathsf{Del}], \Delta) \setminus DSet$
6:   **if** $AnsQ_0 \neq \emptyset$ **then**
7:     **for all** $\varphi_0 \in AnsQ_0$ **do**
8:       $DelSideEffects(i, \varphi_0, DSet)$
9: $D' := D \setminus DSet$
10: **return** $\Delta' := (D', \mathcal{C})$

Algorithm 5. $DelSideEffects(\lambda, \varphi, DSet)$

Input: A line number λ of T-DEL, a fact φ and the current set $DSet$.
Output: The updated set $DSet$.
1: $DSet := DSet \cup \{\varphi\}$
2: $S := \{pos \mid \lambda \text{ occurs in T-DEL}[pos, \mathsf{Ref}]\}$
3: **if** $S \neq \emptyset$ **then**
4:   **for all** $j \in S$ **do**
5:     $AnsQ_1 := eval(\text{T-DEL}[j, \mathsf{Del}], \Delta) \setminus DSet$
6:     **if** $AnsQ_1 \neq \emptyset$ **then**
7:       **for all** $\varphi_1 \in AnsQ_1$ **do**
8:         Let h be a homomorphism such that $h(\text{T-DEL}[j, \mathsf{Del}]) = \varphi_1$
9:         **if** φ is an instance of $h(\text{T-DEL}[j, \mathsf{Query}])$ **then**
10:           $AnsQ_2 := eval(h(\text{T-DEL}[j, \mathsf{Query}]), \Delta) \setminus DSet$
11:           **if** $AnsQ_2 = \emptyset$ **then**
12:             $DelSideEffects(j, h(\text{T-DEL}[j, \mathsf{Del}]), DSet)$
13: **return** $DSet$

$DReq$ for which we successively have $AnsQ_0 = \{B(b, b)\}$ and $AnsQ_0 = \{B(c, b)\}$. Thus, Algorithm 5 is first called with $DSet = \emptyset$, $\lambda = 1$ and $\varphi = B(b, b)$.

After inserting $B(b, b)$ in $DSet$ (line 1), Algorithm 5 computes $S = \{3\}$ and the loop (line 4) is executed. Since T-DEL$[3, \mathsf{Del}] = A(b, N_1)$ (Fig. 2), we have $AnsQ_1 = \{A(b, a)\}$ (line 5). The homomorphism h (line 8) is $h = \{N_1/a\}$. In this case, $AnsQ_2 = \{B(b, a)\}$ (line 10) because $B(b, b)$ is in $DSet$. Since the current values of φ and φ_1 are respectively $B(b, b)$ and $A(b, a)$, the test on line 9 returns **true**. Notice that when the returned value is **false**, the current values of φ and φ_1 do not correspond to the same homomorphism, and the process stops. Since $AnsQ_2 \neq \emptyset$, deleting φ does not require deleting φ_1. The process continues with the loop on line 7 of Algorithm 4, for which we have $\lambda = 2$, $\varphi_0 = B(c, b)$ and $DSet = \{B(b, b)\}$ as input for Algorithm 5. Thus, we obtain $DSet = \{B(b, b), B(c, b)\}$, $S = \{4\}$ and the execution of the loop (line 4) for $j = 4$ implies that $AnsQ_1 = \{A(c, a), A(c, c)\}$.

For $\varphi_1 = A(c,a)$, $AnsQ_2 = \emptyset$ and in the call $DelSideEffects(4, A(c,a),$ $\{B(b,b), B(c,b)\})$ (line 12), $A(c,a)$ is added into $DSet$ (line 1), and S is set to $\emptyset$. Similarly, for $\varphi_1 = A(c,c)$ and $DelSideEffects(4, A(c,c), \{B(b,b), B(c,b),$ $A(c,a)\})$, $DSet = \{B(b,b), B(c,b), A(c,a), A(c,c)\}$ is returned (line 13 of Algorithm 5). Thus, $D' = \{A(b,a), B(b,a)\}$. $\qquad\square$

The following proposition shows that performing deletions according to policy ($P2$) is always possible and satisfies the minimal change requirement.

Proposition 2. *For every finite set of positive constraints C and every finite set of facts DReq, Algorithms 3 and 4 applied to C and DReq always terminate.*
   Moreover, $\Delta' = (D', C)$ is such that: (1) $D' \cap DReq = \emptyset$; (2) $D' \models C$ and (3) for every φ in $DSet \setminus DReq$, $D' \cup \{\varphi\} \not\models C$. $\qquad\square$

3.3   Global Updates

Consider now a database instance with positive constraints (C_P) *and* negative constraints (C_N), and update requests containing *insertions and deletions*. Such updates are processed based on the previous algorithms as follows:

(1) The tableau T-INS is first built up based on C_P and *IReq*, and then we check whether the insertions are possible. If not, the process stops.
(2) Otherwise the consistency of the insertions is checked with respect to negative constraints in C_N, according to policies ($P3$) and ($P4$). If an inconsistency is found, the process stops. Otherwise the deletions necessary to restore consistency (see policy ($P4$)) are computed and added into *DReq*.
(3) Deletions are processed by computing T-DEL and the set *DSet* of all side effects. We check whether these deletions are conflicting with the insertions in step (1). If no conflict is detected the database instance is modified accordingly. Otherwise no modification is performed.

Due to lack of space we do not give details of the corresponding algorithms (see [12]), but we illustrate our approach in the context of our running example.

Example 6. Let $\Delta = (D, C)$ be the database of Fig. 1 and consider *IReq* = $\{\text{Publn}(coolP, KAIS)\}$ and *DReq* = $\{\text{Publn}(coolP, Cool\_J)\}$.

(1) T-INS is defined by C-INS($IReq$) = $\{\text{Publn}(coolP, KAIS), \text{Paper}(coolP),$ $\text{Journal}(KAIS)\}$ and N-INS($IReq$) = $\{\text{Publn}(coolP, N_1), \text{Cites}(coolP, N_2)\}$. The insertion is possible with respect to C_P since $\text{Publn}(coolP, Cool\_J)$, $\text{Publn}(coolP, KAIS)$ and $\text{Cites}(coolP, rdfP)$ are in $D \cup$ C-INS($IReq$).
(2) The consistency with respect to C_N is checked. For $c_6 : (y \neq z) \land$ $\text{Publn}(x,y) \land \text{Publn}(x,z) \Rightarrow \bot)$ and $v = \{x/coolP, y/KAIS\}$, the query $Q : (KAIS \neq z) \land \text{Publn}(coolP, z)$ is processed and returns $\text{Publn}(coolP,$ $Cool\_J)$. As this fact belongs to *DReq*, this set is not modified.

(3) Algorithm 4, called with *DReq* as input, computes $DSet = \{\mathsf{PubIn}(coolP,$ $Cool\_J)\}$. As insertions and deletions are not conflicting, the algorithm outputs: $D' = (D \cup \{\mathsf{PubIn}(coolP, KAIS)\}) \setminus \{\mathsf{PubIn}(coolP, Cool\_J)\}$.   □

Clearly, global updates follow policy (P). Moreover, the following proposition shows that when the database is modified, the updates are deterministically performed and satisfy the minimal change requirement.

Proposition 3. *Given $\Delta = (D, \mathcal{C})$, let $\Delta' = (D', \mathcal{C})$ be the result of the global update processing with Δ, IReq and DReq. Either $\Delta' = \Delta$ or the following properties hold: (1) IReq $\subseteq D'$, DReq $\cap D' = \emptyset$ and $D' \models \mathcal{C}$; (2) For every φ in* C-INS*(IReq)\ IReq, $D'\setminus\{\varphi\} \not\models \mathcal{C}$; (3) For every φ in DSet\DReq, $D'\cup\{\varphi\} \not\models \mathcal{C}$.* □

4   Application Issues

Complexity results. In [8] a general-purpose method dealing with the evolution of a knowledge base with disjunctive embedded dependencies is shown to be NP-complete while its special-purpose version which considers *only* RDF/S semantic constraints leads to a polynomial complexity. We argue that our approach has also a *polynomial* time complexity, while generalizing the work of [8], in the sense that our approach allows both application and RDF/S constraints.

Indeed, denoting by α the maximal arity of predicates and by $|E|$ the cardinality of a set E, it is shown in [12] that the complexities of insertions and deletions are respectively $O(|D| + |IReq|^{2 \cdot \alpha})$ and $O(|D|^2 + |DReq|^{3 \cdot \alpha})$. Thus, the complexity of global updates is $O(|D|^2 + |IReq|^{2 \cdot \alpha} + |DReq|^{3 \cdot \alpha})$, *i.e.*, polynomial with respect to the sizes of D, *IReq* and *DReq*.

Testing our chase algorithms. As our custom versions of chasing are the kernel of our approach, we ran a preliminary JAVA implementation on an Intel(R) Core(TM) i7-6600U CPU@ 2.60 GHz; 16 GB of RAM. Considering the benchmark described in [4], Table 1 illustrates some of our results (chosen out of twenty different tests) using a LUBM [1] scenario (a popular benchmark in the semantic web domain) with 91 positive constraints. In our tests,

Table 1. Chase performance

DReq		T-DEL	Chase
Size	# facts	# atoms	time (in s)
2.1 KB	30	78	0.022
20 KB	290	713	0.075
206 KB	2808	6646	0.169
1.5 MB	19,052	34,685	0.923
9.1 MB	100,543	135,908	10.406

T-DEL is computed using Algorithm 3 and the set *DReq* (seen as a database instance), and the number of new generated atoms is output. These tests show the feasibility of our proposal, regarding the computation time for generating the tableaux. These experimental results should be seen as a first step towards the full implementation of our approach and its assessment. Obtained results are encouraging (apparently close to those of Graal in [4]). Performance comparisons with other chase versions will be conducted shortly, after implementing some optimizations and testing the systems mentioned in [4] in our hardware configuration.

RDF/S semantics as constraints. Besides application constraints, RDF data should be valid with respect to RDF/S semantic constraints. Using the formalism of [8], we classify predicates into two sets: (i) SCHPRED $= \{Cl, Pr, CSub, PSub, Dom, Rng\}$, a set of *schema predicates* standing respectively for classes, properties, sub-classes, sub-properties, property domain and range, and (ii) INSTPRED $= \{CI, PI, Ind, URI\}$, a set of *instance predicates* standing respectively for class and property instances, individuals and URIs.

The RDF/S semantic constraints listed in [8] are split into: *typing constraints* (such as $(\forall x)(Cl(x) \Rightarrow URI(x))$), *schema constraints* involving only predicates in SCHPRED, and *instance constraints* of the form $(\forall \mathbf{X})(\varphi_{Sch}(\mathbf{X}') \Rightarrow \varphi_{Inst}(\mathbf{X}))$ where variables in $\mathbf{X}'$ are in $\mathbf{X}$, $\varphi_{Sch}(\mathbf{X}')$ is an atom over a predicate in SCHPRED, and $\varphi_{Inst}(\mathbf{X})$ is a positive constraint involving instance predicates.

In this setting, an RDF/S database Δ is a pair $(\Delta_{Sch}, \Delta_{Inst})$ where $\Delta_{Sch} = (D_{Sch}, \mathcal{C}_{Sch})$ and $\Delta_{Inst} = (D_{Inst}, \mathcal{C}_{Inst})$ are such that: (i) $\mathcal{C}_{Sch}$ is the set of schema constraints and D_{Sch} is a set of facts over SCHPRED satisfying $\mathcal{C}_{Sch}$; (ii) $\mathcal{C}_{Inst}$ is the set of instance or application constraints and D_{Inst} is a set of facts over INSTPRED such that $D_{Sch} \cup D_{Inst}$ satisfies $\mathcal{C}_{Inst}$; and (iii) all typing constraints are satisfied by $D_{Sch} \cup D_{Inst}$.

Example 7. Applying the previous considerations to our running example, we obtain the database $\Delta = (\Delta_{Sch}, \Delta_{Inst})$ where

- $D_{Sch} = \{Cl(\mathsf{Journal}), Cl(\mathsf{Paper}), Pr(\mathsf{Publn}), Pr(\mathsf{Cites}), Dom(\mathsf{Publn}, \mathsf{Paper}), Rng(\mathsf{Publn}, \mathsf{Journal}), Dom(\mathsf{Cites}, \mathsf{Paper}), Rng(\mathsf{Cites}, \mathsf{Paper})\}$;
- $\mathcal{C}_{Sch}$ expresses the existence and uniqueness of domain and range of properties;
- D_{Inst} contains the facts in D (Fig. 1), expressed with instance predicates ($e.g.$, $CI(dbP, \mathsf{Paper})$ or $PI(dbP, KAIS, \mathsf{Publn})$) and the facts representing individuals and URIs ($e.g.$, $Ind(dbP)$ or $URI(\mathsf{Cites})$);
- $\mathcal{C}_{Inst}$ contains $(PI(x, y, \mathsf{Publn}) \Rightarrow CI(x, \mathsf{Paper}))$, $(PI(x, y, \mathsf{Publn}) \Rightarrow CI(y, \mathsf{Journal}))$, $(PI(x, y, \mathsf{Cites}) \Rightarrow CI(x, \mathsf{Paper}))$, $(PI(x, y, \mathsf{Cites}) \Rightarrow CI(y, \mathsf{Paper}))$, expressing $c_1 - c_4$ of Example 1. Constraints $c_5 - c_8$, which also belong to $\mathcal{C}_{Inst}$, are application constraints. For example, c_5 and c_8 are respectively written: $(CI(x, \mathsf{Paper}) \Rightarrow (PI(x, y, \mathsf{Publn})))$ and $(((x = y) \wedge PI(x, y, \mathsf{Cites})) \Rightarrow \bot)$.

We notice that, to fit the constraint form given earlier, every constraint $(\varphi_{Sch}(\mathbf{X}') \Rightarrow \varphi_{Inst}(\mathbf{X}))$ is replaced by the set of all formulas $h(\varphi_{Inst}(\mathbf{X}''))$ where $\mathbf{X}'' = \mathbf{X} \setminus \mathbf{X}'$ and h is a homomorphism of $\mathbf{X}'$ such that $D_{Sch} \models h(\varphi_{Sch}(\mathbf{X}'))$.□

Application examples. We describe two cases for which our method is well adapted. The first one concerns applications involving the cost or the quality of updates. For instance, extending the database of Example 7 with the hierarchy defined by $CSub(\mathsf{Journal}, \mathsf{Publication})$ and $CSub(\mathsf{DBJournal}, \mathsf{Journal})$, consider the insertions of $CI(anyP, \mathsf{Publication})$, $CI(anyP, \mathsf{Journal})$ and $CI(anyP, \mathsf{DBJournal})$. Due to the constraints, inserting $CI(anyP, \mathsf{DBJournal})$ generates the insertion of the other two facts, while inserting $CI(anyP, \text{Publication})$ has no side effects. Thus, assuming that low quality data are more likely to be deleted and that the three facts are of low quality, the insertion of $CI(anyP, \text{Publication})$ is preferred (as its deletion triggers no other deletions). Notice that this choice can be made because side effects are computed prior to the update. Similar methods are used for ontology revision [17].

A second case is a medical scenario inspired from [10] where an RDF database stores information on patients and their treatments. A clone (virtual machine in the cloud) handles collaborative tasks among users equipped with devices storing a partial copy of the data. When a user requires updates, the clone (where constraints are stored) computes the side effects and send them to the users, to let them update their local database. Suppose the constraints ($CI(X, \text{Injured}) \Rightarrow PI(X, Y, \text{transfusion})$) and ($PI(X, Y, \text{transfusion}) \Rightarrow PI(X, Y, \text{bloodGroup})$), along with the insertion of $CI(Bob, \text{Injured})$. Then, the clone informs users on the potential insertions of $PI(Bob, N_1, \text{transfusion})$ and $PI(Bob, N_1, \text{bloodGroup})$. In this case, if the null N_1 can be instantiated (*i.e.*, Bob's blood group is known), then the insertion is safely accepted, otherwise it is rejected. These examples show the relevance of computing update side effects independently from instances and forbidding null values, as we do in our work.

5   Related Work

Our update strategy is different from proposals such as [3,5,11,15]. Although some of them discuss non determinism, all these approaches consider constraints as inference rules (as in [9,14,16,19]), whereas we deal with constraints in a traditional database viewpoint, as in [8]. In [8] updates are deterministic thanks to a total ordering, which potentially imposes arbitrary choices. Our proposal guarantees determinism by exploiting the linearity of constraints, for which updating has a polynomial time complexity even for RDF/S and application constraints.

According to the update semantics classification in [3], our approach falls in the category Sem_2^{mat} (deletion/insertion of a fact imposes elimination/addition of all its causes/consequences). Approaches in [5,11] address schema updating, but with a more restricted set of RDF/S constraints. While a *three valued* logic framework is used in [13] (which inspired the current proposal), we consider here the standard first-order logic and allow for existential variables in the heads of the rules. Although many works [6,7,20] used TGD and the chase procedure (see [18] for a survey) for data exchange, few of them considered updates. An exception is [20], which studies complexity of the TGD checking problem, including its update variant. Our results can be related to works on repair checking (such as [2]), because the updated database can be seen as a repair of an inconsistent database obtained by performing only updates in *DReq* and *IReq*. However, our minimal change conditions are not necessarily considered in those papers.

6   Concluding Remarks

This paper focuses on consistent updates on a *non-null* database instance satisfying constraints that generalize the commonly used key-foreign key constraints. By observing that update analysis with linear constraints can be done without resorting to analysis of infinite chase, we propose a *deterministic* update strategy based on *specific* chasing algorithms whose time complexity is polynomial.

Our approach, which satisfies *minimal change* requirements, deals with *sets* of insertions and deletions in a two-step procedure where the first step is performed *independently* from the database instance. The motivation comes from RDF applications where nulls cannot be stored and where side effects should be evaluated before effectively updating the database instance. We have shown that the use of our approach in checking validity of RDF documents with respect to RDF/S-semantic together with application constraints is straightforward.

Acknowledgements. The work is partially funded by APR-IA GIRAFON. We sincerely thank Jacques Chabin for his work on implementation and experiments.

References

1. Lubm benchmark. http://swat.cse.lehigh.edu/projects/lubm/
2. Afrati, F.N., Kolaitis, P.G.: Repair checking in inconsistent databases: algorithms and complexity. In: Proceedings of the 12th International Conference on Database Theory - ICDT, Russia, pp. 31–41, 23–25 March 2009
3. Ahmeti, A., Calvanese, D., Polleres, A.: Updating RDFS ABoxes and TBoxes in SPARQL. CoRR, abs/1403.7248 (2014)
4. Benedikt, M., Konstantinidis, G., Mecca, G., et al.: Benchmarking the chase. To appear in Principles of Database Systems (PODS 2017) (2017)
5. Chirkova, R., Fletcher, G.H.L.: Towards well-behaved schema evolution. In: 12th International Workshop on the Web and Databases, WebDB, USA (2009)
6. Fagin, R., Kolaitis, P.G., Miller, R.J., Popa, L.: Data exchange: semantics and query answering. In: Calvanese, D., Lenzerini, M., Motwani, R. (eds.) ICDT 2003. LNCS, vol. 2572, pp. 207–224. Springer, Heidelberg (2003). doi:10.1007/3-540-36285-1_14
7. Fagin, R., Kolaitis, P.G., Popa, L.: Data exchange: getting to the core. ACM Trans. Database Syst. **30**(1), 174–210 (2005)
8. Flouris, G., Konstantinidis, G., Antoniou, G., Christophides, V.: Formal foundations for RDF/S KB evolution. Knowl. Inf. Syst. **35**(1), 153–191 (2013)
9. Gottlob, G., Orsi, G., Pieris, A.: Ontological queries: rewriting and optimization. In: Proceedings of the 27th International Conference on Data Engineering, ICDE, Germany, pp. 2–13 (2011)
10. Guetmi, N.: Design Models for Mobile Collaborative Applications in the Cloud. Ph.D. thesis, École nationale supérieure de mécanique et d'aérotechnique, France (2016). https://tel.archives-ouvertes.fr/tel-01430151
11. Gutierrez, C., Hurtado, C., Vaisman, A.: RDFS update: from theory to practice. In: Antoniou, G., Grobelnik, M., Simperl, E., Parsia, B., Plexousakis, D., Leenheer, P., Pan, J. (eds.) ESWC 2011, Part II. LNCS, vol. 6644, pp. 93–107. Springer, Heidelberg (2011). doi:10.1007/978-3-642-21064-8_7

12. Halfeld-Ferrari, M., Laurent, D.: Updating RDF/S databases under negative and tuple-generating constraints. Technical report, LIFO-Université d'Orléans, RR-2017-05 (2017). https://www.univ-orleans.fr/lifo/rapports.php?lang=en&sub=sub3
13. Halfeld-Ferrari, M., Laurent, D., Spyratos, N.: Update rules in datalog programs. J. Log. Comput. 8(6), 745–775 (1998)
14. Lausen, G., Meier, M., Schmidt, M.: Sparqling constraints for RDF. In: Proceedings of the 11th International Conference on Extending Database Technology, EDBT, France, pp. 499–509 (2008)
15. Lösch, U., Rudolph, S., Vrandečić, D., Studer, R.: Tempus fugit. In: Aroyo, L., Traverso, P., Ciravegna, F., Cimiano, P., Heath, T., Hyvönen, E., Mizoguchi, R., Oren, E., Sabou, M., Simperl, E. (eds.) ESWC 2009. LNCS, vol. 5554, pp. 278–292. Springer, Heidelberg (2009). doi:10.1007/978-3-642-02121-3_23
16. Motik, B., Horrocks, I., Sattler, U.: Bridging the gap between OWL and relational databases. In: Proceedings of the 16th International Conference on World Wide Web, WWW, Canada, pp. 807–816 (2007)
17. Nikitina, N., Rudolph, S., Glimm, B.: Interactive ontology revision. J. Web Sem. 12, 118–130 (2012)
18. Onet, A.: The chase procedure and its applications in data exchange. In: Data Exchange, Integration, and Streams, pp. 1–37 (2013)
19. Patel-Schneider, P.F.: Using description logics for RDF constraint checking and closed-world recognition. In: Proceedings of the Twenty-Ninth AAAI Conference on Artificial Intelligence, USA, pp. 247–253 (2015)
20. Pichler, R., Skritek, S.: The complexity of evaluating tuple generating dependencies. In: Proceedings of the 14th International Conference on Database Theory - ICDT, Sweden, pp. 244–255 (2011)
21. Schätzle, A., Przyjaciel-Zablocki, M., Skilevic, S., Lausen, G.: S2RDF: RDF querying with SPARQL on spark. PVLDB 9(10), 804–815 (2016)

Additional Database
and Information Systems Topics

Migrating Web Archives from HTML4 to HTML5: A Block-Based Approach and Its Evaluation

Andrés Sanoja[1](✉) and Stéphane Gançarski[2]

[1] Escuela de Computación, Universidad Central de Venezuela, Caracas, Venezuela
andres.sanoja@ciens.ucv.ve
[2] Laboratoire d'Informatique de Paris 6, Université Pierre et Marie Curie,
Paris, France
stephane.gancarski@lip6.fr

Abstract. Web archives (and the Web itself) are likely to suffer from format obsolescence. In a few years or decades, future Web browsers will no more be able to properly render Web pages written in HTML4 format. Thus we propose a migration tool from HTML4 to HTML5. This is challenging, because it requires to generate HTML5 semantic elements that do not exist in HTML4 pages. To solve this issue, we propose to use a Web page segmenter. Indeed, blocks generated by a segmenter are good candidates for being semantic elements as both reflect the content structure of the page. We use an evaluation framework for Web page segmentation, that helps defining and computing relevant metrics to measure the quality of the migration process. We ran experiments on a sample of 40 pages. The migrated pages we produce are compared to a ground truth. The automatic labeling of blocks is quite similar to the ground truth, though its quality depends on the type of page we migrate. When comparing the rendering of the original page and the rendering of its migrated version, we note some differences, mainly due to the fact that rendering engines do not (yet) properly render the content of semantic elements.

Keywords: Migration · Web · Segmentation · Blocks · HTML5 · Web archive · Format obsolescence

1 Introduction

Obsolescence, adjustment, and renewal are necessary parts of the development cycle. Improvements usually require changes. In July 2012, the WWW Consortium introduced a recommendation for HTML5, an important evolution with respect to the preceding version of HTML and the XHTML specification. For instance, it introduces the semantic tags allowing browsers to easily access contents, audio and video among others. Laws [5] points that organizations and

© Springer International Publishing AG 2017
M. Kirikova et al. (Eds.): ADBIS 2017, LNCS 10509, pp. 375–393, 2017.
DOI: 10.1007/978-3-319-66917-5_25

publishers need to be ready for this technological change if they want to out-perform their competitors and stay in the technological race. This raises the following question: once publishers switch to HTML5, what happens with the current (on-line or archived) HTML4 content?

As mentioned by Rosenthal [12], eventually modern browsers will no longer be able to render HTML4 (or XHTML) documents in a proper way. Thus, a strategy for preserving HTML4 pages is required. This issue is even more crucial if one consider *Web archives* such as Internet Archive[1], where millions of HTML4 pages are stored and must be preserved for subsequent (possibly long term) retrieval. This archive consists of 2.5 billion resources, crawled between 1996 and 2010 [3]. All those resources are in HTML 2.0, 3.2, 4.0, 4.01 and XHTML 1.0 formats. This gives an example of what is the size of the corpus to work with. To cope with this issue, Web publishers and archivists must decide to perform either an *emulation* or a *migration*.

Emulation is the implementation of functionalities of an obsolete system, but on the new (hardware and software) environment in which the object is rendered [17]. It consists in recreating the environment in which a Web page was originally created. This implies keeping old (versions of) tools and also generates a runtime overhead in response time, due to emulation processing.

Migration refers to transferring data to newer system environments [2]. In our context, this consists in converting a Web page from one file format to another so that the resource (including its functionalities) remains compliant with new rendering engines. Migration may require heavy computing resources such as a cloud or a PC cluster, running map-reduce parallel processing, particularly for large Web archives. However, it has to be performed only once, and migrated pages will be rendered by modern or future browsers as fast as (future) native HTML5 pages. In other words, while emulation is a short-term solution, migration is more adapted to long term preservation. Thus, in this paper, we focus on Web pages migration.

A main goal of HTML5 is to add semantics to the different components of a Web page, called *blocks*, by associating labels, or semantic tags, with blocks. An HTML5 page is composed of *semantic elements* which organize the content. For instance, some blocks are labelled *header* and are processed accordingly by modern rendering engines. As those labels do not exist in HTML4 pages, the migration process must infer them from the structure/content of the existing pages. Discovering the structure of a Web page is called *Web Page Segmentation*, or *segmentation* for short. It consists in dividing a page into (possibly nested) segments, or blocks. A review of the existing approaches to segmentation may be found in [13].

In this article we present how we use Web page segmentation to perform the migration of HTML4 pages to HTML5 format. There are two possible ways to perform such a migration: *tag-by-tag* or *block based*. Tag-by-tag migration means translating the source code of a Web page content, element by element.

---

[1] web.archive.org/.

However, for some tags the heuristic can be very complex, since the HTML5 semantic element can be dependant of the layout of the content, *i.e.* a graphical representation, not deductible from the source code without rendering.

Thus, we choose a block based approach that operates on a rendered version of the page. This rendered version is segmented into blocks, and the geometry of the elements is stored. Blocks are automatically labeled through heuristic rules, which define semantic elements. Then we produce a new version of the page only with the original metadata (*e.g.* style sheets) and semantic elements. Then we migrate each block content and put the result into the corresponding semantic element.

In this work we experiment the benefits of using a block-based segmenter like Block-o-Matic (BoM), that works on the rendered page [15]. Working inside a rendering engine gives access to the DOM (W), the structure (W') result of the segmentation, and the layout (GM) of the page (*c.f.* Sect. 3). This eases the detection of the regions where the semantic elements should be added, and their location in the DOM. Segmenters following other approaches *e.g.* Text-based [4], Tag-based [6] and even Image-based [1] do not produce sufficient information to achieve the same quality.

In theory, semantic tags have no impact in the rendering of the page, but they help to organize the content into coherent regions. Thus, migration can be performed by segmenting HTML4 pages and incorporating semantic tags to the result.

The main contributions of this paper are the following. (1) We propose an enhanced version of BoM, that not only segments an HTML4 page, but also, thanks to a set of heuristic we defined, automatically labels the blocks to produce HTML5 semantics elements, in order to complete the migration process. (2) We propose a framework to measure the quality of a migration and use it to evaluate our migration method. As far as we know, there is no other evaluation framework equivalent to our one in the literature. To measure the correctness of a migration, we compare a set of predefined manual segmentations (ground truth) with automatically migrated versions. The ground truth is built by human assessors using our Manual-design-Of-Blocks tool (MoB). We compare both the structure (blocks) and the semantics (labels) of the migrated version with the corresponding ground truth. We also measure to which extent the rendering of the page is affected by the migration. The results we obtained through experimentation are promising. They help understanding the issues raised by the migration and suggest improvements as future work.

This paper is organized as follow. Section 2 gives a brief summary of the related works. In Sect. 3 we present the Web page segmentation concepts and notations. We introduce BoM, our approach to Web page segmentation, and our evaluation framework. Section 4 details our solution for migrating from HTML4 to HTML5. Section 5 presents the experiments we led and Sect. 6 the results of those experiments. We conclude in Sect. 7 and give hints for future work.

2   Related Works

There are many online references to systems that perform tag-by-tag migration from HTML4 to HTML5[2]. For instance [18] proposes guidelines to manually convert HTML4 documents to HTML5 and points the issues raised by *div* elements. More generally, there is a real difficulty to translate HTML4 elements to HTML5 semantic tags from documents where such tags do not exist. As far as we know, there is no approach that performs a block-based migration comparable with our one.

There are very few systematic and automatic approaches that solve the problem of translating to HTML5 format documents in other formats. As an example, Park [10], present their experience in the migration of ETD (Electronic Theses and Dissertations) from the PDF format to HTML5 format. Most of ETDs have linked multimedia documents and connected by hyperlinks. Storing them in PDF requires to have the corresponding multimedia readers, libraries and plug-ins, as well. HTML5 is a convenient migration format because in this way it is possible to have one single file that has all of the content linked together, including all of the multimedia information in the ETD and metadata available for Web search indexing and other general tasks. In this case, determining which are the semantic elements is easy because the PDF sections are well defined and delimited. This makes possible a full tag-by-tag migration, including semantic elements in the process.

Migration is a crucial issue even for other formats than HTML5. Rosenthal [11] describes the design and implementation of a transparent, on-access format migration capability in the LOCKSS system for preserving Web content. Their implementation is capable of transparently presenting content collected in one Web format to readers in another Web format, without changing the browsers. They present an user case of this type of migration on GIF image format migrated to PNG format.

3   Web Page Segmentation with BOM and Its Evaluation

Our segmenter BoM segments a Web page without *a priori* knowledge of its content. It only uses the heuristic rules defined by the W3C Web standards. For instance, we detect blocks using HTML5 content categories instead of using the tag names or text features.

The segmentation process is composed of two main phases: detecting fine-grained (as small as possible) blocks and then merging them according to a stop condition, so that the segmentation is performed at the desired granularity. It can be expressed by the following function $\Phi(W)$, where W is the rendered DOM of a Web page.

$$\Phi(W, pA, pD) \longrightarrow (W', GM, L)$$

W' is the block graph (a tree) of the segmentation: a block B is the child of block C if C contains B. A flat segmentation of W is obtained by only considering the leaf nodes in W'. GM is the geometric model, pA and pD the stop condition.

---

[2] Googling the term "translating html 4 tag to html5" will give these references.

In few words, GM stores the size and coordinates of the blocks bounding boxes. L is the list of (HTML5) labels produced by the segmentation. This is the main difference with the former version of BoM [14] which was not designed to produce HTML5 compliant segmentations.

pA is the lower bound for the proportional size of a block with respect to the page. So, BoM creates blocks relatively bigger than pA. pD is the minimum separation distance (in pixels) to consider two blocks for merging. More details about the segmentation with BoM can be found in [13].

A segmentation is evaluated by computing its similarity with a reference segmentation. The similarity is defined by using *block correspondence measures* that allow knowing to what extent the generated blocks rectangles match those of the reference segmentation. Consider two segmentations P and G. Based on the geometry of the blocks, we compute correspondences between (sets of) blocks of P and (sets of) blocks of G. A block in P (resp. G) corresponds to a (set of) block(s) in G (resp. P) if they fit in the same rectangle of the page.

If the blocks in BoM segmentation P fits perfectly with the ground-truth blocks G, then there is a perfect matching between G and P. If there are differences between the two segmentations, blocks of P may have zero or several corresponding blocks in G.

Consider that G is the reference segmentation. The metrics for block correspondence of P are the following:

1. **Correct blocks C_c.** The number of one-to-one matches between P and G. C_c is the main metric for measuring the quality of a segmentation.
2. **Oversegmented blocks C_o.** This metric measures how much a segmentation produced too small blocks. However, those small blocks fit inside a block of G.
3. **Undersegmented blocks C_u.** The same as above, but for big blocks, where blocks of G fit in.
4. **Missed blocks C_m.** This metric measures how many blocks of G are not detected by the segmentation P.
5. **False alarm blocks C_f.** This metric measures how many blocks in P are "invented" by the segmentation. There are no corresponding block in G.

C_c is a positive measure, while C_o and C_u count granularity-related errors: found blocks could match with G if they were aggregated or split. C_f and C_m are negative metrics. Formal definitions of those and other metrics can be found in [15] and in [13].

The evaluate function returns a vector made of the computed metrics, *i.e.*

$$evaluate(G, P) = (C_c, C_o, C_u, C_m, C_f) \tag{1}$$

In this paper, we use this evaluation framework twice. First, we use it to test the (visual) quality of a migration operation, by comparing the segmentations obtained before and after the migration (*cf.* Sect. 4.4). In this case, the reference segmentation is the one obtained by segmenting the HTML4 page, before migration. The underlying idea is that if the migration is correct, then the rendering of the migrated version must be similar to the reference version, thus they must lead to similar segmentations with BoM. Second, we use the evaluation framework to compare the labeling computed by BoM with a ground truth. The ground truth is built by human assessors who manually segment and label pages. To ease assessors work, we develop MoB, a tool that allows manually segmenting and labeling pages thanks to a user-friendly interface[3]. In this case, the reference segmentation is the ground truth. By applying the correspondence measures to the segmentation produced by BoM, we can better explain the differences between the labeling obtained with BoM and the labeling of the ground truth (*cf.* Sect. 4.5).

4   Migration and Evaluation

In this section we present the migration process and its evaluation. The goal is to take a Web page in HTML4 format and produce a version of the same page according to the HTML5 format. We evaluate the migration in two ways:

- by comparing the rendering of the migrated version with the rendering of the original version, to measure how much the visual aspect will be preserved by the migration (*cf.* Sect. 4.4);
- comparing the output of our solution with a ground truth, in order to measure the quality of our automatic labeling (*cf.* Sect. 4.5).

Both evaluations help determining the causes and the possible actions to improve the migration method. The process is illustrated on Fig. 1. At the center, the steps necessary for producing the output. At the right side, the evaluation and the manual segmentation (ground truth building) tasks. At the left side, the

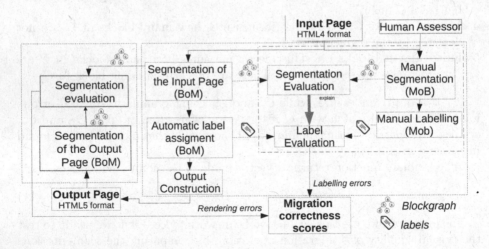

Fig. 1. Migration and its evaluation overview

---

[3] https://github.com/asanoja/web-segmentation-evaluation/tree/master/chrome-extensions/MOB.

evaluation of the rendering of the migrated page with respect to the rendering of the original page. As mentionned before, this evaluation is made by comparing the segmentation of the migrated version with the segmentation of the original version. Each task is described in the rest of the section, with a focus on the blocks labeling and the two evaluation ways above mentioned.

4.1   The Migration Process

The migration process consists of three main tasks:

1. **Segmentation of the input page:** A Web page in HTML4 format is segmented using the BoM segmenter (*c.f.* Sect. 3).
2. **Automatic label assignment:** Based on the properties and characteristics of the blocks found in the segmentation we assign a label (*i.e.* HTML5 semantic elements tags) to each block (*c.f.* Sect. 4.2).
3. **Output construction:** The output is a Web page in HTML5 format. Each block found in the segmentation becomes a semantic element, named by the corresponding block label. The elements associated with the block are included as children of this semantic element. Besides the semantics tags and the original content, no other extra elements are added to the migrated version, *e.g.* style sheets, javascripts or metadata.

4.2   Assigning Labels

BoM creates a block tree including composite and simple blocks (simple blocks are the leaves of the tree) [13]. In our experiments we only consider simple blocks *i.e.* we work on a flat segmentation. The evaluation of labeling composite blocks (in our approach, blocks labelled *page* or *section*) is left for future work.

Heuristics rules are defined in order to determine the label of each block. These rules assign labels depending on the position of a block and its relationship to the others blocks. A block is treated differently if it resides in the visible part of the page (*i.e.* the part of the page visible without using scrolling). For instance, a block is labeled as *header* if it resides in the visible part of the page, and the amount of hyperlinks and content do not exceed certain quantity indicated by a predefined constant. A block with the same characteristics but outside of the visible area and at the bottom of the page is labeled as *footer*.

For the label *nav*, one additional condition is considered. If the proportion of hyperlinks (*i.e.* <A> elements, elements with *onclick* event attribute) a block contains is greater than a predefined constant, it can be considered as a *nav*.

Algorithm 1 describe the label assignment method (the heuristics) for all possible cases. A full version of the algorithm (including composite blocks) can be found at github[4]

---

[4] https://github.com/asanoja/web-segmentation-evaluation/tree/master/ chrome-extensions/BOM.

```
    Data: Block: b
    Result: b.label
1   if b.weight ¿ pA then
2       if b in the visible part of page then
3           if proportion of elements covered by b is greater than a constant then
4               |    b.label=ARTICLE;
5           else if proportion of hyperlinks covered by b is greater than a constant then
6               |    b.label=NAV;
7           else
8               |    b.label=HEADER;
9           end
10      else if b is in the middle/center of the page then
11          if b is at left/right of the page then
12              |    b.label=ASIDE;
13          else if proportion of hyperlinks covered by b is greater than a constant then
14              |    b.label=NAV;
15          else
16              |    b.label=ARTICLE;
17          end
18      else if b is at the bottom of the page then
19          if proportion of hyperlinks covered by b is greater than a constant then
20              |    b.label=NAV;
21          else
22              |    b.label=FOOTER;
23          end
24      else
25          |    b.label=ARTICLE;
26      end
27  end
```

Algorithm 1. Label Assignment Algorithm

4.3 The Evaluation Process

We measure to what extent the migration is correct, and how reliable it is. The evaluation is divided in three tasks, as follows:

1. **Measure of rendering errors:** Using the Web page segmentation evaluation framework (*cf.* Sect. 3) we measure the difference on the rendering both of the original and migrated pages, *cf.* Sect. 4.4.
2. **Manual segmentation and label assignment:** Using the MoB tool (*c.f.* Sect. 3) we produce a *ideal* segmentation of the input page that serves as a ground truth. In the same process the user assigns a label to each block.
3. **Measure of labeling errors:** from both segmentations (*i.e.* the manual -ground truth- and automatic one -BoM-) we compute some measures to determine how different both label assignments are. The metrics are described in detail in Sect. 4.5. In order to better explain the differences, we also compare the structure of the segmentations, using again the measures defined in Sect. 3.

4.4 Dealing with Rendering Errors

In order to measure to what extent the migration affects the rendering of the migrated Web page, we use the correspondence measures defined in Sect. 3.

Consider two rendered DOM, W and $W5$, where W is the rendered DOM of a Web page in HTML4 format and $W5$ is the rendered DOM of the migrated Web page. As mentionned in Sect. 1, W and $W5$ cannot be directly -syntactically-compared tag-by-tag. However, if the migration is correct, W and $W5$ must be similar, *i.e.* they produce the same visual rendering. Thus their respective segmentations must be equal or close. This means that we can indirectly compare W and $W5$ by comparing their respective blocks trees W' and $W5'$ obtained by the BOM segmenter.

If we find only correct blocks then the migration may be perfect. Indeed, if both W and $W5$ produce the same segmentation, *i.e.* $W' = W5'$, there is a high probability that their rendering is the same.

However, in the migrated version some semantic elements may not be rendered correctly. When reaching a semantic element, rendering engines of current Web browsers do not process the user defined CSS style of its content, as it does in the original version. For semantic elements, Web browsers either use a default style, either no style at all. In both cases, they cannot use the CCS style defined for HTML4 elements. Thus, blocks can change their size and position leading to a shifting of the block. Depending on the page design this can affect either the whole page or a part of it. We can interpret the impact of this error by observing the values of the correspondence metrics defined in Sect. 3. Missed and false alarms blocks, corresponding to metrics C_m and C_f, quantify the amount of blocks that are affected by the shifting. In most situations, the shifting of a block generates one missed block and one false alarm, since the evaluation framework do not consider anymore that the same block appears at the same location in both versions. The shifting of a block may also generate over seg-

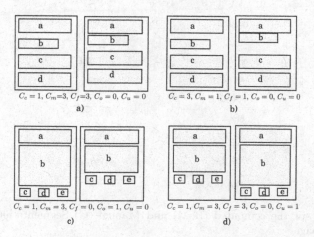

$C_c = 1, C_m=3, C_f=3, C_o = 0, C_u = 0$
a)

$C_c = 3, C_m = 1, C_f = 1, C_o = 0, C_u = 0$
b)

$C_c = 1, C_m = 3, C_f = 0, C_o = 1, C_u = 0$
c)

$C_c = 1, C_m = 3, C_f = 3, C_o = 0, C_u = 1$
d)

Fig. 2. Examples for rendering errors

mentations (resp. under segmentation), thus increasing the value of metric C_o (resp. C_u.): shifted blocks in $W5'$ are considered as part of a different blocks in W', or viceversa.

Figure 2 presents examples of possible rendering errors. In Fig. 2a, block b is shifted up, which causes a cascading shifting up of blocks c and d. Metrics C_m and C_f are equal, then there are three blocks with rendering error. On Fig. 2b only block b is shifted, producing one block with error. In Fig. 2c, block b reduce its size, shifting blocks c, d and e. This lead to an oversegmentation of block b in W' because blocks in b,c,d,e of $W5'$ fit inside. In Fig. 2d, block b in $W5'$ expands its size and covers blocks b and c in W' causing an undersegmentation. Block c in $W5'$ is shifted right, producing a false alarm.

4.5   Measuring Labels Similarity

The manual segmentation Φ_G and the computed segmentation Φ_{BoM} are formally defined in Sect. 3. The manual segmentation, produced by assessors, takes the rendered DOM of a Web page (W) in HTML4 format and produces the W'_G block tree, $i.e.$

$$\Phi_G(W, pA = 5, pD = 30) \longrightarrow (W'_G, GM_G, L_G)$$

The computed segmentation takes the same rendered DOM (W) and produces the W'_{BoM} block tree, $i.e.$

$$\Phi_{BoM}(W, pA = 5, pD = 30) \longrightarrow (W'_{BoM}, GM_{BoM}, L_{BoM})$$

The order in L_G and L_{BoM} is the one in which labels appear on the screen, in a top-down/left-right traversal.

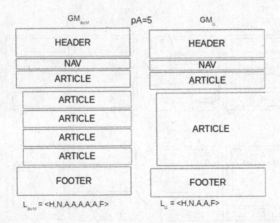

Fig. 3. Labels for the computed (BoM) and Manual (G) Segmentation in a forum example web page

Consider the flat segmentations of a forum page shown in Fig. 3. For the computed segmentation (left) we have a *header* block, the menu as a *nav* block, an *article* block representing the first post, followed by a set of response blocks labeled as *article*. At the end of the page, there is a *footer* block. The figure shows the labels for the manual and computed flat segmentation.

The list of labels from the manual segmentation is: {header, nav, article, article, footer}. The list of labels for the computed segmentation is: {header, nav, article, article, article, article, footer}. For simplicity, we denote the labels with their first letter.

Thus, the list of labels for both segmentations are $L_G =< H, N, A, A, F >$ and $L_{BoM} =< H, N, A, A, A, A, A, F >$.

The assessors who built the manual segmentation (right) considered that all the responses in the forum should be gathered. In other words, the *header, nav, article* and *footer* are equally labeled by BoM and the human assessors, but the responses in BoM version are splitted, which is a granularity error, thus not so serious.

In order to compute to what extent both segmentations are similar, we define sim_D, the inverse proportion of error, based on the Damerau-Levenshtein edit distance [7] as follows:

$$sim_D = 1 - \frac{DamerauLevenshteinDist(L_G, L_{BoM})}{|L_G| + |L_{BoM}|}$$

where $|L_G| + |L_{BoM}|$ is the maximum possible error value for the Edit Distance [8].

$DamerauLevenshteinDist(L_G, L_{BoM})$ represents the minimum number of insertions, deletions, substitutions or transpositions needed to transform L_{BoM} into L_G. sim_D represents a relative similarity between both label sequences. In the example of Fig. 3,

$$sim_D(L_G, L_{BoM}) = 1 - \frac{3}{5 + 8} = 0.61$$

This means that both labelings are 61% similar. Both segmentations have the same content, but one is more detailed than the other. If we merge the *article* blocks corresponding to response in the BoM segmentation, we obtain the same labeling as the manual segmentation. In other words, the sim_D metric value is directly impacted by granularity issues, *i.e.* oversegmentation and undersegmentation (*cf.* Sect. 3).

From the content point of view (without taking granularity into account) both segmentations are totally similar, thus much more than the 61% obtained through sim_D. Thus, we looked for another metric that is not impacted by granularity issues. We use the Jaccard similarity coefficient [9], noted sim_J in this paper. Let SL_G (resp. SL_{BoM}) be the set of elements contained in L_G (resp. L_{BoM}), then

$$sim_J(L_G, L_{BoM}) = \frac{|SL_G \cap SL_{BoM}|}{|SL_G| + |SL_{BoM}| - |SL_G \cap SL_{BoM}|}$$

In the example:
$$sim_J(L_G, L_{BoM}) = 1$$

which means that BoM labeling and manual labeling are 100% similar.

sim_D metric is more affected by small variations in missing/displacement/ repeated labels in the sequence. It is important because it gives an idea on how accurate the labeling assignment algorithm is, in absolute terms.

sim_J minimize the impact of the granularity problems in the segmentation, giving an idea of how good (or similar) the labels were assigned (even with error), in relative terms. This is due to the fact that, when considering SL_G instead of L_G, duplicates are eliminated. However, as SL_G is a set, the order in L_G is lost. This may lead to consider similar couples of lists as $< H, A, F >$ and $< F, A, H >$. Even if such a discrepancy between manual and automatic segmentations is highly unlikely, it justifies that we keep both metrics. Indeed, the sim_D value would be low for such a couple of lists, thus, at the end, the two segmentations $< H, A, F >$ and $< F, A, H >$ would not be considered as very similar. We currently investigate to find a new metric that would decrease the impact of granularity while keeping the labeling order.

To sum up, if sim_J value is greater than sim_D it is because the similarity is affected by the granularity. If sim_J value is lesser than sim_D it is because the similarity is affected either by the correctness of labels assignments or the quality of the segmentation. Both metrics allow us to interpret the results obtained from the experience. As the quality of the labeling may depend on the quality of the segmentation, we also apply the correspondance measure defined in Sect. 3 in order to better explain the labeling differences (*cf.* Sect. 6.2).

5   Experiments

In this section we describe the setup of experiments we led to evaluate our migration method.

5.1   Experimentation Design

We present the Web pages we used and a brief description of the experiments we led on those pages.

The MIG5 Collection. The dataset holds the offline version of Web pages, together with their segmentations created manually by human assessors (ground truth) or automatically with BoM (with a predefined stop condition), organized in categories.

Web pages of the MIG5 collection are taken from the GOSH (GOogle SearcH) collection that we built. Web pages in this collection are selected with respect to their category. This selection is based in the categorization made by Brian Solis [16], "The Conversation Prism". It depicts the social media landscape from ethnography point of view. In this work, we considered the five most common of these categories, namely Blog, Forum, Picture, Enterprise and Wiki. For each category, a set of 25 sites have been selected using Google search to find the pages with the highest *PageRank*. Within each of those sites, one page is crawled[5]. The GOSH collection contains 125 pages.

The MIG5 collection is a subset of the GOSH collection which only contains Web pages in HTML 4.01 format, but from the same categories (Blog, Enterprise, Forum, Picture and Wiki).

Experiments. The first experiment aims at measuring if including the semantic elements (*i.e.* the migration process) affects the rendering of the page. The block correspondence method, as presented in Sect. 3, is used for evaluating the correctness of the migration. The segmentation of the original Web page is used as a ground truth, while the segmentation of the migrated Web page is the evaluated segmentation.

The second experiment is devoted to measure to what extent the labels found with the BoM segmentation algorithm match to those in a ground truth of manually labeled blocks.

5.2   Manual-design-Of-Blocks Tool

In order to help human assessors building a Web page segmentation (ground truth) we developed the tool MoB (Manual-design-Of-Blocks). It is designed as a browser extension and offers functionalities to expert users for creating manual segmentations (See footnote 3).

Users can create blocks based on Web page elements. They can merge blocks, navigate into the element hierarchy to produce a block graph[6] (*c.f* Sect. 3), or produce a flat segmentation (*i.e.* leaves in the block tree). These segmentations are stored in a repository[7] for the evaluation.

MoB is implemented as an extension for Google Chrome. We are currently working on producing an equivalent extension for Firefox.

5.3   Ground Truth Building

Table 1 shows the organization of the MIG5 collection. It is composed of 40 pages organized by category.

The MoB tool (*cf.* Sect. 5.2) is used by human assessors to annotate the blocks. Besides specifying the blocks, assessors assign a label to each block. Labels correspond to a subset of the semantic elements defined in the HTML5 specification (header, footer, section, article, nav, aside). We ask them to produce a flat segmentation. The stop condition for all the experiments are set to $pA = 5$ and $pD = 30$ (*cf.* Sect. 3). Indeed, through experiments, we noticed that this pA

---

[5] https://github.com/asanoja/web-segmentation-evaluation/tree/master/dataset.
[6] Usually a tree.
[7] http://www-poleia.lip6.fr/~sanojaa/BOM/inventory.

Table 1. MIG5 pages by categories

Category	Pages
Blog	5
Enterprise	9
Forum	14
Picture	7
Wiki	5
Total	40

value generates blocks likely to correspond to elements in the layout of the page. The separation is set to $pD = 30$ because usually these regions can be very close one to each other.

6 Results

We present now the results of the experiments described in the previous section. We measure the quality (rendering and labeling) of our approach to migrate HTML4 pages to HTML5 format.

6.1 Measuring Rendering Errors

In this Section, we compare the segmentation of an HTML4 page and the segmentation of its migrated HTML5 version.

Table 2 shows the average correspondence metrics (defined in Sect. 3) that measure rendering errors as described in Sect. 4.4, by category, for the MIG5 collection. C_c represents the correct blocks, C_o the oversegmentations, C_u the undersegmentations, C_m the missed blocks and C_f the false alarms.

Table 2. Correspondence metrics for the MIG5 collection for comparing rendering errors between W' and $W5'$

Category	C_c	C_o	C_u	C_m	C_f
Blog	6.50	0	0.50	0	0.50
Enterprise	4.00	0.33	0.33	1.11	2.77
Forum	3.41	0.59	0.41	2.11	1.29
Picture	2.71	1.00	0.29	2.00	0.71
Wiki	6.00	0	0	0.60	0.60

For the Blog category, on average, there are 0.5 false alarms blocks and 0.50 blocks undersegmented. In some Blog pages the main content expands horizontally by the rendering of the article element of the main content. It covers the main content and the menu on the right side of the original version, producing an undersegmentation. The menu in the migrated version shifts right so that there is no correspondence in the original segmentation, which produces false alarms. This situation is depicted on Fig. 2d.

Wiki pages have blocks shifting vertically, producing some missed blocks and false alarms. No granularity issues are raised by this category of pages.

Forum pages are affected by the shifting of block at the top of the page affecting all the blocks below. Some forum responses which size is bigger than the rest can cover shifted blocks, producing oversegmentations. On the other hand, responses in the migrated version can also expand, producing undersegmentations. Pages in categories Picture and Enterprise have a similar situation. These pages have complex Web design making them more sensitive to size and position changes. Pages in these three categories are the most affected by rendering errors in the collection.

Blog and Wiki categories have the best results. The regions in these type of pages are simple and the position and order of blocks are standard. The regions are well separated, and rendering errors have little impact.

6.2   Measuring Labels

In this Section, we compare the labeling obtained with BoM with the labeling defined by human assessor in the ground truth. As the labeling can be impacted by the rendering of the migrated page, we also use the correspondence metrics between the migrated version and the ground truth, in order to better explain the results for labeling.

Table 3. Average values for sim_J and sim_D for the MIG5 collection

Category	sim_J	sim_D
Blog	0.79	0.73
Enterprise	0.58	0.59
Forum	0.87	0.74
Picture	0.55	0.60
Wiki	0.88	0.66

Table 3 shows the average values, by category, of the metrics defined in Sect. 4.5 for the MIG5 collection. Column sim_J represents the Jaccard-based similarity measure. The sim_D column represents the DamerauLevenshtein-based similarity measure.

Table 4 shows the average correspondence metrics values per category, when comparing the manual segmentation W'_G and the automatic segmentation W'_{BoM}. C_c represents the correct blocks, C_o the oversergmentations, C_u the

Table 4. Correspondence metrics for the MIG5 collection for comparing labeling between W'_G and W'_{BoM}

Category	C_c	C_o	C_u	N
Blog	6,95	0,05	0	7
Enterprise	6,24	0,01	0,20	6,45
Forum	5,81	0,66	0,12	6,59
Picture	4,21	0,02	0,67	6,71
Wiki	6,45	0	0,15	6,60

undersegmentations and N is the number of blocks in the ground truth. This table helps interpreting the results obtained in Table 3.

As seen in Table 3, Blog, Forum and Wiki categories have the best results for labelling. The structure of those types of pages comes, in most of the cases, from templates or CMS, which make them uniform and thus easy to segment and label. However, if we compare the metric values, sim_J value is greater than sim_D, which means that the labeling is affected by the granularity. We observed in the dataset that the common issue for the labeling method is at the top of the page, where the *header* is sometimes divided in two or three blocks, because navigation links and other aesthetic information sometimes are included in the header. The Forum category is the one most affected by granularity (sim_D significantly lower than sim_J). This is explained by the fact that, in Forum pages, blocks in the middle of the page (*e.g.* responses) are frequently oversegmented (*cf.* Fig. 3). This is consistent with the oversegmented metric value $C_o = 0.66$ for forum category, which is the highest among the five categories. We consider that those blocks in the center and middle of the page should be treated differently, with respect to granularity. In order to improve the accuracy of BoM, we plan to study this situation and replace the scalar pA parameter by a vector (one value per region in the page). This is a challenge, because we must determine a set of stop conditions for each category. Machine-Learning techniques could be used to better estimate the components of this vector.

The Enterprise and Picture categories get the worst values. The occurrence of undersegmentations are the main reasons of the low values for labeling in Picture and Enterprise categories. Due to the complex design of some of these pages, BoM usually creates a big block covering the blocks in the ground truth. This causes the undersegmentation. Picture category is the most affected by this problem, with correct blocks $C_c = 4.21$, which means that it has 2.5 blocks in average of difference than the ground truth. This has an influence in the labeling, that is why $sim_D = 0.60$ and $sim_J = 0.55$.

7   Conclusion and Future Works

In this paper we presented our approach to block-based migration of Web pages from HTML4 format to HTML5 format, and its evaluation. Thanks to Web page segmentation, we produce a migrated version that complies with the HTML5 specification. Using our evaluation framework we measure the quality of the migration process. We analyzed how the algorithm assigned labels to blocks in comparison to a ground truth made of manually labeled segmentations. The rendering errors were measured using the block correspondence metrics defined in Sect. 3. The results show that, in the context of digital preservation, migrating Web pages from one format to another is possible using the BoM Web page segmentation algorithm, even if the results are not perfect. This allows for avoiding emulation in Web archives, which slows down the retrieval of pages, and requires to keep old tools or versions of tools. In its current version, our algorithm is rather simple, as it only takes into account the spatial features of the blocks. To get better results, we plan to consider other features, such as the textual content and the text properties.

The shifting of blocks is an issue that should be considered. The segmentation is affected by the semantic tags. For instance, some browsers have no default style for these elements, thus shifting their position. Even if they do have one, it may not render the semantic element at the same position as the corresponding HTML4 element which may be handled by a user defined CCS style.

Also, unexpected margins and spacing may appear. This clearly impacts the correspondence metrics. A solution to this issue would be to include some tolerance in the rectangles comparison during the BCG construction (*c.f.* Sect. 3) and consider default styles for the semantic elements.

Our approach gives insights of the upcoming issues raised by the migration of Web content in the context of Web preservation. Migrating the HTML source is the main goal, but we need to assure that all its dependencies are also migrated (*i.e* JavaScript, CSS and images).

We plan to add more pages to the MIG5 dataset, and also include more categories. To this end we think that a good strategy to follow is to enhance the functionality and usability of MoB in order to include non-experts assessors in the annotation process.

There is also an issue with the stop condition, that impacts all phases of the segmentation. Using only one stop condition (*i.e.* pA parameter, *c.f.* Sect. 3) may negatively impact the quality of the segmentation in some regions of the page. Indeed, the whole content of a page is not designed with a uniform granularity in mind: some regions are more detailed than other ones. Thus, we need different granularity parameters, for different regions of a Web page, one for the main content, another for the menus and so on. In other words, we plan replace the scalar pA parameter by a vector.

The metrics sim_D and sim_J (*c.f.* Sect. 4.5) allow us to better understand the behavior of data in the migration process. We are currently investigating a new metric that would decrease the impact of granularity while keeping labeling order into account. Another improvement is related with the ground truth. Human assessors produced a labeled segmentation, *i.e.* an HTML5 document, used to evaluate the segmentation and the labeling at the same time, which renders the interpretation of the results quite complex. In future experimentation, we plan to separate both concerns (by giving already segmented pages to assessors) so that the labeling can be evaluated independently from the segmentation.

References

1. Cao, J., Mao, B., Luo, J.: A segmentation method for web page analysis using shrinking and dividing. Int. J. Parallel Emergent Distrib. Syst. **25**(2), 93–104 (2010)
2. Garret, J.: Preserving digital information. Technical report, Commission on Preservation and Access and the Research Libraries Group (1996)
3. Jackson, A.N.: Formats over time: exploring UK web history. CoRR, pp. 1210–1714 (2012)
4. Kohlschütter, C., Nejdl, W.: A densitometric approach to web page segmentation. In: Proceedings of the 17th ACM Conference on Information and Knowledge Management, pp. 1173–1182, New York, NY, USA. ACM (2008)
5. Laws, B.: Seriously, another format? You must be kidding. CSE News **36**(2), 41 (2013)
6. Lin, S.-H., Ho, J.-M.: Discovering informative content blocks from web documents. In: Proceedings of the Eighth ACM SIGKDD International Conference on Knowledge Discovery and Data Mining, KDD 2002, pp. 588–593, Edmonton, Alberta, Canada. ACM (2002). ISBN: 1-58113-567-X. doi:10.1145/775047.775134
7. Moreau, E., Yvon, F., Cappé, O.: Robust similarity measures for named entities matching. In: Proceedings of the 22Nd International Conference on Computational Linguistics, vol. 1, COLING 2008, pp. 593–600, Stroudsburg, PA, USA. Association for Computational Linguistics (2008). ISBN: 978-1-905593-44-6, http://dl.acm.org/citation.cfm?id=1599081.1599156
8. Navarro, G.: A guided tour to approximate string matching. ACM Comput. Surv. **33**(1), 31–88 (2001). doi:10.1145/375360.375365. ISSN: 0360-0300
9. Niwattanakul, S., Singthongchai, J., Naenudorn, E., Wanapu, S.: Using of Jaccard coefficient for keywords similarity. In: Proceedings of the International MultiConference of Engineers and Computer Scientists, vol. 1, pp. 13–15 (2013)
10. Park, S.H., Lynberg, N., Racer, J., McElmurray, P., Fox, E.A.: Html5 etds. In: Proceedings of International Symposium on Electronic thesis and Dissertations, Austin, TX, USA (2010)
11. Rosenthal, D.S.H., Lipkis, T., Robertson, T., Morabito, S.: Transparent format migration of preserved web content. D-Lib Mag. **11**(1) (2005). http://dblp.uni-trier.de/db/journals/dlib/dlib11.html#RosenthalLRM05
12. Rosenthal, D.S.H.: Format obsolescence: assessing the threat and the defenses. Libr. Hi Tech **28**(2), 195–210 (2010)
13. Sanoja, A.: Web page segmentation, evaluation and applications. PhD thesis, Université Pierre et Marie Curie-Paris VI (2015). https://hal.inria.fr/tel-01128002/
14. Sanoja, A., Gançarski, S.: Block-o-matic: a web page segmentation framework. In: International Conference on Multimedia Computing and Systems (ICMCS), pp. 595–600, Marrakesh, Moroco, April 2014
15. Sanoja, A., Gançarski, S.: Web page segmentation evaluation. In: Proceedings of the 30th Annual ACM Symposium on Applied Computing, pp. 753–760. ACM (2015)
16. Solis, B.: The conversation prism (2014). https://conversationprism.com

17. Van der Hoeven, J.: Emulation for digital preservation in practice: the results. Int. J. Dig. Curation **2**(2), 123–132 (2007)
18. W3Schools.com. HTML5 Migration: Migration from HTML4 to HTML5. W3Schools (2016). http://www.w3schools.com/html/html5_migration.asp

A Tool for Design-Time Usability Evaluation of Web User Interfaces

Jevgeni Marenkov$^{(\boxtimes)}$, Tarmo Robal, and Ahto Kalja

Tallinn University of Technology, Tallinn, Estonia
jevgeni.marenkov@gmail.com, tarmo.robal@ati.ttu.ee,
ahto.kalja@ttu.ee

Abstract. The diversity of smartphones and tablet computers has become intrinsic part of modern life. Following usability guidelines while designing web user interface (UI) is an essential requirement for each web application. Even a minor change in UI could lead to usability problems, e.g. changing background or foreground colour of buttons could cause usability problems especially for people with disabilities. Empirical evaluation methods such as questionnaires and Card Sorting are effective in finding such problems. Nevertheless, these methods cannot be used widely when time, money and evaluators are scarce. The purpose of our work is to deliver a tool for design-time automatic evaluation of UI conformance to category-specific usability guidelines. The main contribution of this solution is enabling immediate cost-efficient and automatic web UI evaluation that conforms to available and set standards. This approach is being integrated into the Estonian eGovernment authority in order to automate usability evaluation of web applications.

Keywords: Web usability · Usability guidelines · Web user interface

1 Introduction

The diversity of smartphones, tablet computers and smartwatches have become an intrinsic part of modern life and culture. Therefore, web user interface (UI) compatibility with mobile platforms is an essential requirement for each web application. Furthermore, UI should also be compatible with the diversity of platforms (including Android, iOS, Windows, Linux, etc.) and different browsers (Safari, Chrome, Firefox, etc.) regardless of their version. Notwithstanding, device and platform compatibility covers only a minor part of UI requirements. In fact, UI should be consistent between pages, attractive, user-friendly, easy to use and navigate. All such characteristics are included in the definition of usability. Usability is the extent to which a product can be used by specified users to achieve specified goals with effectiveness, efficiency and satisfaction in a specified context of use [1]. Usability covers many areas such as accessibility referring to UI requirement for people experiencing disabilities, learnability assuring web application functionality to be complete and correctly displayed.

A demand for usable web applications has led to a variety of approaches and methods helping to achieve a high level of usability. Many studies provide usability guidelines, best design practices, recommendations and patterns to follow when

© Springer International Publishing AG 2017
M. Kirikova et al. (Eds.): ADBIS 2017, LNCS 10509, pp. 394–407, 2017.
DOI: 10.1007/978-3-319-66917-5_26

designing web UI [2, 3]. Usability guideline is a usability criterion advising on how certain UI element should be designed. In a context of our research, we are focusing on those usability guidelines that could be evaluated automatically - without the involvement of potential users to the UI evaluation process.

A crucial responsibility of designers and UI quality assurance specialists is to verify that designed solution satisfies all business requirements, predefined usability guidelines and its overall information architecture is clear for potential users. There are two major groups of methods for evaluating usability: empirical and inspection methods. Usability Inspection methods require the expertise of usability inspectors to detect usability problems in user interface design. They include such methods as Heuristic Evaluation, Formal Usability Inspection, Pluralistic and Cognitive Walkthrough [4]. In its turn, Empirical Testing methods require the participation of real users and include Card Sorting, Eye tracking and Questionnaires with usability tests participants [5]. Empirical Testing is efficient in discovering key issues in information architecture, identifying the flaws and misplacements in web application design. Despite the fact that Empirical Testing methods are commonly more efficient than inspection methods, there are many obstacles preventing from applying these methods widely:

- Organising and conducting user tests is relatively expensive because it requires a high demand for human and time resources [6, 7];
- Small software companies do not have the funds to pay for complete consultancy or involving usability specialist as they are expensive to hire [8];
- Difficulty to get potential users participating in usability evaluations [7] [9];
- It is not always possible to increase the coverage of evaluated features by evaluating every single aspect of UI [6];

Applying empirical evaluation methods is beneficial at the designing and prototyping stages of UI design and development. Evaluating minor changes in UI structure is not always feasible with empirical methods due to a high demand for human and financial resources. In fact, usability inspection methods can assess usability fast and at a lower cost [6, 7]

There are multiple sources of guidelines for inspection methods like Web Content Accessibility Guidelines (WCAG) [10] and Sect. 508 [11] standards, multiple design recommendations and best practices to create a good web experience [12, 13]. The evaluation of UI conformance to guidelines using inspection methods can be done without involving potential users of web applications. Moreover, it is feasible to improve the manual usability inspection via automation. There are multiple tools for UI automatic evaluation like Ocawa[1], Magenta[2], Evaluera[3] and the list of tools provided by World Wide Web Consortium[4]. All aforementioned tools are limited by finding deviations in the HTML code and finding certain accessibility issues; evaluating visual aspects of web application like measuring contrast rate of UI elements, evaluating position of elements

---

[1] http://www.ocawa.com.

[2] http://giove.isti.cnr.it/accessibility/magenta.

[3] http://www.evaluera.co.uk.

[4] https://www.w3.org/WAI/ER/tools/.

on the screen and many other is not possible with that approach. Their integration into the UI development process is extremely complicated and very often not possible at all because solutions are distributed as standalone applications without a possibility to be extended and integrated into the process of continuous delivery.

Due to the rapidly changing business requirements, UI design and structure should also be continuously developed to meet new requirements. Changing UI should be done with extreme caution as every even minor change of UI or the content of page could lead to severe usability problems [14]. For instance, according to usability guidelines, the contrast ration between the letters and the background that is immediately behind the letter should be kept above 4:5:1. Violating the guideline leads to the lower usability for the target users including people with disabilities. Thus, usability is extremely dependent on every modification of UI and, as a result, the immediate evaluation of UI conformance to usability guidelines and subsequent feedback to UI developer becomes another critical demand. That is important because finding usability problems early in the development stage makes the fix less costly than found later.

The purpose of our study is to propose a fully-functional tool for the design-time automatic evaluation of UI conformance to the category-specific usability guidelines during UI design and implementation stage. The main contribution of this solution is enabling immediate cost-efficient and automatic web UI evaluation and feedback to developers to ensure the UI under development conforms to set guidelines. Hence, this approach will assist developers and UI testers in finding out usability problems automatically in early stages of UI development. The value of current study with respect to the previous achievements [14] is that we are focusing on the technical implementation of the solution for immediate usability evaluation of web UI, based on the earlier proposed framework.

The rest of the paper is organized as follows. In Sect. 2 we discuss related works of the research area. Section 3 provides an architecture of proposed framework, while Sect. 4 presents the evaluation of the tool itself. In Sect. 5 we present the work in progress, and, finally, Sect. 6 draws conclusions.

2 Related Works

Essential part of every automated usability evaluation tool (including our solution) is a set of guidelines against what the UI is to be evaluated. Web Content Accessibility Guidelines (WCAG) [10] and Sect. 508 Standards for Electronic and Information Technology [11] are technical standards providing guidelines that explain how to make web content more accessible to target users including people with disabilities. In fact, WCAG and Sect. 508 standards contain quite similar and partly covering accessibility guidelines. Web accessibility as an attribute through which people with disabilities can perceive, understand, navigate, and interact with the web, and they can contribute to the web [15]. Nevertheless, accessibility is only a certain subset of usability. Many other categories like home page, navigation, content organization guidelines are not covered by standards. That is the reason why many researches aim to establish usability guidelines covering certain elements of UI [2, 3] [12, 13]

Automated usability evaluation is very perspective area of research having multiple advantages over approaches involving potential users in usability evaluation like reducing the costs of usability evaluation and increasing the coverage of UI areas [16, 17]. Dingli proposed a framework that is effective in identifying usability aspects that violate usability guidelines [16]. The main drawback of the approach is that it cannot evaluate visual aspects of web application. Another disadvantage of the proposed solution is the sophisticated way of adding new usability guidelines using Guideline Definition Table.

Multiple solutions are concentrating on checking compliance of UI to accessibility standards [18, 19]. The main disadvantage of these tools is that it is not possible to evaluate visual aspects of a web application like measuring the contrast rate of UI element and the positions of elements on the screen. The solutions are based on validation of HTML syntax against WCAG guidelines. In order to evaluate fully functional UI, it is needed to apply CSS styles and Javascript scripts to the parsed HTML.

Analyzing user's behavior and reusing the knowledge of user with the purpose of providing more usable UI is also a promising research direction. There is a category of tools predicting the usage of the UI based on the knowledge discovery approach [20, 21]. Boza et al. presented a heuristic approach based on data mining techniques with the purpose to determine relationships between UI components and discover possible problems [21]. Using data mining in combination with mathematical algorithms, they generated rules based on the analysis of test reports. E.g. when "the site prevents users from making mistakes" then "error messages are written in the user language". Preliminary results indicate that the approach is viable to discover patterns and relationship between different UI components. In common, such approaches cannot guarantee high accuracy of evaluation results, having misleading results because of the used algorithms [22].

An essential output of each evaluation tool is a report providing feedback about UI compliance to predefined usability guidelines. There are multiple types of research analyzing the structure of reports containing usability defects with the purpose to improve existing format [23–25]. Yusop et al. surveyed practitioners in industrial software organizations and in open source communities about their usability defect reporting practices [23]. Their research showed that usability reports should contain at least the next information: title/summary, steps to reproduce, observed result and expected result.

In terms of our previous research in the field of UI usability, we have formulated the problem of assessing application UI conformance to usability guidelines and designed a framework that could be used to solve the addressed problem [14]. This framework enables to tackle a large set of usability issues at once already during UI development and saves cost and resources in later system development phases, especially in testing. The value of the current study with respect to the previous achievements is that we propose a fully functional tool for design-time usability evaluation of web UI. Our purpose is to provide overview of all developed components emphasizing technical implementation details.

3 Tool for Immediate Usability Evaluation

3.1 Tool Overview

We designed a high-level architecture to separate different aspects of software functionality into self-sufficient software modules. Figure 1 presents the high-level architecture of the tool consisting of four primary components: ontology, ontology processing engine, UI evaluation engine and client-side application. All the components will be described in details in next Sections.

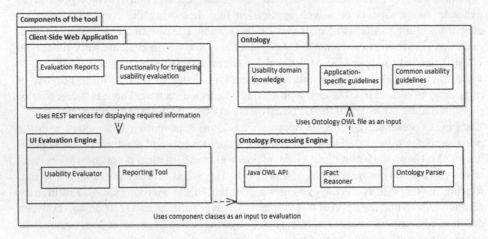

Fig. 1. High-level architecture of the tool

A core element of the tool is an ontology containing descriptions of usability guidelines that can be machine-processed on a particular web user interface. Such approach simplifies sharing of metrics and concepts between various guidelines and between different groups. The ontology processing engine is a mediator component responsible for errorless communication between the ontology and user interface evaluation engine. The UI evaluation engine is a component that assesses UI conformance to usability guidelines containing reporting component responsible for generating usability evaluation report. Client side web application contains easy to use UI for managing guidelines and triggering the evaluation process.

Initially, usability guideline is created by the mean of ontology. Then, Ontology Processing Engine transforms the guideline to the format understandable to UI Evaluation Engine. Afterwards, the process of automatic evaluation is triggered checking whether UI is accessible from the browser and whether there is a predefined set of guidelines to be processed. Then, the evaluation of UI is performed according to the set of guidelines. Finally, a report is provided to the developer containing the conformance of UI to the usability guidelines.

The proposed solution supports HTML 4 and 5, CSS 3 (providing backwards compatibility with previous versions) and Javascript based user interfaces. It does not require additional adoptions for web user interfaces allowing evaluation of any web UI without any extra configurations. The only limitation is that UIs requiring specific environments to run (e.g. Flash and Java Applets) are not supported by the tool.

3.2 Ontology-Based UI Guidelines Description

An ontology is a description (like a formal specification of a program) of the concepts and relationships that can exist for an agent or a community of agents [26]. Ontologies describe classes, attributes, relations presenting knowledge formally as a set of concepts within a domain and their relations. In order to design usability domain knowledge, it is necessary to understand, which domain concepts usability contains and how the hierarchy between concepts could be designed.

The use of ontology provides powerful means to capture usability domain knowledge including various sets of guidelines. In fact, ontology allows automated reasoning over the domain knowledge allowing automatic categorization. Ontology provides an effective way to describe and store usability domain knowledge through various aspects, describing the types of concepts that exist in the domain, and their properties and relations. All things considered, the application of ontology turns our tool to be more flexible and coherent in re-using designed domain knowledge where is needed.

An alternative approach to the ontology would be adopting custom metalanguage. In a first prototype of the tool we used custom metalanguage, but it turned out that it is very limited in expressing the guidelines and definition of the guidelines eventually became very complex as there is no good way of having an overview of all defined elements and their relations.

We used Web Ontology Language (OWL) as a knowledge representation language and open source feature rich Protégé ontology editor 5.1.[5] for creating the ontology.

Our ontology defines only those usability guidelines that can be automatically evaluated. The ontology contains WCAG and Sect. 508 guidelines including guidelines involved with people with disabilities as well as common usability guidelines. Initially, the guideline should be expressed as thorough and detailed as possible, so the conversion them into an ontology could be done with little effort. The guideline should contain the element, attribute or property being tested and the acceptance condition. The appropriate example of the guideline defined in a human readable way is: *The visual presentation of link text and background colour behind it should have a contrast ratio of at least 4.5:1*. To evaluate that guideline, the actual contrast rates of all link on UI will be compared to the contrast rate defined by the guideline.

Presently the ontology is used only for storing usability domain knowledge. It does not perform any kind of evaluations of UI conformance to the guidelines, however, based on available descriptions this could be achieved. The evaluation process of UI is carried out by the UI evaluation engine (see Sect. 3.3 for more details).

Figure 2 presents a segment of the ontology structure showing a *Link* class including its subclasses and dependencies with other concepts. *Link* is a *UIElement*

---

[5] Protégé, http://protege.stanford.edu/.

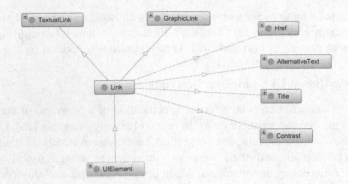

Fig. 2. A segment of the ontology structure: excerpt from the Protégé ontology editor

having attributes *Title*, *Href*, *Contrast*, *AlternativeText* and *Unit*. There are two possible *Link* types: *TextualLink* and *GraphicalLink*. Similar connections are defined for various elements of UI.

Usability guidelines are defined in ontology as subclasses of class *Guideline*. For instance, to compose a new usability guideline for defining contrast rate of links, we should create a new subclass of *UsabilityGuideline*. *UsabilityGuideline* has object property *hasGuidelineElement* which defines the element being evaluated. Statement '*hasGuidelineElement only Link*' shows that the defined guideline is used to evaluate only links. Figure 2 demonstrates that *Link* element has object property *Contrast*. Contrast in its turn is defined as '*hasContrast some xsd:integer*' showing that we can set a value of *Contrast* to an integer. Every guideline has a number of required annotations to be filled: guideline - contains a brief name of usability guidelines; reference - contains the URL or path to the source of the guideline; rdfs: comment – contains the full description of guideline to be evaluated. The design of this ontology has been described in our previous paper [27], and thereby we limit its discussion within this paper to the above.

3.3 User Interface Evaluation Engine

The UI evaluation engine is a component that assesses UI conformance to usability guidelines. Figure 3 shows the relation between components of the tool. The ontology described in Sect. 3.2 is used as an input for Ontology Processing Engine. Before evaluating UI, we should transform usability guidelines (individuals) defined in ontology by the mean of OWL Web ontology language (in XML format) to format understandable for the UI evaluation engine. The transformation is required because processing guidelines in native OWL format is not trivial due to the complicated API of OWL language changing the application structure to unclear and opaque. OWL API [28] library has been used for serializing usability ontology into appropriate Java classes. We used JFact for reasoning over the domain - a Java port of FaCT ++ reasoner [29] having full compatibility with the OWL API library.

Ontology Processing Engine provides API for retrieving usability guidelines in a format understandable to Guideline Evaluator (see Fig. 3). Guideline Evaluator

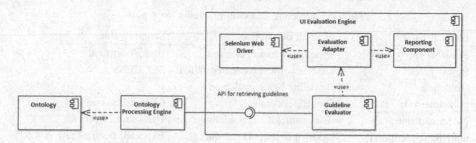

Fig. 3. UML Component diagram representing components of the UI Evaluation Engine and dependencies to other components

retrieves guideline to evaluate, processes guideline identifying the element being evaluated (e.g. Link, Button) and calls the corresponding evaluation adaptor (e.g. LinkAdapter, ButtonAdapter) providing the properties of the element as a parameter to the adapter (e.g. contrast $> = 4.5$). The corresponding adapter retrieves the actual values using Selenium Web Driver and checks if the evaluated values of properties correspond to the value defined in guideline. In other words, if we are evaluating the contrast rate of Link then the LinkAdapter asks Selenium Web Driver to process all Links on the Web Page and provide their contrast rates. Afterwards, the adapter checks if the values returned by Selenium Web Driver correspond to the value defined in guideline. If the contrast rate of all Links corresponds to the contrast rate in the guideline then the success response is generated, otherwise failure response is generated and added to the report.

Thereby, the core mechanism of processing the Web User Interface is Selenium WebDriver API[6]. Selenium is used for automation of UI tests providing simple and concise programming interface. Selenium has full support for most programming languages (Java, C#, Python, Javascript and others), being compatible with most popular browsers (Chrome, Firefox, Internet Explore and others). Selenium WebDriver provides rich API commands and operations containing an interface for fetching a page, locating UI elements on the screen, filling in forms and many other operations.

An essential part of the evaluation engine is the reporting component responsible for generating usability evaluation results. Table 1 presents the structure of usability evaluation report containing all relevant information. Figure 4 and 5 demonstrate the screenshots of reports containing evaluation results.

3.4   Client Side Web Application

Client side web application contains easy to use UI for managing guidelines and triggering the evaluation process. The potential users of proposed solution are technical (including developers and quality assurance specialists) and business users (including analysts and product owners). The primary deliverable for UI developers is a library providing the API for evaluating UI conformance to the guidelines on a local or remote machine.

---

[6] http://www.seleniumhq.org/projects/webdriver.

Table 1. Structure of usability evaluation report

Element name	Element description	Sample value
elementType	Type of evaluated element	Link, Text field
Result	Identifies if test passed or failed	SUCCESS
Guideline	Guideline being evaluated	
guideline.code	Guideline identifier	03-FluidLayout
guideline.name	Brief name of the guideline	Use Fluid Layout
guideline.description	Detailed description of the Guideline	Expected contrast rate is 4.5
failedElements	Elements violating guideline.	
failedElements.path	Path for downloading screenshot.	screenshot42.jpg
failedElements.description	Reason of failure	Actual contrast rate is 3.7

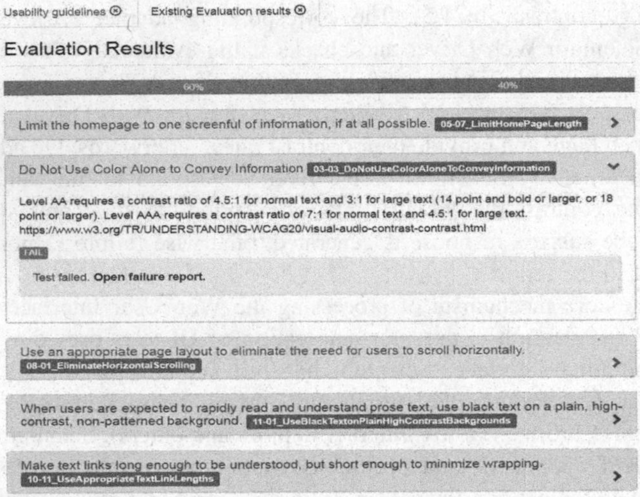

Fig. 4. Screenshot of the view containing usability evaluation results

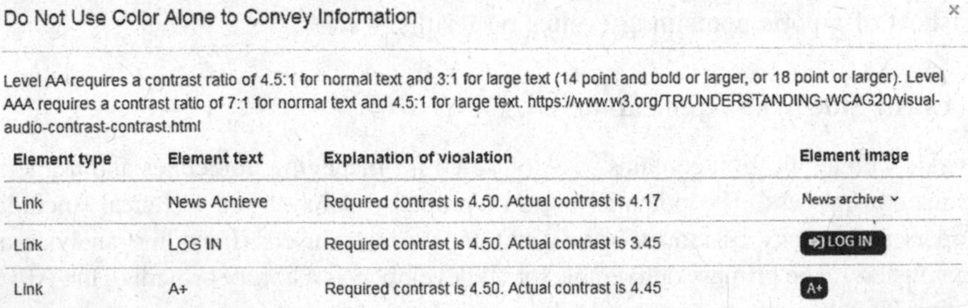

Fig. 5. Screenshot of dialog presenting additional information about a failed guideline

Business users commonly do not have required technical background to run UI tests from the code. That is why a web application has been designed containing visual functionality for managing guidelines and triggering the evaluation process. Figure 4 presents a screenshot of the view containing usability evaluation results. Violated guidelines are highlighted in red color; passed guidelines in green color. Evaluation results provide full information of evaluated guideline including name, code and description. By clicking the link Open Failure Report opens dialog containing image, text and description of elements violating the guideline (see Fig. 5).

The Client-Side Web Application has been build based on Angular[7] Javascript framework using Bootstrap[8] stylesheets containing an extensive list of components for designing client web applications.

4   Tool Evaluation

The purpose of the section is to validate the ontology design, and compare how the UI evaluation engine we developed stands out against existing automated usability evaluation tools - USEFul [16] and Mauve [18]. These tools are not suitable for design-time usability evaluation but are rather used in pre-release testing. Nevertheless, these tools were selected for comparison as they are capable of evaluating Web UI conformance to many usability guidelines with certain limitations discussed further. In fact, there is a detailed description of UI evaluation approaches used in both tools.

4.1   Evaluation of Ontology Design

In order to verify the ontology design and to check that it fulfills its intended purpose within then tool, we selected 115 automatically testable usability guidelines from different categories. The test suite has been composed analyzing recommendations in scientific publications [2, 3], WCAG [10] and Sect. 508 [11] standards and guidelines provided by U.S. Department of health and human services (HHS)[9]. Moreover, multiple evidence-based user experience and usability researches have been inspected [12, 13]. Table 2 contains categories of combined usability guidelines with the number of guidelines for each category. All guidelines were defined with proposed ontology proving that proposed ontology provides enough concepts like object, object properties, data properties to define custom usability guidelines of various types. Users can combine application-specific test suites including guidelines from different categories and standards.

---

[7] Angular, https://angular.io/.
[8] Bootstrap, http://getbootstrap.com/.
[9] HHS.gov U.S. Department of Health & Human Services, http://www.hhs.gov/.

Table 2. Categories of usability guidelines with number of guidelines defined

Category of guideline	Number of guidelines
Accessibility and Compatibility	15
The home page and Search	18
Page layout and Navigation	22
Scrolling, paging and Links	14
Heading, titles and labels	4
Text appearance and Lists	14
Screen-based controls	12
Graphics, Images, and Multimedia	6
Organisation of information and Content	10

4.2   Comparison of UI Evaluation Engines

The UI evaluation engine is a component that assesses UI conformance to usability guidelines. The UI evaluation components of existing similar solutions (like USEFul [16] and Mauve [18]) are based on the parsing of HTML code of Web page verifying that the certain value of HTML tag or element is the same as defined in a guideline. Let's compare, how the evaluation is performed in these tools taking a simple guideline as an example: *Images that are presented to the user have a text alternative that serves the equivalent purpose.* Technically, we should verify that every *img* tag contains attribute *alt*. The sample image element in HTML code looks like:

```
< img src = "/img/logo.png" alt = "Eesti.ee">
```

Both USEFul and Mauve are parsing HTML code as a String extracting all elements with tag *img* and, afterwards, checking if each element has an *alt* attribute. We are using Selenium Web Driver (discussed in Sect. 3.3) that has also support for processing HTML code. For instance, in order to get all *img* elements we call the next code `seleniumDriver.findElements(By.tagName("img"))`. Afterwards, we check if all *img* elements contain the *alt* attributes `img.getAttribute ("alt")`. Overall, Selenium covers all possible variations of parsing HTML making the parsing process simpler by the mean of clear API.

In common, aforementioned tools are limited by finding deviations only in HTML code; evaluating visual aspects of a web application like measuring the contrast rate of UI element or evaluating the positions of elements on the screen and many other assessments are not possible with that approach. As far as our solution is based on Selenium Web Driver, the evaluation of such aspects is also possible with our tool. For instance, the next guideline could not be evaluated by any of aforementioned tools: *Horizontal scrolling is not allowed,* as it is not possible to verify the existence of horizontal scrolling only by processing HTML code. Our tool is capable of evaluating the guideline using Selenium *JavascriptExecutor* that helps to detect if scrolling is allowed and scroll the page horizontally if it is possible.

To sum up, the UI evaluation engine is capable of evaluating various guidelines that require parsing of HTML code. Moreover, the guidelines evaluating the element position, color scheme or scrolling possibility are also supported by the tool covering more usability guidelines than other discussed usability evaluating tools.

5 Work in Progress

The proposed tool has been introduced to the Estonian Information Systems Authority[10] (RIA) with the purpose to integrate the solution established based on the framework to evaluate the usability of Estonian eGovernment components. RIA coordinates the development and administration of Estonian National Portal[11]. Estonian National Portal contains approximately 100 pages for testing including various components such as pages for citizens, entrepreneurs. Following usability guidelines for public service portals is extremely important as they provide services for all inhabitants of Estonia, but not only, having a different experience in using user interfaces and various devices.

In cooperation with RIA quality assurance team and designers, we concluded that integration of the framework for the evaluation of UI compliance to usability guidelines (especially WCAG guidelines) is beneficial for RIA. Based on the feedback provided we extended designed ontology by dividing all guidelines into subgroups like WCAG and Sect. 508 guidelines. Such approach allows evaluating UI conformance only to the selected group of guidelines, not to all of them.

6 Conclusions

Usability is very capricious to any modifications of UI elements because even a minor change of UI could lead to severe usability problems. The appropriate example is that appearance of horizontal scrolling on desktop devices is one of the few interactions generating negative responses from users decreasing the usability of UI. Another example could be that changing the foreground or background color of any UI element like link or button potentially could reduce the accessibility for users with low vision or cognitive impairments. All things considered, it is important to provide feedback to developers concerning usability problems as soon as possible.

The paper is addressing the gap between design and assessment phases of UI development proposing the tool for design-time automatic usability evaluation of UI conformance to the category-specific usability guidelines during UI design and implementation stage. Currently, there are no any other tools providing design-time automatic usability evaluation.

Existing solutions for automated usability evaluation cannot be integrated to the implementation stage of UI, structuring their solutions in a way that could be used only in pre-release testing. The advantage of our solution is that it fits into the implementation stage of UI development and allows automatic validation of introduced changes and their conformance to guidelines. Afterwards, the tool provides immediate feedback informing UI developer, how the changes affected UI conformance to the usability guidelines.

---

[10] RIA, https://www.ria.ee/en/.

[11] Estonian National Portal, https://www.eesti.ee/eng.

The proposed tool is capable of evaluating HTML and Javascript based web UIs without additional configurations. Based on the analysis of WCAG, Sect. 508 standards as well as on the UI best practices and recommendations introduced in scientific publications and various usability researches, the test suite containing 115 usability guidelines was prepared. All guidelines were described by means of ontology. In fact, the tool is not limited to the predefined set of guidelines, it also allows creating custom application-specific usability guidelines.

Existing solutions for automated usability evaluation mostly include the guidelines evaluating the structure of HTML code. Evaluating the visual aspects like color and contrast of UI elements, position of content within the page or the use of horizontal or vertical scrolling is not possible with these tools. We overcome these limitations by introducing UI evaluation engine based on Selenium allowing evaluating HTML code based guidelines as well as guidelines checking visual consistency of UI elements.

Acknowledgements. This research was supported by the Estonian Ministry of Research and Education institutional research grant no. IUT33-13. Authors are very thankful to Estonian Information Systems Authority team for governmental portal www.eesti.ee consulting.

References

1. International Organization for Standardization: ISO 9241-210:2010 Ergonomics of human-system interaction Part 210: human-centred design process for interactive systems (2010)
2. Smith, S.L., Mosier, J.N.: Guidelines for Designing User Interface Software. Bed-ford, Mitre (1986)
3. Borges, J.A., Morales, I., Rodriguez, N.J.: Guidelines for designing usable World Wide Web pages. In: Conference on Human Factors in Computing Systems, pp. 277–278. ACM, New York (1996)
4. Mack, R.L., Nielsen, J.: Usability Inspection Methods. Wiley, New York (1994)
5. Kock, E., Biljon, J., Pretorius, M.: Usability evaluation methods: mind the gaps. In: Proceedings of the 2009 Annual Research Conference of the South African Institute of Computer Scientists and Information Technologists, pp. 122–131. ACM, New York (2009)
6. Ivory, M.Y., Hearst, M.: The state of the art in automating usability evaluation of user interfaces. ACM Comput. Surv. **33**, 470–516 (2001)
7. Bak, J.O., Nguyen, K., Risgaard, P, Stage, J.: Obstacles to usability evaluation in practice: a survey of software development organizations. In: Proceedings of the 5th Nordic conference on Human-computer interaction, pp 23–32. ACM, New York (2008)
8. Häkli, A.: Introducing user-centered design in a small-size software development organization. Helsinki University of Technology, Helsinki (2005)
9. Lizano, F., Sandoval, M.M., Bruun, A., Stage, J.: Is usability evaluation important: the perspective of novice software developers. In: The 27th International BCS Human Computer Interaction Conference, Article 31, British Computer Society, Swinton (2013)
10. Web Content Accessibility Guidelines. http://www.w3.org/WAI/intro/wcag
11. Section 508. https://www.section508.gov/
12. User Experience for Mobile Applications and Websites. https://www.nngroup.com/reports/mobile-website-and-application-usability/

13. Navigation and Page Layout. https://www.nngroup.com/reports/intranet-navigation-layout-and-text/
14. Marenkov, J., Robal, T., Kalja, A.: A framework for improving web application user interfaces through immediate evaluation. In: Databases and Information Systems, pp. 283–296. IOS Press, Amsterdam (2016)
15. World Wide Web Consortium W3C, 2010b. Web Accessibility Initiative (WAI). http://www.w3.org/WAI/intro/accessibility.php
16. Dingli, A.: USEFul: a framework to mainstream web site usability. Int. J. Hum. Comput. Interact. **2**, 10–30 (2011)
17. Dingli, A., Cassar, S.: An intelligent framework for website usability. Adv. Hum Comput Interact. (2014). Article 5
18. Schiavone, A.G., Paterno, F.: An extensible environment for guideline-based accessibility evaluation of dynamic Web applications. J. Univ. Access Inf. Soc. **14**, 111–132 (2015)
19. Leporini, B., Paterno, F., Scorcia, A.: Flexible tool support for accessibility evaluation. Interact. Comput. **18**(5), 869–890 (2006)
20. Davis, P.A., Shipman, F.M.: Learning usability assessment models for web sites. In: Proceedings of the 16th International Conference On Intelligent User Interfaces, pp. 195–204. ACM, New York (2011)
21. Boza, B.C., Schiaffino S., Teyseyre A., Godoy D.: An approach for knowledge discovery in a web usability context. In: Proceedings of the 13th Brazilian Symposium on Human Factors in Computing Systems, pp 393–396, Porto Alegre (2014)
22. Winckler, M.A.A., Freitas, C.M.D.S., De Lima, J.V.: Usability remote evaluation for WWW. In: CHI 2000 Extended Abstracts on Human Factors in Computing Systems (CHI EA 2000), pp. 131–132. ACM, New York (2000)
23. Yusop, N.S.M.Y., Grundy, J., Vasa, R.:Reporting usability defects: do reporters report what software developers need? In: Proceedings of the 20th International Conference on Evaluation and Assessment in Software Engineering, pp. 1–10. ACM, New York (2016)
24. Davies, S., Roper, M.: What's in a bug report? In: Proceedings of the 8th ACM/IEEE International Symposium on Empirical Software Engineering and Measurement, Article 26. ACM, Torino (2014)
25. Yusop, N.S.M., Grundy, J., Vasa, R.: Reporting Usability Defects: Limitations of open source defect repositories and suggestions for improvement. In: Proceedings of ASWEC Australasian Software Engineering Conference, pp. 38–43. ACM, New York (2015)
26. OWL Web Ontology Language Guide, https://www.w3.org/TR/owl-guide/
27. Robal, T., Marenkov, J., Kalja, A.: Ontology design for automatic evaluation of web user interface usability. In: PICMET 2017 Conference: "Technology Management for Interconnected World", Portland, USA. 9–13 July 2017 (2017, in press)
28. Horridge, M., Bechhofer, S.: The OWL API: A Java API for OWL ontologies. Semantic Web **2**, 11–21 (2011)
29. Tsarkov, D., Horrocks, I.: FaCT ++ description logic reasoner: system description. In: Proceedings of the Third International Joint Conference, pp. 292–297. Springer, Heidelberg (2006)

Genotypic Data in Relational Databases: Efficient Storage and Rapid Retrieval

Ryan N. Lichtenwalter[(✉)], Katerina Zorina-Lichtenwalter,
and Luda Diatchenko

McGill University, Montreal, QC H3A0G4, Canada
ryan.lichtenwalter@mcgill.ca
https://mcgill.ca

Abstract. As technologies to produce genotypic data have become less expensive, the widths and depths of such data have sharply increased. Relational databases have performed poorly in this domain. Data storage and retrieval is now mostly conducted by highly coupled and specialized software packages and file formats, but relational databases offer advantages if the domain challenges can be overcome. We revisit their feasibility as a tool for efficiently storing and querying extremely large genotypic data sets. We describe a technique for managing genotypic data in the PostgreSQL relational database, compare it to common existing techniques for storing and querying genotypic data, and demonstrate that it can greatly reduce both query times and storage requirements.

1 Introduction

Genomics is among the fastest-growing scientific fields in terms of data production [13]. One complete human genome represents many gigabytes of data, and in the five years between 2007 and 2012, the cost to sequence an individual genome decreased by a factor of one thousand [4]. Data integration is a significant challenge [5,6,12], and while relational databases offer expressive power, it is difficult to employ them successfully to manage genotypic data at current scales [12,15,17]. Genotypic data sets are dense matrices, with each element representing a small unit of information. Normalizing these matrices results in difficult design and maintenance, an explosion in storage requirements, and potentially long query times. Despite the limitations of relational databases in this domain, there is significant interest in managing genotypic data in them [1,5,8,12,14,15,17,18]. Nonetheless, when relational databases are used to manage genotypic data, it is very often either normalized or stored as long strings. We are only aware of one system that uses denormalized array-based storage [15].

We propose and explore the array-based technique, show that it scales to far greater volumes than existing techniques, and extend it to incorporate probabilistic genotype data. The technique supports fast queries within hundreds of billions of data points and scales to tens of trillions of data points with modest hardware available to the typical research lab. It reduces storage requirements by several orders of magnitude versus normalized representations.

© Springer International Publishing AG 2017
M. Kirikova et al. (Eds.): ADBIS 2017, LNCS 10509, pp. 408–421, 2017.
DOI: 10.1007/978-3-319-66917-5_27

2   Domain Background

The genetic information for an individual is biologically stored as a long sequence with an alphabet of four nucleotide bases that combine to form two strands of DNA. Each human has a genome of about 3 billion base pairs, but we are typically interested in analyzing only those that differ between individuals. These differing sequences are called *variants*, and approximately 100 million are known to exist. Figure 1 illustrates two copies of the same segment of a chromosome with 5 base pairs and a single-nucleotide variant.

The genotypes of individuals are determined either by sequencing and variant calling their complete genomes or by genotyping a selection of variants with a microarray. The resulting information for an individual is called a *sample*. The specific base pair or sequence of base pairs that an individual possesses for a given variant is called an *allele*. At most variant loci, humans carry two alleles, one on each homologous chromosome. Therefore, some analyses are chromosome-specific

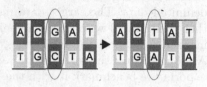

Fig. 1. Two copies of the same segment of a chromosome, including 5 base pairs and one single-nucleotide variant.

and take into account *phase*, which is a property that respects the allele pair permutation. Variant allele determination is a biochemical procedure that occasionally fails, resulting in a *missing call*.

Accounting for missingness and phase, there are 17 possible genotype values, and thus $\log_2(17) \approx 4.08$ bits are required to encode single-nucleotide alleles for one variant in the genome of an individual. We can instead store a *reference* allele and an *alternate* allele with other variant characteristics and store a genotype as the count of either reference or alternate alleles. Preserving phase, $\log_2(5) \approx 2.32$ bits are required, and 2 bits are required without phase.

3   Existing Storage Solutions

In this work, we use domain-specific tools as a frame of reference and focus on relational databases. We also examine design tradeoffs regarding whether the underlying genotype storage is row-oriented (sample-major) or columnar (variant-major).

3.1   File-Based Solutions

Many special-purpose file formats have been devised to manage genomic data. These formats are commonly used for storage, transmission, query, and analysis.

gzip and Shell Scripting. Storing gzip-compressed flat files and querying with shell utilities is a common practice and an informative baseline. We compress a sample-major tab-delimited flat file sorted row-wise by sample and column-wise by variant with `gzip -6`. We then perform sample-based filtering using `head`, `tail`, and `sed -n 's_1p;s_2p;...s_np;'`, where s is a sample vector, and variant-based filtering with `cut -f 'v_1,v_2,...,v_n'`, where v is a variant vector. Indices in s and v come from `grep -w '...'` operations on dictionary files.

tabix Indexing. The `tabix` utility [10] is a tool for indexing and querying tab-delimited data. The target data must first be sorted by genomic position and then compressed with the companion `bgzip` utility. The resulting file is then indexed to support variant queries, so filtering samples must be done with `cut`. Although `tabix` supports a wide range of tab-delimited formats, including SQL exports, we benchmark it with the VCFv4.2 file format.

PLINK Data Management. The `PLINK` utility [16] is a staple of genetic association studies. It does various statistical analyses, performs format conversions, and supports basic sample and variant querying. Its primary on-disk and in-memory format is a compact binary representation. Despite its power and efficiency, PLINK is limited in the types of data that it is designed to store and integrate. Each invocation reads all data into memory from disk before performing the requested operations and writing output to one or more files.

3.2   Relational Database Solutions

Several solutions have been deployed for storing genomic data in relational databases. One uses the database as a container to store bioinformatic file formats and uses specialized functions to operate on these objects [17]. Some are mainly used for storage and wholesale exports of data and are not suitable for analytical queries. We are interested in a solution amenable to both efficient storage and querying. We chose the open-source PostgreSQL 9.4 relational database management system because it is freely accessible to researchers, possesses a powerful and informative query planner, supports compressed storage of large, variable-length attributes outside of primary table storage, and provides for easy extension with user-defined types and C functions. It is a common choice for users of relational databases in this domain [5,12,14].

We avoid duplicating the sequences that correspond to reference and alternate alleles in the encoding of each genotype by storing them separately. Genotypes are stored as the count of either reference or alternate alleles and, optionally, information about phase. We employ the PostgreSQL `TINYINT` extension, which provides 1-byte integers, to represent the count of alternate alleles compactly in the coordinate list and array-based strategies below. We also tried using the 4-byte PostgreSQL `ENUM` type and a 1-byte user-defined type written in C to store allele nucleotides instead of alternate allele counts, but these types result in larger disk footprints (334% and 218% as large, respectively) and slower queries.

Normalized Coordinate Lists. The pure 5NF relational method of storing genotypic data associates each genotype with a composite key comprising the sample and variant identifiers. In sparse matrix terms, this is called the coordinate list representation, and it is the standard 5NF approach for storing matrices. This method allows foreign key integrity checks, rapid DML write operations, and conventional joins against both sample and variant tables. Unfortunately, it produces long, narrow tables with row counts proportional to the Cartesian product of samples and variants. This causes several problems. First, it incurs a significant per-row overhead. Assuming 4-byte foreign keys for the sample and variant, and 1 byte to represent the genotype, then overhead for this table in PostgreSQL 9.4 is 35 bytes per row: a 4-byte pointer in the page header, 23 bytes for the heap tuple header, a 1-byte NULL bit mask or pad, and a 7-byte row tuple alignment pad to achieve 8-byte alignment. Storing a modest data set of just 1000 samples and 1 million variants requires 41 GB of space discounting indices. Second, even excluding row overhead, 89% of the space is occupied with duplicated foreign keys. Third, sample and variant locality may be lost, so even fortuitous queries requesting subsets of the table require full table scans unless specific clustering and partitioning is carefully maintained. Fourth, increased data size decreases file system cache effectiveness. Fifth, indices become too unwieldy to perform well without table partitioning. Sixth, it is more difficult to take advantage of the trivial compressibility that exists in this domain when the data is interspersed with key values.

Despite these limitations, this is a common approach employed when using relational databases to manage genotypic data, including several published methods and scientific research tools [5,12,17]. In our benchmark tests with this storage method, we range-partitioned tables by variant, indexed with the composite variant-sample primary key, and clustered on the index. This is contrary to the sample-based partitioning used in other systems [12] but consistent with our strategy of prioritizing rapid filtering of the variant space, which facilitates variant-level and gene-level association studies. This decision comes at the cost of expensive sample-based operations. Finally, although PostgreSQL does not perform row-level or page-level compression for fixed-length inline table data, using such compression may reduce the size requirements of the coordinate list representation by roughly 50% [17].

PostgreSQL HSTORE Storage. In the HSTORE organizational scheme, one of the dimensions is stored in an associative data structure. Like coordinate lists, this is a reasonable representation for sparse matrices, and the PostgreSQL HSTORE type conveniently supports efficient, native multiple-indexing operations via the HSTORE->TEXT[] operator. Nonetheless, genotype data is not sparse. Although collapsing one dimension into an HSTORE reduces the per-row overhead of a coordinate list table, it suffers two serious drawbacks. First, it must store both indices and values, so it avoids key replication in only one dimension. Second, it stores both of these as strings, greatly inflating key footprint.

SQL VARCHAR Storage. It is fairly common in genomic data warehouses that employ relational databases to store the data as one genotype string per sample. For instance, genotypes [AC,TG,AA] for a sample would be stored as the string "ACTGAA". This obviates the per-row overhead of coordinate list storage but requires 16 bits of storage per single-nucleotide variant, which is almost 8 times the theoretical minimum requirement for storing phased genotypes. Storing genotypes as long strings also has the disadvantage of either denying the possibility of variants involving sequences of different length or denying the use of consistent variant addressing. Finally, although the string allows for random access once it has been read into memory, the syntax is cumbersome and non-contiguous selections require separate invocations of the SUBSTRING function.

4   The Array Storage Strategy

Each of the preceding representations has serious limitations, whether a lack of relational database capabilities with file-based tools and analytical suites, or crippling performance issues with common relational database strategies. We explore a relational database organization similar to that of [15], which stores and queries genomic data far more efficiently than overwhelmingly employed relational database methods. It requires only 0.003% as much space as the often-used coordinate list method for our benchmark data, offers superior performance across a broad spectrum of query types, and stores and queries data as efficiently as the best of the file-based tools while providing the full expressive power of SQL.

4.1   Non-contiguous Multiple Indexing

PostgreSQL arrays allow organization of data into an arbitrary number of dimensions and come with sophisticated primitives to manage this complexity. Given an array A of n elements, indexing is achievable with A[i], where i is a 1-indexed address in the array. Furthermore, it is possible to select slices of arrays as A[i:j], where i and j specify the lower and upper bounds of the slice. PostgreSQL provides no efficient way to access multiple non-contiguous elements in its arrays. The standard way to access non-contiguous elements is as multiple fields in the SELECT clause (i.e. SELECT A[i1], A[i2], A[i3], ...). Each element selected in this manner causes the implementation to rescan the array. For non-contiguous subsets of indices approaching the length of the array, this results in $O\left(n^2\right)$ behavior.

Another method for selecting multiple elements is performing a set-returning UNNEST WITH ORDINALITY operation on the array. This provides the array elements with their associated indices. We can then select an arbitrary subset of the elements by joining and filtering the indices with other data. Finally, we can use the `array_agg` function to recompose the result into an array that contains only the elements to select. The entire procedure can be neatly wrapped in a SQL function, and it operates in $O(n)$ time. Despite the superior asymptotic performance, this method is slow, and it is easily outperformed by selecting array elements as multiple fields when the number of fields to access is small.

Fortunately, it is possible to write a C function, `array_multi_index`, to overcome this difficulty. It accepts an array into which indexing should be performed and a second array of indices to use. We strove to maintain consistency with other PostgreSQL array primitives and allow as much generality as possible: the function returns NULL for a NULL value array or index array, returns NULL array values corresponding to NULL or out-of-bounds indices, returns repeated values corresponding to repeated indices, and supports arbitrary index ordering. Since the underlying back-end implementation of PostgreSQL arrays is a C array, once the array is read from storage and unpacked, the only penalties involved in accessing elements are related to cache size and data locality. This function is therefore only slightly slower than native single-indexing via `A[i]`.

4.2   Array Organization

With array storage of genotypic data, each row contains a foreign key and one or more arrays. Structuring the `genotype` table so that each row references the `sample` table and contains all variant genotypes for a single sample gives sample-major organization. Structuring it so that each row references the `variant` table and contains all sample genotypes for a single variant gives variant-major organization, as in Fig. 2. The planner can take advantage of indices on whichever dimension is supplied with a foreign key, and this dimension supports standard SQL join syntax. Each method offers advantages and disadvantages, some of which depend on whether there is roughly the same number of samples as variants, and some of which exist irrespective of the relative size of the dimensions.

Genetic studies often have many variants and relatively few samples. Since PostgreSQL arrays only offer random access after the complete array is loaded into memory, array-based organizations only allow true random access by row. As a result, we often stand to achieve significant performance gains by choosing variant-major storage. The `genotype` table is longer and the arrays are narrower, which offers greater benefits from indices and faster operations over arrays. Further, genetic studies are often performed as either genome-wide association studies, thus requiring all variants for analysis, or per-gene or per-variant studies, thus requiring only a miniscule percentage of variants. Indices on the `variant` table can quickly winnow the set of variants, and the join can take advantage of the foreign key index on the `genotype` table so that it is only necessary to read a tiny fraction of the genotypic data. While genetic studies also frequently analyze a subset of samples according to demographic characteristics or phenotype availability, these subsets typically represent a significantly higher percentage along that dimension, and it is efficient to accomplish this join virtually through random access in the narrow arrays of samples. Finally, regardless of the `sample` and `variant` table cardinalities, many variants have a low alternate allele frequency, so many variant-major arrays contain homogeneous data, which allows for superior data compression and a smaller disk footprint. For these reasons, variant-major organization is used by several genotypic data representations [3, 9, 16].

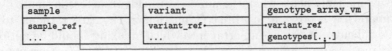

Fig. 2. Variant-major array organization.

Though making use of variant-ordering may slightly increase compressibility [2], for most genotypic data sets, sample-major storage is not ideal, because it results in wide arrays that are slower to access. We can minimize this penalty by dividing the single array of genotypes into multiple arrays, or when table column limitations are exceeded, even into multiple tables, but this undesirably and vastly increases schema complexity and requires dynamic SQL or long duplicated SQL expressions. Computing genomic statistics per sample over a set of variants is fast but is an infrequent operation. Sample-major organization has one significant advantage: it is better aligned with the natural and intuitive way of organizing and analyzing research data sets. It is convenient to join sample-major genotypic data with its corresponding subject-major phenotypic data to obtain one wide subject-major matrix. Fortunately it is possible to perform matrix transpose operations on variant-major query results efficiently with domain-aware tools such as PLINK, generic tools such as GNU datamash, or even inside the database.

Listing 1. Coordinate List Query

```
SELECT variant_ref, array_agg( genotype ORDER BY sample_ref )
FROM genotype_normalized
WHERE sample_ref IN (SELECT sample_ref FROM sample WHERE ...) AND
variant_ref IN (SELECT variant_ref FROM variant WHERE ...)
GROUP BY variant_ref
```

Listing 2. Variant-Major Array Query

```
SELECT variant_ref, array_multi_index( genotypes,
(SELECT array_agg(sample_ref) FROM sample WHERE ...) )
FROM genotype_array_vm
WHERE variant_ref IN (SELECT variant_ref FROM variant WHERE ...)
```

4.3 Alternative Querying

The array strategy requires only minimal changes to coordinate list queries. Listings 1 and 2 show queries that return identical results. We use the IN operator instead of a JOIN only to explicitly show the location of the filter condition for each dimension. The difference in the array organization query is that the result of applying the selection criteria for the dimension stored in the array must be aggregated into an array itself and passed as a parameter to the array_multi_index function. Note that we will typically want the two dimensions to be projected into rows and columns, so the coordinate list query incurs the additional cost of aggregation, which is an incidental advantage of array organization.

The principle is the same for arbitrarily complex queries. The subsets of samples and variants may be obtained with sub-queries, from SQL variables, using common table expressions, or through the use of joins with no restrictions on the broader structure of the query. The only requirement is that an implied foreign key relationship exists between array indices and some column in the related entity table. The queries in the listings above assume that this relationship exists in terms of the primary key of the entity table, but it could just as easily exist over a column established specifically to index into the array.

4.4 Sample and Variant Data Flexibility

Most genotypic data for genome-wide and targeted analyses are write-once [1], so traditional UPDATE operations are usually unnecessary. Sometimes additional samples or variants join the data set due to continued genotyping. The array-based storage scheme supports both insertions and updates. Insertions to the row dimension are accomplished with a SQL INSERT, but insertions and updates in the array dimension are more complicated. PostgreSQL arrays support SQL UPDATE and various concatenation and element assignment operations. Adding new data in the array dimension requires performing an array concatenation, and the indices of the new elements implicitly correspond to data within newly added rows in the parent table. The database cannot enforce referential integrity on the array in the traditional sense, but the indices of the array act as implicit foreign keys into the parent table. We can choose or create a prime attribute to act as a candidate key in the parent table and additionally specify that it should take the form of consecutive integers starting with 1. We can enforce referential integrity procedurally, and we can use typical SQL join and filter facilities in the form of Listing 2 to perform all the standard DML operations.

4.5 Array Operations

We must often filter variants based on information generated from an analysis over the sample space, such as *variant call rate* or *alternate allele frequency*. This a primary driver of our choice of variant-major organization. Variant call rate is the percentage of non-missing genotypes in a set of samples. As an indicator of data quality, it is often relevant only over the entire corpus of variants and so can be precomputed [15]. Alternate allele frequency, however, is sometimes useful to know for sample subsets too. Precomputation is infeasible, so it is desirable for the database to perform this computation efficiently with arbitrary sample subsets.

Listing 3. Alternate Allele Frequency Query

```
SELECT SUM(unnest)/(2*COUNT(unnest))::FLOAT
FROM UNNEST( genotypes )
```

To do this, we can return to the `UNNEST` technique highlighted in Listing 2. The necessary SQL code is simple and concise but slow. Because this is a common operation, we offer another C function that computes call rate and alternate allele frequency for both the complete set of samples and for the requested subset. It returns this information in a composite user-defined type, which is conveniently accessible by field.

We might also like to convert genotypes stored in arrays between alternate allele count and allele nucleotide sequence representations. We can do this with SQL, but when we must apply it over an entire array it again requires a SQL function involving the `UNNEST` and `array_agg` sequence of operations. We instead provide another small C function that performs this operation over the array and returns a `VARCHAR[]`.

5   Comparative Performance

We performed all experiments on a dedicated machine running CentOS 6.7 containing an Intel Xeon E5-2630v2 CPU with 2 cores enabled, 8 GB of RAM, and an 8-disk RAID-10 array of 7200 RPM hard disks. Our benchmark data consists of 3104 subjects with 2,567,845 variants for a total of 8.0 billion genotypic data points. Imputation generated probability tuples for 34,985,077 variants, which totals 108.6 billion tuples or 325.8 billion imputed genotypic data points.

Figure 3 shows the disk requirements for each storage scheme including necessary indices. For relational database representations, `VACUUM FULL ANALYZE` was performed after loading, and indices were generated with `FILLFACTOR=100`. This aligns with the typical write-once, read-only nature of this data.

The `gzip` size provides a baseline for space requirements. The PLINK textual entry indicates the size of a typical uncompressed representation of the data set. Storage requirements of the database representations vary widely. The coordinate list representation suffers from row overhead, key duplication, and lack of compression, and the index for each partition requires nearly half as much space as the table. The array-based representations are clearly smallest among the relational database storage options.

5.1   Query Times

We tested several queries representative of tasks necessary for downstream analysis. All random values for queries are identical across storage schemes. To simulate a production system with a warm file system cache, we ran queries against each representation prior to benchmarking it. For the relational database tests, we used output from the PostgreSQL query planner (i.e. `EXPLAIN ANALYZE`) to optimize each of the SQL queries for their respective storage representations. We note that many of the queries were CPU-bound, particularly in the case of the compact storage representations.

All experiments were repeated 10 times to obtain the arithmetic mean. We used the `bash time` shell keyword to record wall time in the same way for both the file-based utilities and the `psql` invocations. A variety of `psql` configuration options can greatly affect query performance as measured by the client. We employed the following command line for our timing runs:

```
<<< $QUERY psql -v FETCH_COUNT=1 -P pager=off -A -t -F $'\t' -q
```

The baseline time for any `psql` call on the system was, on average, 12 ms. Query planning times were typically around 5 ms, though query planning for the partitioned table with constraint exclusion required roughly 200 ms. Output was sent to the null device, `/dev/null`, to remove output device performance from consideration. Since this was not possible with PLINK, we provided PLINK with output storage on an unswapped `tmpfs` RAM disk. The result of these benchmarks appears in Fig. 3.

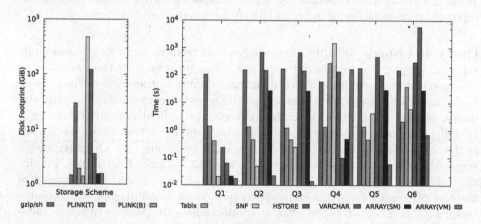

Fig. 3. Space requirements (left, logarithmic y-axis) and query performance (right, logarithmic y-axis) of various storage schemes.

Query 1: Atomic Lookup. The most basic query benchmark retrieves a specific genotype given a randomly selected sample and variant. This is not a common production query, but it is an instructive indicator of performance at the atomic scale. Most storage methods perform well on this query. All of the database representations effectively leverage their indices. The HSTORE performs badly due to data structure bloat. The variant-major array representation is fastest, minimizing the total data processing requirement to perform the query. Its smaller index outperforms the coordinate list representation and its narrower array outperforms the sample-major representation.

Query 2: Alternate Allele Frequency. Computing the alternate allele frequency in a set of samples is a common operation in genetic analysis and underlies many larger, more complex queries. It is an interesting variant descriptor and is useful for filtering variants that are unsuitable for downstream analysis. It must be repeated once per variant to filter variants by alternate allele frequency over a set of samples, so times above a fraction of a second are problematic. This query computes alternate allele frequency for a randomly selected variant in a random subset of 50% of samples. The result demonstrates the efficiency of the C procedure in performing single-pass computations over the array. The variant-major array representation performs best, with database server timing showing that calculations require less than 0.1 ms per variant.

Query 3: A Single Variant. Researchers are often interested in considering the quality of statistical models relating a single genetic variant with an observed phenotype. This query extracts a randomly selected variant for all samples. The sample-major representations must perform a table scan, but the coordinate list representation takes advantage of its partitioning, indexing, and clustering. While both array structures are fast due to compactness, the variant-major array representation exploits its index to quickly access the requisite row.

Query 4: A Single Sample. This query extracts all variants for one randomly selected sample. Our benchmark data has many more variants than samples, so this query must assemble and return far more data than the single-variant query. The coordinate list representation cannot use constraint exclusion to satisfy this query, because sample data is scattered evenly throughout the partitions. Additionally, three factors induce a sequential table scan. First, the sample foreign key is the second column in the multi-column index, so using that index requires a sequential index scan. Second, the indices are half the size of the tables. Finally, the tables are clustered by the index, so variants are stored contiguously at the cost of samples being distributed evenly. Adding an additional multi-column index with `sample_ref` as the leftmost column allows the planner to use an index effectively, but it must still perform a bitmap index scan over each partition, and the time requirement is only reduced from 1471 s to 594 s at a cost of another 149.3 GB for the second index. Further improvements require altering the dimension of partitioning or clustering and decrease efficiency of variant-centric queries. Sample major representations are ideal for this query. The `VARCHAR` and `ARRAY` both use the same PostgreSQL `varlena` implementation, so we directly observe the benefit of factoring the allele nucleotide sequences out of the genotypes. Though the sample-major array is 348 times as fast as the variant-major array for this query and soundly the fastest, the variant-major array is 1952 times as fast for the single-variant query. This is an artifact of the relative cardinalities of the dimensions but is representative of typical data.

Query 5: All Variants on a Gene. This query selects subsets of variants that reside on the same randomly selected gene and extracts genotypes for a randomly generated set of 75% of samples, which represents a common task in genetic association studies. Variant-based partitioning and clustering allow the coordinate list representation to remain competitive, but the small index and superior compactness of data in the variant-major array representation minimize the total quantity of data that it must read and process.

Query 6: Complex Subsets. This query is representative of the complex set of criteria that arise in exploratory association studies. Researchers are interested in subsets of samples based on demographic and phenotypic qualities and in subsets of variants based on their genomic properties, such as location and functional effect. We randomly select between 1 and 200 variants from anywhere on the genome for a random subset of samples. This query tests the performance of constructing results of varying sizes across non-contiguous selections of both sample and variant dimensions.

The `VARCHAR` data structure, for which we did not implement a multiple indexing operator, demonstrates the cost of non-contiguous selection. The coordinate list representation benefits from partitioning, indexing, and clustering, because these queries filter both dimensions to require a sufficiently small percentage of the data. Performance deteriorates as result set size increases, and ill-fated queries trigger sequential table scans. The variant-major array offers by far the best performance.

6 Probabilistic Genotypes

Increasingly projects such as 1000genomes are used to impute genotypes that are not directly measured in an individual based on the correlation of measured genotypes with genotypes in a reference dataset [11]. While current genotyping microarrays typically measure alleles for between 0.5 and 2.5 million variants [7], imputation may output tens of millions of probabilistic genotypes per sample. The output of the statistical procedures that perform this imputation is often a tuple $T : \{\mathcal{P}(AA), \mathcal{P}(AB), \mathcal{P}(BB)\}$, where $\mathcal{P}(x) \in [0,1]$ is a probability and A and B are potential alleles.

Native database floating point types are potential storage media for tuple elements, but we know that $\mathcal{P}(x) \in [0,1]$ and that $\sum_{\mathcal{P}(x)\in T} \mathcal{P}(x) \leq 1$, where the probability sum need not be unity to allow for missing genotype calls. Widely used software for performing imputation by default emits only 3 decimal digits in $[0.000, 1.000]$, and the utility of precision beyond this is dubious. With three digits of precision, each probability fits into a 10-bit unsigned integral representation. Three such entities can be packed into a 4-byte PostgreSQL custom data type to be queried as a unit. This increases data locality and reduces query times versus native floating point types, and it decreases post-compression storage requirements by 70%. For our data set, there is only a 58% storage penalty per variant for maintaining the 4-byte bit-packed imputed probability tuples versus the 1-byte `TINYINT` unimputed alternate allele counts. We implemented this extension to manage 108.6 billion imputed genotype probability tuples.

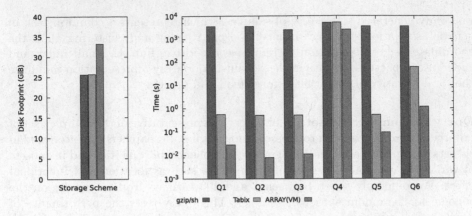

Fig. 4. Space requirements (left, linear y-axis) and query performance (right, logarithmic y-axis) of various storage schemes.

Of the storage schemes previously offered, most are intractable at this scale. For example, the coordinate list representation requires 4.3 TB of space. We provide only the best-performing methods for comparison. Figure 4 shows storage requirements and query performance. Alternate allele frequency is not relevant, so query 2 is absent.

7 Conclusion

We have presented a contemporary study of storing genotypic data in relational databases and proposed a scalable array-based solution in PostgreSQL that has escaped serious consideration. We further extended it to efficiently support probabilistic data. Benchmarks using a large research data set of genotypes demonstrate that array-based organization reduces storage requirements compared to existing relational database methods by several orders of magnitude while improving query speeds significantly. It competes well with domain-aware tools that sacrifice many of the benefits of relational databases. This technique is of similar utility in other domains with dense data matrices containing small, homogeneous units of rarely written information.

User-defined types, C and SQL functions, and benchmark code are available on GitHub at https://github.com/rlichtenwalter/pgsql_genomics. A production database using this code is heavily used by multiple researchers in the Human Pain Genetics Lab at McGill University, where it backs an analytical web portal supporting complex queries integrating genotypes, variant annotations, genomic intervals, phenotypes, quality metrics, and other diverse data. Researchers across disciplines have used this portal to generate many terabytes worth of throughput via large, complex queries with sub-second median completion times. Research was sponsored by the NIH/NIDCR under grants U01DE017018 and R01DE023846.

References

1. Bloom, T., Sharpe, T.: Managing data from high-throughput genomic processing: a case study. In: Proceedings of the Thirtieth International Conference on Very Large Data Bases, pp. 1198–1201. VLDB Endowment (2004)
2. Chanda, P., Elhaik, E., Bader, J.S.: HapZipper: sharing HapMap populations just got easier. Nucleic Acids Res. **40**(20), e159–e159 (2012)
3. Danecek, P., Auton, A., Abecasis, G., Albers, C.A., Banks, E., DePristo, M.A., Handsaker, R.E., Lunter, G., Marth, G.T., Sherry, S.T., et al.: The variant call format and VCFtools. Bioinformatics **27**(15), 2156–2158 (2011)
4. Davies, K.: The $1,000 Genome: The Revolution in DNA Sequencing and the New Era of Personalized Medicine. Simon and Schuster, New York (2015)
5. Fong, C., Ko, D.C., Wasnick, M., Radey, M., Miller, S.I., Brittnacher, M.: GWAS analyzer: integrating genotype, phenotype and public annotation data for genome-wide association study analysis. Bioinformatics **26**(4), 560–564 (2010)
6. Gabetta, M., Limongelli, I., Rizzo, E., Riva, A., Segagni, D., Bellazzi, R.: BigQ: a NoSQL based framework to handle genomic variants in i2b2. BMC Bioinform. **16**(1), 1 (2015)
7. Ha, N.-T., Freytag, S., Bickeboeller, H.: Coverage and efficiency in current SNP chips. Europ. J. Hum. Genet. **22**(9), 1124–1130 (2014)
8. Jolley, K.A., Maiden, M.C.: BIGSdb: scalable analysis of bacterial genome variation at the population level. BMC Bioinform. **11**(1), 595 (2010)
9. Layer, R.M., Kindlon, N., Karczewski, K.J., Quinlan, A.R., et al.: Efficient genotype compression, analysis of large genetic-variation data sets. Nat. Methods **13**(1), 63–65 (2016)
10. Li, H.: Tabix: fast retrieval of sequence features from generic TAB-delimited files. Bioinformatics **27**(5), 718–719 (2011)
11. Marchini, J., Howie, B.: Genotype imputation for genome-wide association studies. Nat. Rev. Genet. **11**(7), 499–511 (2010)
12. Mitha, F., Herodotou, H., Borisov, N., Jiang, C., Yoder, J., Owzar, K.: SNPpy-database management for SNP data from Genome wide association studies. PLOS ONE **6**(10), e24982 (2011)
13. O'Driscoll, A., Daugelaite, J., Sleator, R.D.: 'Big data', hadoop and cloud computing in genomics. J. Biomed. Inform. **46**(5), 774–781 (2013)
14. Orro, A., Guffanti, G., Salvi, E., Macciardi, F., Milanesi, L.: SNPLims: a data management system for genome wide association studies. BMC Bioinform. **9**(Suppl 2), S13 (2008)
15. Paila, U., Chapman, B.A., Kirchner, R., Quinlan, A.R.: GEMINI: integrative exploration of genetic variation and genome annotations. PLoS Comput. Biol. **9**(7), e1003153 (2013)
16. Purcell, S., Neale, B., Todd-Brown, K., Thomas, L., Ferreira, M.A.R., Bender, D., Maller, J., Sklar, P., De Bakker, P.I.W., Daly, M.J., et al.: PLINK: a tool set for whole-genome association and population-based linkage analyses. Am. J. Hum. Genet. **81**(3), 559–575 (2007)
17. Röhm, U., Blakeley, J.: Data management for high-throughput genomics. arXiv preprint arXiv:0909.1764 (2009)
18. Yeung, J.M.Y., Sham, P.C., Chan, A.S.W., Cherny, S.S.: OpenADAM: an open source genome-wide association data management system for Affymetrix SNP arrays. BMC Genomics **9**(1), 1–4 (2008)

Author Index

Printed in the United States
By Bookmasters